DIGITAL IMAGE WARPING

Digital Image Warping

George Wolberg

IEEE Computer Society Press Monograph

DIGITAL IMAGE WARPING

George Wolberg

Department of Computer Science
Columbia University
New York

IEEE Computer Society Press
Los Alamitos, California

Washington • Brussels • Tokyo

Published by

IEEE Computer Society Press
10662 Los Vaqueros Circle
P.O. Box 3014
Los Alamitos, CA 90720-1264

Cover credit: Raziel's Transformation Sequence from *Willow*.
Courtesy of Industrial Light & Magic, a Division of Lucasfilm Ltd.
© 1988 Lucasfilm Ltd. All Rights Reserved.

Cover layout by Vic Grenrock

IEEE Computer Society Press Order Number 1944
IEEE Catalog Number EH0322-8
ISBN 0-8186-8944-7 (case)
ISBN 0-8186-5944-0 (microfiche)
SAN 264-620X

Additional copies can be ordered from:

IEEE Computer Society Press	IEEE Computer Society	IEEE Computer Society	IEEE Service Center
Customer Service Center	13, Avenue de l'Aquilon	Ooshima Building	445 Hoes Lane
10662 Los Vaqueros Circle	B-1200 Brussels	2-19-1 Minami-Aoyama,	P.O. Box 1331
P.O. Box 3014	BELGIUM	Minato-Ku	Piscataway, NJ 08855-1331
Los Alamitos, CA 90720-1264		Tokyo 107, JAPAN	

PREFACE

Digital image warping is a growing branch of image processing that deals with geometric transformation techniques. Early interest in this area dates back to the mid-1960s when it was introduced for geometric correction applications in remote sensing. Since that time it has experienced vigorous growth, finding uses in such fields as medical imaging, computer vision, and computer graphics. Although image warping has traditionally been dominated by results from the remote sensing community, it has recently enjoyed a new surge of interest from the computer graphics field. This is largely due to the growing availability of advanced graphics workstations and increasingly powerful computers that make warping a viable tool for image synthesis and special effects. Work in this area has already led to successful market products such as real-time video effects generators for the television industry and cost-effective warping hardware for geometric correction. Current trends indicate that this area will have growing impact on desktop video, a new technology that promises to revolutionize the video production market in much the same way as desktop publishing has altered the way in which people prepare documents.

Digital image warping has benefited greatly from several fields, ranging from early work in remote sensing to recent developments in computer graphics. The scope of these contributions, however, often varies widely owing to different operating conditions and assumptions. This state is reflected in the image processing literature. Despite the fact that image processing is a well-established subject with many textbooks devoted to its study, image warping is generally treated as a peripheral subject with only sparse coverage. Furthermore, these textbooks rarely present image warping concepts as a single body of knowledge. Since the presentations are usually tailored to some narrow readership, different components of the same conceptual framework are emphasized. This has left a noticeable gap in the literature with respect to a unified treatment of digital image warping in a single text. This book attempts to redress this imbalance.

The purpose of this book is to introduce the fundamental concepts of digital image warping and to lay a foundation that can be used as the basis for further study and research in this field. Emphasis is given to the development of a single coherent framework. This serves to unify the terminology, motivation, and contributions of many disciplines that have each contributed in significantly different ways. The coherent framework puts the diverse aspects of this subject into proper perspective. In this manner, the needs and goals of a diverse readership are addressed.

This book is intended to be a practical guide for eclectic scientists and engineers who find themselves in need of implementing warping algorithms and comprehending the underlying concepts. It is also geared to students of image processing who wish to apply their knowledge of that subject to a well-defined application. Special effort has been made to keep prerequisites to a minimum in the hope of presenting a self-contained treatment of this field. Consequently, knowledge of elementary image processing is helpful, although not essential. Furthermore, every effort is made to reinforce the discussion with an intuitive understanding. As a result, only those aspects of supporting theory that are directly relevant to the subject are brought to bear. Interested readers may consult the extensive bibliography for suggested readings that delve further into those areas.

This book originally grew out of a survey paper that I had written on geometric transformation techniques for digital images. During the course of preparing that paper, the large number of disparate sources with potential bearing on digital image warping became strikingly apparent. This writing reflects my goal to consolidate these works into a self-contained central repository. Since digital image warping involves many diverse aspects, from implementation considerations to the mathematical abstractions of sampling and filtering theory, I have attempted to chart a middle path by focusing upon those basic concepts, techniques, and problems that characterize the geometric transformation of digital images, given the inevitable limitations of discrete approximations. The material in this book is thus a delicate balance between theory and practice. The practical segment includes algorithms which the reader may implement. The theory segment is comprised of proofs and formulas derived to motivate the algorithms and to establish a standard of comparison among them. In this manner, theory provides a necessary context in which to understand the goals and limitations of the collection of algorithms presented herein.

The organization of this book closely follows the components of the conceptual framework for digital image warping. Chapter 1 discusses the history of this field and presents a brief overview of the subsequent chapters. A review of common terminology, mathematical preliminaries, and digital image acquisition is presented in Chapter 2. As we shall see later, digital image warping consists of two basic operations: a spatial transformation to define the rearrangement of pixels and interpolation to compute their values. Chapter 3 describes various common formulations for spatial transformations, as well as techniques for inferring them when only a set of correspondence points are known. Chapter 4 provides a review of sampling theory, the mathematical framework used to describe the filtering problems that follow. Chapter 5 describes image resampling, including several common interpolation kernels. They are applied in the discussion of antialiasing in Chapter 6. This chapter demonstrates several approaches used to avoid artifacts that manifest themselves to the discrete nature of digital images. Fast warping techniques based on scanline algorithms are presented in Chapter 7. These results are particularly useful for both hardware and software realizations of geometric transformations. Finally, the main points of the book are recapitulated in Chapter 8. Source code, written in C, is scattered among the chapters and appendices to demonstrate implementation details for various algorithms.

It is often difficult to measure the success of a book. Ultimately, the effectiveness of this text can be judged in two ways. First, the reader should appreciate the difficulties and subtleties in actually warping a digital image. This includes a full understanding of the problems posed due to the discrete nature of digital images, as well as an awareness of the tradeoffs confronting an algorithm designer. There are valuable lessons to be learned in this process. Second, the reader should master the key concepts and techniques that facilitate further research and development. Unlike many other branches of science, students of digital image warping benefit from the direct visual realization of mathematical abstractions and concepts. As a result, readers are fortunate to have images clarify what mathematical notation sometimes obscures. This makes the study of digital image warping a truly fascinating and enjoyable endeavor.

George Wolberg

ACKNOWLEDGEMENTS

This book is a product of my doctoral studies at Columbia University. I consider myself very fortunate to have spent several exciting years among a vibrant group of faculty and students in the Computer Science department. My deepest thanks go to Prof. Terry Boult, my advisor and good friend. He has played an instrumental role in my professional growth. His guidance and support have sharpened my research skills, sparked my interest in a broad range of research topics, set high standards for me to emulate, and made the whole experience a truly special one for me. I am very grateful for our many fruitful discussions that have influenced the form and content of this book and my related dissertation.

I also wish to thank Profs. Steven Feiner and Gerald Maguire for many scintillating conversations and for their meticulous review of the manuscript. Their suggestions helped me refine my ideas and presentation. Special thanks are owed to Henry Massalin for his invaluable insights and thoughtful discussions. He innovated the ideas for the exponential filters described in Chapter 5. His original implementation of these filters for audio applications prompted me to suggest their usage in video processing where a different set of operating assumptions make them particularly cost-effective and robust (Qua!). I also thank David Kurlander for his help and support, including his assistance with Figs. 4.1, 4.2, 5.2, 6.8, and my photograph on the back cover.

The source of my inspiration for this book, and indeed for my decision to pursue doctoral studies, stems from my friendship with Dr. Theo Pavlidis. While still an undergraduate electrical engineering student at Cooper Union, I was priviledged to work for him at AT&T Bell Laboratories over the course of two summers. During that time, I experienced a great deal of professional growth, an enthusiasm towards research, and a fascination with image processing, pattern recognition, and computer graphics. I am greatly indebted to him for his long-standing inspiration and support.

My early interest in digital image warping is rooted in my consulting work with Fantastic Animation Machine, a computer animation company in New York City. I wish to thank Jim Lindner and his staff of software gurus and artists for making the hectic pace of television production work a lot of fun. It was an excellent learning experience.

The people at Industrial Light and Magic (ILM) were very helpful in providing images for this book, including the image that appears on the front cover. Thanks go to Douglas Smythe for sharing the details of his mesh warping algorithm. Lincoln Hu deserves special mention for expediently coordinating the transfer of images and for his meticulous attention to detail. Doug Kay helped make all of this possible by pushing this past the lawyers and red tape.

The contributions from Pixar were handled by Rick Sayre. Thanks also go to Ed Catmull and Alvy Ray Smith for various discussions on the subject, and for their seminal paper that sparked my original interest in image warping. Tom Brigham contributed the image in Fig. 1.2. Generous contributions in content and form were made by Paul Heckbert, Ken Turkowski, Karl Fant, and Norman Chin. Thanks are owed to Profs. Peter Allen and John Kender for their advice and encouragement. I wish to thank Margaret Brown and Jon Butler for their support in the production of this book. They handled the project most professionally, and their prompt and courteous attention made my job a lot easier.

I gratefully acknowledge the U.S. National Science Foundation (NSF) for funding my graduate work via an NSF Graduate Fellowship. Further support was provided by the U.S. Defense Advanced Research Projects Agency (DARPA), and by the Center for Telecommunications Research (CTR) at Columbia University. Most of my software development was done on HP 9000/370 graphics workstations that were generously donated by Hewlett-Packard.

Finally, I am sincerely thankful and appreciative to my dear mother for her love, understanding, and constant support. My persistance during the writing of this book was largely a product of her example. This book is dedicated to my mother and to the memory of my beloved father.

TABLE OF CONTENTS

1

INTRODUCTION

1.1. BACKGROUND

Digital image warping is a growing branch of image processing that deals with the geometric transformation of digital images. A *geometric transformation* is an operation that redefines the spatial relationship between points in an image. Although image warping often tends to conjure up notions of highly distorted imagery, a warp may range from something as simple as a translation, scale, or rotation, to something as elaborate as a convoluted transformation. Since all warps do, in fact, apply geometric transformations to images, the terms ''warp'' and ''geometric transformation'' are used interchangeably throughout this book.

It is helpful to interpret image warping in terms of the following physical analogy. Imagine printing an image onto a sheet of rubber. Depending on what forces are applied to that sheet, the image may simply appear rotated or scaled, or it may appear wildly distorted, corresponding to the popular notion of a warp. While this example might seem to portray image warping as a playful exercise, image warping does serve an important role in many applied sciences. Over the past twenty years, for instance, image warping has been the subject of considerable attention in remote sensing, medical imaging, computer vision, and computer graphics. It has made its way into many applications, including distortion compensation of imaging sensors, decalibration for image registration, geometrical normalization for image analysis and display, map projection, and texture mapping for image synthesis.

Historically, geometric transformations were first performed on continuous (analog) images using optical systems. Early work in this area is described in [Cutrona 60], a landmark paper on the use of optics to perform transformations. Since then, numerous advances have been made in this field [Horner 87]. Although optical systems offer the distinct advantage of operating at the speed of light, they are limited in control and flexibility. Digital computer systems, on the other hand, resolve these problems and potentially offer more accuracy. Consequently, the algorithms presented in this book deal exclusively with digital (discrete) images.

1

The earliest work in geometric transformations for digital images stems from the remote sensing field. This area gained attention in the mid-1960s, when the U.S. National Aeronautics and Space Administration (NASA) embarked upon aggressive earth observation programs. Its objective was the acquisition of data for environmental research applicable to earth resource inventory and management. As a result of this initiative, programs such as Landsat and Skylab emerged. In addition, other government agencies were supporting work requiring aerial photographs for terrain mapping and surveillance.

These projects all involved acquiring multi-image sets (i.e., multiple images of the same area taken either at different times or with different sensors). Immediately, the task arises to align each image with every other image in the set so that all corresponding points match. This process is known as *image registration*. Misalignment can occur due to any of the following reasons. First, images may be taken at the same time but acquired from several sensors, each having different distortion properties, e.g., lens aberration. Second, images may be taken from one sensor at different times and at various viewing geometries. Furthermore, sensor motion will give rise to distortion as well.

Geometric transformations were originally introduced to invert (correct) these distortions and to allow the accurate determination of spatial relationships and scale. This requires us to first estimate the distortion model, usually by means of reference points which may be accurately marked or readily identified (e.g., road intersections and land-water interface). In the vast majority of cases, the coordinate transformation representing the distortion is modeled as a bivariate polynomial whose coefficients are obtained by minimizing an error function over the reference points. Usually, a second-order polynomial suffices, accounting for translation, scale, rotation, skew, and pincushion effects. For more local control, affine transformations and piecewise polynomial mapping functions are widely used, with transformation parameters varying from one region to another. See [Haralick 76] for a historical review of early work in remote sensing.

An example of the use of image warping for geometric correction is given in Figs. 1.1 and 1.2. Figure 1.1 shows an example of an image distorted due to viewing geometry. It was recorded after the Viking Lander 2 spacecraft landed on Mars in September 1976. A cylindrical scanner was used to acquire the image. Since the spacecraft landed with an 8° downward tilt, the level horizon appears curved. This problem is corrected in Fig. 1.2, which shows the same image after it was rectified by a transformation designed to remove the tilt distortion.

The methods derived from remote sensing have direct application in other related fields, including medical imaging and computer vision. In medical imaging, for instance, geometric transformations play an important role in image registration and rotation for digital radiology. In this field, images obtained after injection of contrast dye are enhanced by subtracting a mask image taken before the injection. This technique, known as digital subtraction angiography, is subject to distortions due to patient motion. Since motion causes misalignment of the image and its subtraction mask, the resulting produced images are degraded. The quality of these images is improved with transformation algorithms that increase the accuracy of the registration.

Figure 1.1: Viking Lander 2 image distorted due to downward tilt [Green 89].

Figure 1.2: Viking Lander 2 image after distortion correction [Green 89].

Image warping is a problem that arises in computer graphics as well. However, in this field the goal is not geometric correction, but rather inducing geometric distortion. Graphics research has developed a distinct repertoire of techniques to deal with this problem. The primary application is texture mapping, a technique to map 2-D images onto 3-D surfaces, and then project them back onto a 2-D viewing screen. Texture mapping has been used with much success in achieving visually rich and complicated imagery. Furthermore, additional sophisticated filtering techniques have been promoted to combat artifacts arising from the severe spatial distortions possible in this application. The thrust of this effort has been directed to the study and design of efficient spatially-varying low-pass filters. Since the remote sensing and medical imaging fields have generally attempted to correct only mild distortions, they have neglected this important area. The design of fast algorithms for filtering fairly general areas remains a great challenge.

Image warping is commonly used in graphics design to create interesting visual effects. For instance, Fig. 1.3 shows a fascinating sequence of warps that depicts a transformation between two faces, a horse and rider, two frogs, and two dancers. Other examples of such applications include the image sequence shown on the front cover, as well as other effects described in [Holzmann 88].

Figure 1.3: Transformation sequence: faces → horse/rider → frogs → dancers. Copyright © 1983 Tom Brigham / NYIT-CGL. All rights reserved.

The continuing development of efficient algorithms for digital image warping has gained impetus from the growing availability of fast and cost-effective digital hardware. The ability to process high resolution imagery has become more feasible with the advent of fast computational elements, high-capacity digital data storage devices, and improved display technology. Consequently, the trend in algorithm design has been towards a more effective match with the implementation technology. This is reflected in the recent surge of warping products that exploit scanline algorithms.

It is instructive at this point to illustrate the relationship between the remote sensing, medical imaging, computer vision, and computer graphics fields since they all have ties to image warping. As stated earlier, image warping is a subset of image processing. These fields are all connected to image warping insofar as they share a common usage for image processing. Figure 1.4 illustrates these links as they relate to images and mathematical scene descriptions, the two forms of data used by the aforementioned fields.

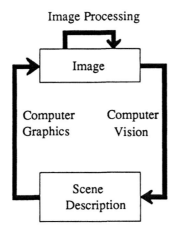

Figure 1.4: Underlying role of image processing [Pavlidis 82].

Consider the transition from a scene description to an image, as shown in Fig. 1.4. This is a function of a renderer in computer graphics. Although image processing is often applied after rendering, as a postprocess, those rendering operations requiring proper filtering actually embed image processing concepts directly. This is true for warping applications in graphics, which manifests itself in the form of texture mapping. As a result, texture mapping is best understood as an image processing problem.

The transition from an input image to an output image is characteristic of image processing. Image warping is thereby considered an image processing task because it takes an input image and applies a geometric transformation to yield an output image. Computer vision and remote sensing, on the other hand, attempt to extract a scene description from an image. They use image registration and geometric correction as preliminary components to pattern recognition. Therefore, image warping is common to these fields insofar as they share images which are subject to geometric transformations.

1.2. OVERVIEW

The purpose of this book is to describe the algorithms developed in this field within a consistent and coherent framework. It centers on the three components that comprise all geometric transformations in image warping: spatial transformations, resampling, and antialiasing. Due to the central importance of sampling theory, a review is provided as a preface to the resampling and antialiasing chapters. In addition, a discussion of efficient scanline implementations is given as well. This is of particular importance to practicing scientists and engineers.

In this section, we briefly review the various stages in a geometric transformation. Each stage has received a great deal of attention from a wide community of people in many diverse fields. As a result, the literature is replete with varied terminologies, motivations, and assumptions. A review of geometric transformation techniques, particularly in the context of their numerous applications, is useful for highlighting the common thread that underlies their many forms. Since each stage is the subject of a separate chapter, this review should serve to outline the contents of this book. We begin with some basic concepts in spatial transformations.

1.2.1. Spatial Transformations

The basis of geometric transformations is the mapping of one coordinate system onto another. This is defined by means of a *spatial transformation* — a mapping function that establishes a spatial correspondence between all points in the input and output images. Given a spatial transformation, each point in the output assumes the value of its corresponding point in the input image. The correspondence is found by using the spatial transformation mapping function to project the output point onto the input image.

Depending on the application, spatial transformation mapping functions may take on many different forms. Simple transformations may be specified by analytic expressions including affine, projective, bilinear, and polynomial transformations. More sophisticated mapping functions that are not conveniently expressed in analytic terms can be determined from a sparse lattice of control points for which spatial correspondence is known. This yields a spatial representation in which undefined points are evaluated through interpolation. Indeed, taking this approach to the limit yields a dense grid of control points resembling a 2-D spatial lookup table that may define any arbitrary mapping function.

In computer graphics, for example, the spatial transformation is completely specified by the parameterization of the 3-D object, its position with respect to the 2-D projection plane (i.e., the viewing screen), viewpoint, and center of interest. The objects are usually defined as planar polygons or bicubic patches. Consequently, three coordinate systems are used: 2-D texture space, 3-D object space, and 2-D screen space. The various formulations for spatial transformations, as well as methods to infer them, are discussed in Chapter 3.

1.2.2. Sampling Theory

In the continuous domain, a geometric transformation is fully specified by the spatial transformation. This is due to the fact that an analytic mapping is bijective — one-to-one and onto. However, in our domain of interest, complications are introduced due to the discrete nature of digital images. Undesirable artifacts can arise if we are not careful. Consequently, we turn to sampling theory for a deeper understanding of the problem at hand.

Sampling theory is central to the study of sampled-data systems, e.g., digital image transformations. It lays a firm mathematical foundation for the analysis of sampled signals, offering invaluable insight into the problems and solutions of sampling. It does so by providing an elegant mathematical formulation describing the relationship between a continuous signal and its samples. We use it to resolve the problems of image reconstruction and aliasing. Note that reconstruction is an interpolation procedure applied to the sampled data and that aliasing simply refers to the presence of unreproducibly high frequencies and the resulting artifacts.

Together with defining theoretical limits on the continuous reconstruction of discrete input, sampling theory yields the guidelines for numerically measuring the quality of various proposed filtering techniques. This proves most useful in formally describing reconstruction, aliasing, and the filtering necessary to combat the artifacts that may appear at the output. The fundamentals of sampling theory are reviewed in Chapter 4.

1.2.3. Resampling

Once a spatial transformation is established, and once we accommodate the subtleties of digital filtering, we can proceed to resample the image. First, however, some additional background is in order.

In digital images, the discrete picture elements, or *pixels*, are restricted to lie on a sampling grid, taken to be the integer lattice. The output pixels, now defined to lie on the output sampling grid, are passed through the mapping function generating a new grid used to resample the input. This new resampling grid, unlike the input sampling grid, does not generally coincide with the integer lattice. Rather, the positions of the grid points may take on any of the continuous values assigned by the mapping function.

Since the discrete input is defined only at integer positions, an interpolation stage is introduced to fit a continuous surface through the data samples. The continuous surface may then be sampled at arbitrary positions. This interpolation stage is known as *image reconstruction*. In the literature, the terms "reconstruction" and "interpolation" are used interchangeably. Collectively, image reconstruction followed by sampling is known as *image resampling*.

Image resampling consists of passing the regularly spaced output grid through the spatial transformation, yielding a resampling grid that maps into the input image. Since the input is discrete, image reconstruction is performed to interpolate the continuous input signal from its samples. Sampling the reconstructed signal gives us the values that are assigned to the output pixels.

The accuracy of interpolation has significant impact on the quality of the output image. As a result, many interpolation functions have been studied from the viewpoints of both computational efficiency and approximation quality. Popular interpolation functions include cubic convolution, bilinear, and nearest neighbor. They can exactly reconstruct second-, first-, and zero-degree polynomials, respectively. More expensive and accurate methods include cubic spline interpolation and convolution with a sinc function. Using sampling theory, this last choice can be shown to be the ideal filter. However, it cannot be realized using a finite number of neighboring elements. Consequently, the alternate proposals have been given to offer reasonable approximations. Image resampling and reconstruction are described in Chapter 5.

1.2.4. Aliasing

Through image reconstruction, we have solved the first problem that arises due to operating in the discrete domain — sampling a discrete input. Another problem now arises in evaluating the discrete output. The problem, related to the resampling stage, is described below.

The output image, as described earlier, has been generated by *point sampling* the reconstructed input. Point (or zero-spread) sampling refers to an ideal sampling process in which the value of each sampled point is taken independently of its neighbors. That is, each input point influences one and only one output point.

With point sampling, entire intervals between samples are discarded and their information content is lost. If the input signal is smoothly varying, the lost data is recoverable through interpolation, i.e., reconstruction. This statement is true only when the input is a member of a class of signals for which the interpolation algorithm is designed. However, if the skipped intervals are sufficiently complex, interpolation may be inadequate and the lost data is unrecoverable. The input signal is then said to be *undersampled*, and any attempt at reconstruction gives rise to a condition known as *aliasing*. Aliasing distortions, due to the presence of unreproducibly high spatial frequencies, may surface in the form of jagged edges and moire patterns.

Aliasing artifacts are most evident when the spatial mapping induces large-scale changes. As an example, consider the problem of image magnification and minification. When magnifying an image, each input pixel contributes to many output pixels. This one-to-many mapping requires the reconstructed signal to be densely sampled. Clearly, the resulting image quality is closely tied to the accuracy of the interpolation function used in reconstruction. For instance, high-degree interpolation functions can exactly reconstruct a larger class of signals than low-degree functions. Therefore, if the input is poorly reconstructed, artifacts such as jagged edges become noticeable at the output grid. Note that the computer graphics community often considers jagged edges to be synonymous with aliasing. As we shall see in Chapter 4, this is sometimes a misconception. In this case, for instance, jagged edges are due to inadequate reconstruction, *not* aliasing.

Under magnification, the output contains at least as much information as the input, with the output assigned the values of the densely sampled reconstructed signal. When minifying (i.e., reducing) an image, the opposite is true. The reconstructed signal is sparsely sampled in order to realize the scale reduction. This represents a clear loss of data, where many input samples are actually skipped over in the point sampling. It is here where aliasing is apparent in the form of moire patterns and fictitious low-frequency components. It is related to the problem of mapping many input samples onto a single output pixel. This requires appropriate filtering to properly integrate all the information mapping to that pixel.

The filtering used to counter aliasing is known as *antialiasing*. Its derivation is grounded in the well established principles of sampling theory. Antialiasing typically requires the input to be blurred *before* resampling. This serves to have the sampled points influenced by their discarded neighbors. In this manner, the extent of the artifacts is diminished, but not eliminated.

Completely undistorted sampled output can only be achieved by sampling at a sufficiently high frequency, as dictated by sampling theory. Although adapting the sampling rate is more desirable, physical limitations on the resolution of the output device often prohibit this alternative. Thus, the most common solution to aliasing is smoothing the input prior to sampling.

The well understood principles of sampling theory offer theoretical insight into the problem of aliasing and its solution. However, due to practical limitations in implementing the ideal filters suggested by the theory, a large number of algorithms have been proposed to yield approximate solutions. Chapter 6 details the antialiasing algorithms.

1.2.5. Scanline Algorithms

The underlying theme behind many of the algorithms that only approximate ideal filtering is one recurring consideration: speed. Fast warping techniques are critical for numerous applications. There is a constant struggle in the speed/accuracy tradeoff. As a result, a large body of work in digital image warping has been directed towards optimizing special cases to obtain major performance gains. In particular, the use of scanline algorithms has reduced complexity and processing time. Scanline algorithms are often based on separable geometric transformations. They reduce 2-D problems into a sequence of 1-D (scanline) resampling problems. This makes them amenable to streamline processing and allows them to be implemented with conventional hardware. Scanline algorithms have been shown to be useful for affine and perspective transformations, as well as for mappings onto bilinear, biquadratic, bicubic, and superquadric patches. Recent work has also shown how it may be extended to realize arbitrary spatial transformations. The dramatic developments due to scanline algorithms are described in Chapter 7.

1.3. CONCEPTUAL LAYOUT

Figure 1.5 shows the relationship between the various stages in a geometric transformation. It is by no means a strict recipe for the order in which warping is achieved. Instead, the purpose of this figure is to convey a conceptual layout, and to serve as a roadmap for this book.

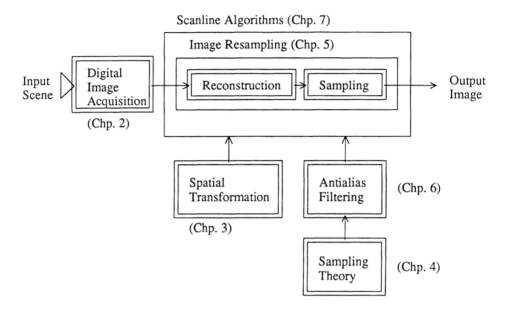

Figure 1.5: Conceptual layout.

An image is first acquired by a digital image acquisition system. It then passes through the image resampling stage, consisting of a reconstruction substage to compute a continuous image and a sampling substage that samples it at any desired location. The exact positions at which resampling occurs is defined by the spatial transformation. The output image is obtained once image resampling is completed.

In order to avoid artifacts in the output, the resampling stage must abide by the principles of digital filtering. Antialias filtering is introduced for this purpose. It serves to process the image so that artifacts due to undersampling are mitigated. The theory and justification for this filtering is derived from sampling theory. In practice, image resampling and digital filtering are collapsed into efficient algorithms which are tightly coupled. As a result, the stages that contribute to image resampling are depicted as being integrated into scanline algorithms.

2

PRELIMINARIES

In this chapter, we begin our study of digital image warping with a review of some basic terminology and mathematical preliminaries. This shall help to lay our treatment of image warping on firm ground. In particular, elements of this chapter comprise a formulation that will be found to be recurring throughout this book. After the definitions and notation have been clarified, we turn to a description of digital image acquisition. This stage is responsible for converting a continuous image of a scene into a discrete representation that is suitable for digital computers. Attention is given to the imaging components in digital image acquisition systems. The operation of these devices is explained and an overview of a general imaging system is given. Finally, we conclude with a presentation of input images that will be used repeatedly throughout this book. These images will later be subjected to geometric transformations to demonstrate various warping and filtering algorithms.

2.1. FUNDAMENTALS

Every branch of science establishes a set of definitions and notation in which to formalize concepts and convey ideas. Digital image warping borrows its terminology from its parent field, digital image processing. In this section, we review some basic definitions that are fundamental to image processing. They are intended to bridge the gap between an informal dialogue and a technical treatment of digital image warping. We begin with a discussion of signals and images.

2.1.1. Signals and Images

A *signal* is a function that conveys information. In standard signal processing texts, signals are usually taken to be one-dimensional functions of time, e.g., $f(t)$. In general, though, signals can be defined in terms of any number of variables. Image processing, for instance, deals with two-dimensional functions of space, e.g., $f(x,y)$. These signals are mathematical representations of *images*, where $f(x,y)$ is the brightness value at spatial coordinate (x,y).

Images can be classified by whether or not they are defined over all points in the spatial domain, and by whether their image values are represented with finite or infinite precision. If we designate the labels "continuous" and "discrete" to classify the spatial domain as well as the image values, then we can establish the following four image categories: continuous-continuous, continuous-discrete, discrete-continuous, and discrete-discrete. Note that the two halves of the labels refer to the spatial coordinates and image values, respectively.

A *continuous-continuous image* is an infinite-precision image defined at a continuum of positions in space. The literature sometimes refers to such images as *analog images*, or simply *continuous images*. Images from this class may be represented with finite-precision to yield *continuous-discrete images*. Such images result from discretizing a continuous-continuous image under a process known as *quantization* to map the real image values onto a finite set (e.g., a range that can be accommodated by the numerical precision of the computer). Alternatively, images may continue to have their values retained at infinite-precision, however these values may be defined at only a discrete set of points. This form of spatial quantization is a manifestation of *sampling*, yielding *discrete-continuous images*. Since digital computers operate exclusively on finite-precision numbers, they deal with *discrete-discrete images*. In this manner, both the spatial coordinates *and* the image values are quantized to the numerical precision of the computer that will process them. This class is commonly known as *digital images*, or simply *discrete images*, owing to the manner in which they are manipulated. Methods for converting between analog and digital images will be described later.

We speak of *monochrome* images, or *black-and-white* images, when f is a single-valued function representing shades of gray, or *gray levels*. Alternatively, we speak of *color* images when f is a vector-valued function specifying multiple color components at each spatial coordinate. Although various color spaces exist, color images are typically defined in terms of three color components: red, green, and blue (RGB). That is, for color images we have

$$f(x,y) = \left\{ f_{red}(x,y), \ f_{green}(x,y), \ f_{blue}(x,y) \right\} \qquad (2.1.1)$$

Such vector-valued functions can be readily interpreted as a stack of single-valued images, called *channels*. Therefore, monochrome images have one channel while RGB color images have three (see Fig. 2.1). Color images are instances of a general class known as *multispectral images*. This refers to images of the same scene that are acquired in different parts of the electromagnetic spectrum. In the case of color images, the scene is passed through three spectral filters to separate the image into three RGB components. Note that nothing requires image data to be acquired in spectral regions that fall in the visible range. Many applications find uses for images in the ultraviolet, infrared, microwave, and X-ray ranges. In all cases, though, each channel is devoted to a particular spectral band or, more generally, to an image attribute.

Depending on the application, any number of channels may be introduced to an image. For instance, a fourth channel denoting opacity is useful for image compositing

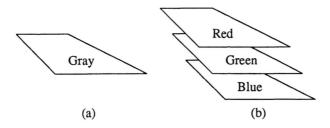

Figure 2.1: Image formats. (a) monochrome; (b) color.

operations which must smoothly blend images together [Porter 84]. In remote sensing, many channels are used for multispectral image analysis in earth science applications (e.g., the study of surface composition and structure, crop assessment, ocean monitoring, and weather analysis). In all of these cases, it is important to note that the number of variables used to index a signal is independent of the number of vector elements it yields. That is, there is no relationship between the number of dimensions and channels. For example, a two-dimensional function $f(x,y)$ can yield a 3-tuple color vector, or a 4-tuple (color, transparency) vector. Channels can even be used to encode spatially-varying signals that are not related to optical information. Typical examples include population and elevation data.

Thus far, all of the examples referring to images have been two-dimensional. It is possible to define higher-dimensional signals as well, although in these cases they are not usually referred to as images. An animation, for instance, may be defined in terms of function $f(x,y,t)$ where (x,y) again refers to the spatial coordinate and t denotes time. This produces a stack of 2-D images, whereby each slice in the stack is a snapshot of the animation. Volumetric data, e.g., CAT scans, can be defined in a similar manner. These are truly 3-D "images" that are denoted by $f(x,y,z)$, where (x,y,z) are 3-D coordinates. Animating volumetric data is possible by defining the 4-D function $f(x,y,z,t)$ whereby the spatial coordinates (x,y,z) are augmented by time t.

In the remainder of this book, we shall deal almost exclusively with 2-D color images. It is important to remember that although warped output images may appear as though they lie in 3-D space, they are in fact nothing more than 2-D functions. A direct analogy can be made here to photographs, whereby 3-D world scenes are projected onto flat images.

Our discussion thus far has focused on definitions related to images. We now turn to a presentation of terminology for filters. This proves useful because digital image warping is firmly grounded in digital filtering theory. Furthermore, the elements of an image acquisition system are modeled as a cascade of filters. This review should help put our discussion of image warping, including image acquisition, into more formal terms.

2.1.2. Filters

A *filter* is any system that processes an input signal $f(x)$ to produce an output signal, or a *response*, $g(x)$. We shall denote this as

$$f(x) \rightarrow g(x) \qquad (2.1.2)$$

Although we are ultimately interested in 2-D signals (e.g., images), we use 1-D signals here for notational convenience. Extensions to additional dimensions will be handled by considering each dimension independently.

Filters are classified by the nature of their responses. Two important criteria used to distinguish filters are *linearity* and *spatial-invariance*. A filter is said to be *linear* if it satisfies the following two conditions:

$$\alpha f(x) \rightarrow \alpha g(x) \qquad (2.1.3)$$
$$f_1(x) + f_2(x) \rightarrow g_1(x) + g_2(x)$$

for all values of α and all inputs $f_1(x)$ and $f_2(x)$. The first condition implies that the output response of a linear filter is proportional to the input. The second condition states that a linear filter responds to additional input independently of other signals present. These conditions can be expressed more compactly as

$$\alpha_1 f_1(x) + \alpha_2 f_2(x) \rightarrow \alpha_1 g_1(x) + \alpha_2 g_2(x) \qquad (2.1.4)$$

which restates the following two linear properties: scaling and superposition at the input produces equivalent scaling and superposition at the output.

A filter is said to be *space-invariant*, or *shift-invariant*, if a spatial shift in the input causes an identical shift in the output:

$$f(x-a) \rightarrow g(x-a) \qquad (2.1.5)$$

In terms of 2-D images, this means that the filter behaves the same way across the entire image, i.e., with no spatial dependencies. Similar constraints can be imposed on a filter in the temporal domain to qualify it as *time-variant* or *time-invariant*. In the remainder of this discussion, we shall avoid mention of the temporal domain although the same statements regarding the spatial domain apply there as well.

In practice, most physically realizable filters (e.g., lenses) are not entirely linear or space-invariant. For instance, most optical systems are limited in their maximum response and thus cannot be strictly linear. Furthermore, brightness, which is power per unit area, cannot be negative, thereby limiting the system's minimum response. This precludes an arbitrary range of values for the input and output images. Most optical imaging systems are prevented from being strictly space-invariant by finite image area and lens aberrations.

Despite these deviations, we often choose to approximate such systems as linear and space-invariant. As a byproduct of these modeling assumptions, we can adopt a rich set of analytical tools from linear filtering theory. This leads to useful algorithms for processing images. In contrast, nonlinear and space-variant filtering is not well-understood by many engineers and scientists, although it is currently the subject of much active research [Marvasti 87]. We will revisit this topic later when we discuss nonlinear image warping.

2.1.3. Impulse Response

In the continuous domain, we define

$$\delta(x) = \begin{cases} \lim_{\varepsilon \to 0} \int_{-\varepsilon}^{\varepsilon} \delta(x')\,dx' = 1, & x = 0 \\ \\ 0, & x \neq 0 \end{cases} \tag{2.1.6}$$

to be the *impulse function*, known also as the *Dirac delta function*. The impulse function can be used to sample a continuous function $f(x)$ as follows

$$f(x_0) = \int_{-\infty}^{\infty} f(\lambda)\,\delta(x_0 - \lambda)\,d\lambda \tag{2.1.7}$$

If we are operating in the discrete (integer) domain, then the *Kronecker delta function* is used:

$$\delta(x) = \begin{cases} 1, & x = 0 \\ 0, & x \neq 0 \end{cases} \tag{2.1.8}$$

for integer values of x. The two-dimensional versions of the Dirac and Kronecker delta functions are obtained in a separable fashion by taking the product of their 1-D counterparts:

$$\text{Dirac:} \quad \delta(x,y) = \delta(x)\,\delta(y) \tag{2.1.9}$$

$$\text{Kronecker:} \quad \delta(m,n) = \delta(m)\delta(n)$$

When an impulse is applied to a filter, an altered impulse, referred to as the *impulse response*, is generated at the output. The first direct outcome of linearity and spatial-invariance is that the filter can be uniquely characterized by its impulse response. The significance of the impulse and impulse response function becomes apparent when we realize that any input signal can be represented in the limit by an infinite sum of shifted and scaled impulses. This is an outcome of the *sifting integral*

$$f(x) = \int_{-\infty}^{\infty} f(\lambda)\,\delta(x - \lambda)\,d\lambda \tag{2.1.10}$$

which uses the actual signal $f(x)$ to scale the collection of impulses. Accordingly, the output of a linear and space-invariant filter will be a superposition of shifted and scaled impulse responses.

For an imaging system, the impulse response is the image in the output plane due to an ideal point source in the input plane. In this case, the impulse may be taken to be an infinitesimally small white dot upon a black background. Due to the limited accuracy of the imaging system, that dot will be resolved into a broader region. This impulse response is usually referred to as the *point spread function* (PSF) of the imaging system. Since the inputs and outputs represent a positive quantity (e.g., light intensity), the PSF is restricted to be positive. The term impulse response, on the other hand, is more general and is allowed to take on negative and complex values.

As its name suggests, the PSF is taken to be a bandlimiting filter having blurring characteristics. It reflects the physical limitations of a lens to accurately resolve each input point without the influence of neighboring points. Consequently, the PSF is typically modeled as a low-pass filter given by a bell-shaped weighting function over a finite aperture area. A PSF profile is depicted in Fig. 2.2.

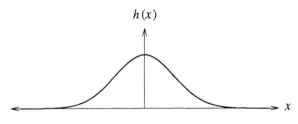

Figure 2.2: PSF profile.

2.1.4. Convolution

The response $g(x)$ of a digital filter to an arbitrary input signal $f(x)$ is expressed in terms of the impulse response $h(x)$ of the filter by means of the *convolution integral*

$$g(x) = f(x) * h(x) = \int_{-\infty}^{\infty} f(\lambda) h(x-\lambda) \, d\lambda \qquad (2.1.11)$$

where * denotes the convolution operation, $h(x)$ is used as the *convolution kernel*, and λ is the dummy variable of integration. The integration is always performed with respect to a dummy variable (such as λ) and x is a constant insofar as the integration is concerned. Kernel $h(x)$, also known as the *filter kernel*, is treated as a sliding window that is shifted across the entire input signal. As it makes its way across $f(x)$, a sum of the pointwise products between the two functions is taken and assigned to output $g(x)$. This process, known as *convolution*, is of fundamental importance to linear filtering theory.

The convolution integral given in Eq. (2.1.11) is defined for continuous functions $f(x)$ and $h(x)$. In our application, however, the input and convolution kernel are discrete. This warrants a discrete convolution, defined as the following summation

$$g(x) = f(x) * h(x) = \sum_{-\infty}^{\infty} f(\lambda) h(x-\lambda) \, d\lambda \qquad (2.1.12)$$

where x may continue to be a continuous variable, but λ now takes on only integer values. In practice, we use the discrete convolution in Eq. (2.1.12) to compute the output for our discrete input $f(x)$ and impulse response $h(x)$ at only a limited set of values for x.

If the impulse response is itself an impulse, then the filter is ideal and the input will be untampered at the output. That is, the convolution integral in Eq. (2.1.11) reduces to the sifting integral in Eq. (2.1.10) with $h(x)$ being replaced by $\delta(x)$. In general, though, the impulse response extends over neighboring samples; thus several scaled values may overlap. When these are added together, the series of sums forms the new filtered signal values. Thus, the output of any linear, space-invariant filter is related to its input by convolution.

Convolution can best be understood graphically. For instance, consider the samples shown in Fig. 2.3a. Each sample is treated as an impulse by the filter. Since the filter is linear and space-invariant, the input samples are replaced with properly scaled impulse response functions. In Fig. 2.3b, a triangular impulse response is used to generate the output signal. Note that the impulse responses are depicted as thin lines, and the output (summation of scaled and superpositioned triangles) is drawn in boldface. The reader will notice that this choice for the impulse response is tantamount to linear interpolation. Although the impulse response function can take on many different forms, we shall generally be interested in symmetric kernels of finite extent. Various kernels useful for image reconstruction are discussed in Chapter 5.

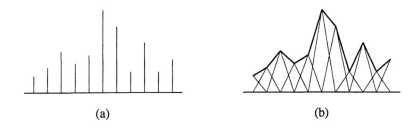

$$(a) \qquad\qquad\qquad\qquad (b)$$

Figure 2.3: Convolution with a triangle filter. (a) Input; (b) Output.

It is apparent from this example that convolution is useful to derive continuous functions from a set of discrete samples. This process, known as *reconstruction*, is fundamental to image warping because it is often necessary to determine image values at noninteger positions, i.e., locations for which no input was supplied. As an example, consider the problem of magnification. Given a unit triangle function for the impulse response, the output $g(x)$ for the input $f(x)$ is derived below in Table 2.1. The table uses a scale factor of four, thereby accounting for the .25 increments used to index the input. Note that $f(x)$ is only supplied for integer values of x, and the interpolation makes use of the two adjacent input values. The weights applied to the input are derived from the value of the unit triangle as it crosses the input while it is centered on the output position x.

x	$f(x)$	$g(4x)$
0.00	150	150
0.25		$(150)(.75) + (78)(.25) = 132$
0.50		$(150)(.50) + (78)(.50) = 114$
0.75		$(150)(.25) + (78)(.75) = 96$
1.00	78	78
1.25		$(78)(.75) + (90)(.25) = 81$
1.50		$(78)(.50) + (90)(.50) = 84$
1.75		$(78)(.25) + (90)(.75) = 87$
2.00	90	90

Table 2.1: Four-fold magnification with a triangle function.

In general, we can always interpolate the input data as long as the centered convolution kernel passes through zero at all the input sample positions but one. Thus, when the kernel is situated on an input sample it will use that data alone to determine the output value for that point. The unit triangle impulse response function complies with this interpolation condition: it has unity value at the center from which it linearly falls to zero over a single pixel interval.

The Gaussian function shown in Fig. 2.4a does not satisfy this interpolation condition. Consequently, convolving with this kernel yields an *approximating* function that passes near, but not necessarily through, the input data. The extent to which the impulse response function blurs the input data is determined by its region of support. Wider kernels can potentially cause more blurring. In order to normalize the convolution, the scale factor reflecting the kernel's region of support is incorporated directly into the kernel. Therefore, broader kernels are also shorter, i.e., scaled down in amplitude.

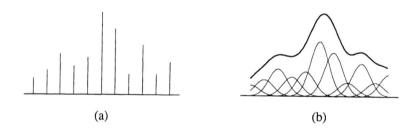

(a)   (b)

Figure 2.4: Convolution with a Gaussian filter. (a) Input; (b) Output.

2.1.5. Frequency Analysis

Convolution is a process which is difficult to visualize. Although a graphical construction is helpful in determining the output, it does not support the mathematical rigor that is necessary to design and evaluate filter kernels. Moreover, the convolution integral is not a formulation that readily lends itself to analysis and efficient computation. These problems are, in large part, attributed to the domain in which we are operating.

Thus far, our entire development has taken place in the *spatial domain*, where we have represented signals as plots of amplitude versus spatial position. These signals can just as well be represented in the *frequency domain*, where they are decomposed into a sum of sinusoids of different frequencies, with each frequency having a particular amplitude and phase shift. While this representation may seem alien for images, it is intuitive for audio applications. Therefore, we shall first develop the rationale for the frequency domain in terms of audio signals. Extensions to visual images will then follow naturally.

2.1.5.1. An Analogy To Audio Signals

Most modern stereo systems are equipped with graphic equalizers that permit the listener to tailor the frequency content of the sound. An equalizer is a set of filters that are each responsible for manipulating a narrow frequency band of the input frequency spectrum. In this instance, manipulation takes the form of attenuation, emphasis, or merely allowing the input to pass through untampered. This has direct impact on the richness of the sound. For instance, the low frequencies can be enhanced to compensate for inadequate bass in the music. We may simultaneously attenuate the high frequencies to eliminate undesirable noise, due perhaps to the record or tape. We may, alternatively, wish to emphasize the upper frequencies to enhance the instruments or vocals in that range.

The point to bear in mind is that sound is a sum of complex waveforms that each emanate from some contributing instrument. These waveforms sum together in a linear manner, satisfying the superposition principle. Each waveform is itself composed of a wide range of sinusoids, including the fundamental frequency and overtones at the harmonic frequencies [Pohlmann 89]. Graphic equalizers therefore provide an intuitive interface in which to specify the manipulation of the audio signal.

An alternate design might be one that requests the user for the appropriate convolution kernels necessary to achieve the same results. It is clear that this approach would overwhelm most users. The primary difficulty lies in the unintuitive connection between the shape of the kernel and its precise filtering effects on the audio signal. Moreover, considering audio signals in the frequency domain is more consistent with the signal formation process.

Having established that audio signals are readily interpreted in the frequency domain, a similar claim can be made for visual signals. A direct analogy holds between the frequency content in music and images. In music, the transition from low- to high-frequencies corresponds to the spectrum between baritones and sopranos, respectively.

In visual data, that same transition corresponds to the spectrum between blurred imagery and images rich in visual detail. Note that high frequencies refer to wild intensity excursions. This tends to correspond to visual detail like edges and texture in high contrast images. High frequencies that are subjectively determined to add nothing to the information content of the signal are usually referred to as noise. Since blurred images have slowly varying intensity functions, they lack significant high frequency information. In either case, music and images are time- and spatially-varying functions whose information content is embedded in their frequency spectrum. The conversion between the spatial and frequency domains is achieved by means of the Fourier transform.

We are familiar with other instances in which mathematical transforms are used to simplify a solution to a problem. The logarithm is one instance of such a transform. It simplifies problems requiring products and quotients by substituting addition for multiplication and subtraction for division. The only tradeoff is the accuracy and time necessary to convert the operands into logarithms and then back again. Similar benefits and drawbacks apply to the Fourier transform, a method introduced by the French physicist Joseph Fourier nearly two centuries ago. He derived the method to transform signals between the spatial (or time) domain and the frequency domain. As we shall see later, using two representations for a signal is useful because some operations that are difficult to execute in one domain are relatively easy to do in the other domain. In this manner, the benefits of both representations are exploited.

2.1.5.2. Fourier Transforms

Fourier transforms are central to the study of signal processing. They offer a powerful set of analytical tools to analyze and process single and multidimensional signals and systems. The great impact that Fourier transforms has had on signal processing is due, in large part, to the fundamental understanding gained by examining a signal from an entirely different viewpoint.

We had earlier considered an arbitrary input function $f(x)$ to be the sum of an infinite number of impulses, each scaled and shifted in space. In a great leap of imagination, Fourier discovered that an alternate summation is possible: $f(x)$ can be taken to be the sum of an infinite number of sinusoidal waves. This new viewpoint is justifiable because the response of a linear, space-invariant system to a complex exponential (sinusoid) is another complex exponential of the same frequency but altered amplitude and phase. Determining the amplitudes and phase shifts for the sinusoids is the central topic of *Fourier analysis*. Conversely, the act of adding these scaled and shifted sinusoids together is known as *Fourier synthesis*. Fourier analysis and synthesis are each made possible by the Fourier transform pair:

$$F(u) = \int_{-\infty}^{\infty} f(x) e^{-i 2\pi u x} dx \qquad (2.1.13)$$

$$f(x) = \int_{-\infty}^{\infty} F(u) e^{+i 2\pi u x} du \qquad (2.1.14)$$

where $i = \sqrt{-1}$, and

$$e^{\pm i 2\pi u x} = \cos 2\pi u x \pm i \sin 2\pi u x \qquad (2.1.15)$$

is a succinct expression for a complex exponential at frequency u.

The definition of the *Fourier transform*, given in Eq. (2.1.13), is valid for any integrable function $f(x)$. It decomposes $f(x)$ into a sum of complex exponentials. The complex function $F(u)$ specifies, for each frequency u, the amplitude and phase of each complex exponential. $F(u)$ is commonly known as the signal's *frequency spectrum*. This should not be confused with the Fourier transform of a filter, which is called the *frequency response* (for 1-D filters) or the *modulation transfer function* (for 2-D filters). The frequency response of a filter is computed as the Fourier transform of its impulse response.

It is important to realize that $f(x)$ and $F(u)$ are two different representations of the same function. In particular, $f(x)$ is the signal in the spatial domain and $F(u)$ is its counterpart in the frequency domain. One goes back and forth between these two representations by means of the Fourier transform pair. The transformation from the frequency domain back to the spatial domain is given by the *inverse Fourier transform*, defined in Eq. (2.1.14).

Although $f(x)$ may be any complex signal, we are generally interested in real functions, i.e., standard color images. The Fourier transform of a real function is usually complex. This is actually a clever encoding of the orthogonal basis set, which consists of sine and cosine functions. Together, they specify the amplitude and phase of each frequency component, i.e., a sine wave. Thus, we have $F(u)$ defined as a complex function of the form $R(u) + iI(u)$, where $R(u)$ and $I(u)$ are the real and imaginary components, respectively. The *amplitude*, or *magnitude*, of $F(u)$ is defined as

$$|F(u)| = \sqrt{R^2(u) + I^2(u)} \qquad (2.1.16)$$

It is often referred to as the *Fourier spectrum*. This should not be mistaken with the Fourier transform $F(u)$, which is itself commonly known as the spectrum. In order to avoid confusion, we shall refer to $|F(u)|$ as the magnitude spectrum. The *phase spectrum* is given as

$$\phi(u) = \tan^{-1}\left[\frac{I(u)}{R(u)}\right] \qquad (2.1.17)$$

This specifies the *phase shift*, or *phase angle*, for the complex exponential at each frequency u.

The Fourier transform of a signal is often plotted as magnitude versus frequency, ignoring phase angle. This form of display has become conventional because the bulk of the information relayed about a signal is embedded in its frequency content, as given by the magnitude spectrum $|F(u)|$. For example, Figs. 2.5a and 2.5b show a square wave and its spectrum, respectively. In this example, it just so happens that the phase function $\phi(u)$ is zero for all u, with the spectrum being defined in terms of the following infinite series:

$$F(u) = \cos u - \frac{1}{3}\cos 3u + \frac{1}{5}\cos 5u - \frac{1}{7}\cos 7u + \cdots \qquad (2.1.18)$$

$$= \sin u + \frac{1}{3}\sin 3u + \frac{1}{5}\sin 5u + \frac{1}{7}\sin 7u + \cdots$$

Consequently, the spectrum is real and we display its values directly. Note that both positive and negative frequencies are displayed, and the amplitudes have been halved accordingly. An application of Fourier synthesis is shown in Fig. 2.5c, where the first five nonzero components of $F(u)$ are added together. With each additional component, the reconstructed function increasingly takes on the appearance of the original square wave. The ripples that persist are a consequence of the oscillatory behavior of the sinusoidal basis functions. They remain in the reconstruction unless all frequency components are considered in the reconstruction. This artifact is known as *Gibbs phenomenon* which predicts an overshoot/undershoot of about 22% near edges that are not fully reconstructed [Antoniou 79].

A second example is given in Fig. 2.6. There, an arbitrary waveform undergoes Fourier analysis and synthesis. In this case, $F(u)$ is complex and so only the magnitude spectrum $|F(u)|$ is shown in Fig. 2.6b. Since the spectrum is defined over infinite frequencies, only a small segment of it is shown in the figure. The results of Fourier synthesis with the first ten frequency components are shown in Fig. 2.6c. As before, incorporating the higher frequency components adds finer detail to the reconstructed function.

The two examples given above highlight an important property of Fourier transforms that relate to periodic and aperiodic functions. Periodic signals, such as the square wave shown in Fig. 2.5a, can be represented as the sum of phase-shifted sine waves whose frequencies are integral multiples of the signal's lowest nonzero frequency component. In other words, a periodic signal contains all the frequencies that are *harmonics* of the *fundamental frequency*. We normally associate the analysis of periodic signals with *Fourier series* rather than Fourier transforms. The Fourier series can be expressed as the following summation

$$f(x) = \sum_{n=-\infty}^{\infty} c(nu_0) e^{i2\pi nu_0 x} \qquad (2.1.19)$$

where $c(nu_0)$ is the nth *Fourier coefficient*

$$c(nu_0) = \int_{-x_0/2}^{x_0/2} f(x) e^{-i2\pi nu_0 x} dx \qquad (2.1.20)$$

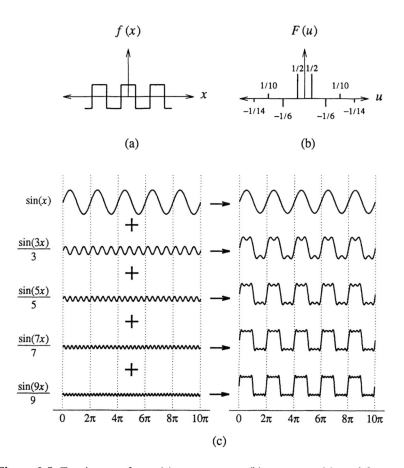

Figure 2.5: Fourier transform. (a) square wave; (b) spectrum; (c) partial sums.

and u_0 is the fundamental frequency. Note that since $f(x)$ is periodic, the integral in Eq. (2.1.20) that is used to compute the Fourier coefficients must only integrate over period x_0.

Aperiodic signals do not enjoy the same compact representation as their periodic counterparts. Whereas a periodic signal is expressed in terms of a sum of frequency components that are integer multiples of some fundamental frequency, an aperiodic signal must necessarily be represented as an integral over a continuum of frequencies, as in Eq. (2.1.13). This is reflected in the spectra of Figs. 2.5b and Fig. 2.6b. Notice that the square wave spectrum consists of a discrete set of impulses in the frequency domain, while the spectrum of the aperiodic signal in Fig. 2.6 is defined over all frequencies. For this reason, we distinguish Eq. (2.1.13) as the *Fourier integral*. It can be shown that the Fourier series is a special case of the Fourier integral. In summary, periodic signals have discrete Fourier components and are described by a Fourier series. Aperiodic signals

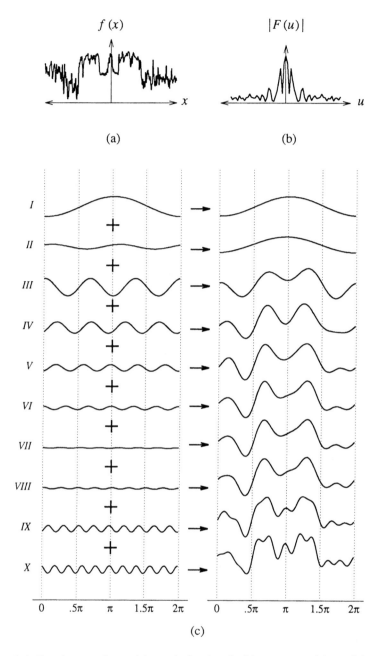

Figure 2.6: Fourier transform. (a) aperiodic signal; (b) spectrum; (c) partial sums.

have continuously varying Fourier components and are described by a Fourier integral.

There are several other symmetries that apply between the spatial and frequency domains. Table 2.2 lists just a few of them. They refer to functions being real, imaginary, even, and odd. A function is real if the imaginary component is set to zero. Similarly, a function is imaginary if its real component is zero. A function $f(x)$ is even if $f(-x)=f(x)$. If $f(-x)=-f(x)$, then $f(x)$ is said to be an odd function. Finally, $F^*(u)$ refers to the complex conjugate of $F(u)$. That is, if $F(u)=R(u)+iI(u)$, then $F^*(u)=R(u)-iI(u)$.

Spatial Domain, $f(x)$	Frequency Domain, $F(u)$
Real	$F(-u)=F^*(u)$
Imaginary	$F(-u)=-F^*(u)$
Even	Even
Odd	Odd
Real and Even	Real and Even
Real and Odd	Imaginary and Odd
Imaginary and Even	Imaginary and Even
Imaginary and Odd	Real and Odd
Periodic	Disrete
Periodic Sampling	Periodic Copies

Table 2.2: Symmetries between the spatial and frequency domains.

The last two symmetries listed above are particularly notable. By means of the Fourier series, we have already seen periodic signals produce discrete Fourier spectra (Fig. 2.5). We shall see later that periodic spectra correspond to sampled signals in the spatial domain. The significance of this symmetry will become apparent when we discuss discrete Fourier transforms and sampling theory.

In addition to the symmetries given above, there are many properties that apply to the Fourier transform. Some of them are listed in Table 2.3. Among the most important of these properties is linearity because it reflects the applicability of the Fourier transform to linear system analysis. Spatial scaling is of significance, particularly in the context of simple warps that consist only of scale changes. This property establishes a reciprocal relationship between the spatial domain and the frequency domain. Therefore, expansion (compression) in the spatial domain corresponds to compression (expansion) in the frequency domain. Furthermore, the frequency scale not only contracts (expands), but the amplitude increases (decreases) vertically in such a way as to keep the area constant. Finally, the property that establishes a correspondence between convolution in one domain and multiplication in the other is of great practical significance. The proofs of these properties are left as an exercise for the reader. A detailed exposition can be found in [Bracewell 86], [Brigham 88], and most signal processing textbooks.

Property	Spatial Domain, $f(x)$	Frequency Domain, $F(u)$
Linearity	$\alpha_1 f_1(x) + \alpha_2 f_2(x)$	$\alpha_1 F_1(u) + \alpha_2 F_2(u)$
Spatial Scaling	$f(ax)$	$\dfrac{1}{a} F(\dfrac{u}{a})$
Frequency Scaling	$\dfrac{1}{a} f(\dfrac{x}{a})$	$F(au)$
Spatial Shifting	$f(x-a)$	$F(u) e^{-i 2\pi u a}$
Frequency Shifting (Modulation)	$f(x) e^{i 2\pi a x}$	$F(u-a)$
Convolution	$g(x) = f(x) * h(x)$	$G(u) = F(u) H(u)$
Multiplication	$g(x) = f(x) h(x)$	$G(u) = F(u) * H(u)$

Table 2.3: Fourier transform properties.

The Fourier transform can be easily extended to multidimensional signals and systems. For 2-D images $f(x,y)$ that are integrable, the following Fourier transform pair exists:

$$F(u,v) = \int \int f(x,y) e^{-i 2\pi(ux + vy)} \, dx \, dy \qquad (2.1.21)$$

$$f(x,y) = \int \int F(u,v) e^{+i 2\pi(ux + vy)} \, du \, dv \qquad (2.1.22)$$

where u and v are frequency variables. Extensions to higher dimensions are possible by simply adding exponent terms to the complex exponential, and integrating over the additional space and frequency variables.

2.1.5.3. Discrete Fourier Transforms

The discussion thus far has focused on continuous signals. In practice, we deal with discrete images that are both limited in extent and sampled at discrete points. The results developed so far must be modified to be useful in this domain. We thus come to define the discrete Fourier transform pair:

$$F(u) = \frac{1}{N} \sum_{x=0}^{N-1} f(x) e^{-i 2\pi ux/N} \qquad (2.1.23)$$

$$f(x) = \sum_{u=0}^{N-1} F(u) e^{i 2\pi ux/N} \qquad (2.1.24)$$

for $0 \leq u,x \leq N-1$, where N is the number of input samples. The $1/N$ factor that appears in front of the forward transform serves to normalize the spectrum with respect to the length of the input. There is no strict rule which requires the normalization to be applied to $F(u)$. In some sources, the $1/N$ factor appears in front of the inverse transform instead. For reasons of symmetry, other common formulations have the forward and inverse transforms each scaled by $1/\sqrt{N}$. As long as a cumulative $1/N$ factor is applied somewhere along the transform pair, the final results will be properly normalized.

The *discrete Fourier transform* (DFT), defined in Eq. (2.1.23), assumes that $f(x)$ is an input array consisting of N regularly spaced samples. It maps these N complex numbers into $F(u)$, another set of N complex numbers. Since the frequency domain is now discrete, the DFT must treat the input as a periodic signal (from Table 2.2). As a result, we have let the limits of summation change from $(-N/2, N/2)$ to $(0, N-1)$, bearing in mind that the negative frequencies now occupy positions $N/2$ to $N-1$. Although negative frequencies have no physical meaning, they are a byproduct of the mathematics in this process. It is noteworthy to observe, though, that the largest reproducible frequency for an N-sample input is $N/2$. This corresponds to a sequence of samples that alternate between black and white, in which the smallest period for each cycle is two pixels.

The data in $f(x)$ is treated as one period of a periodic signal by replicating itself indefinitely, thereby tiling the input plane with copies of itself. This makes the opposite ends of the signal adjacent by virtue of wraparound from $f(N-1)$ to $f(0)$. Note that this is only a model of the infinite input, given only the small (aperiodic) segment $f(x)$, i.e., no physical replication is necessary. While this permits $F(u)$ to be defined for discrete values of u, it does introduce artifacts. First, the transition across the wraparound border may be discontinuous in value or derivative(s). This has consequences in the high frequency components of $F(u)$. One solution to this problem is windowing, in which the actual signal is multiplied by another function which smoothly tapers off at the borders (see Chapter 5). Another consideration is the value of N. A small N produces a coarse approximation to the continuous Fourier transform. However, by choosing a sufficiently high sampling rate, a good approximation to the continuous Fourier transform is obtained for most signals.

The 1-D discrete Fourier transform pairs given in Eqs. (2.1.23) and (2.1.24) can be extended to higher dimensions by means of the separability property. For an $N \times M$ images, we have the following DFT pair:

$$F(u,v) = \frac{1}{MN} \sum_{x=0}^{N-1} \sum_{y=0}^{M-1} f(x,y) e^{-i2\pi(ux/N + vy/M)} \qquad (2.1.25)$$

$$f(x,y) = \sum_{u=0}^{N-1} \sum_{v=0}^{M-1} F(u,v) e^{i2\pi(ux/N + vy/M)} \qquad (2.1.26)$$

The DFT pair given above can be expressed in the separable forms

$$F(u,v) = \frac{1}{N} \sum_{x=0}^{N-1} f(x,y) e^{-i2\pi ux/N} \frac{1}{M} \sum_{y=0}^{M-1} f(x,y) e^{-i2\pi uy/M} \qquad (2.1.27)$$

$$f(x,y) = \sum_{u=0}^{N-1} F(u,v) e^{i2\pi ux/N} \sum_{v=0}^{M-1} F(u,v) e^{i2\pi vx/M} \qquad (2.1.28)$$

for $u,x = 0, 1, ..., N-1$, and $v,y = 0, 1, ..., M-1$.

The principal advantage of this reformulation is that $F(u,v)$ and $f(x,y)$ can each be obtained by successive applications of the 1-D DFT or its inverse. By regrouping the operations above, it becomes possible to compute the transforms in the following manner. First, transform each row independently, placing the results in intermediate image I. Then, transform each column of I independently. This yields the correct results for either the forward or inverse transforms. In Chapter 7, we will show how separable algorithms of this kind have been used to greatly reduce the computational cost of digital image warping.

Although the DFT is an important tool that is amenable for computer use, it does so at a high price. For an N-sample input, the computational cost of the DFT is $O(N^2)$. This accounts for N summations, each requiring N multiplication operations. Even with high-speed computers, the cost of such a transform can be overwhelming for large N. Consequently, the DFT is often generated with the *fast Fourier transform* (FFT), a computational algorithm that reduces the computing time to $O(N \log_2 N)$. The FFT achieves large speedups by exploiting the use of partial results that combine to produce the correct output. This increase in computing speed has completely revolutionized many facets of scientific analysis. A detailed description of the FFT algorithm, and its variants, are given in Appendix 1. In addition to this review, interested readers may also consult [Brigham 88] and [Ramirez 85] for further details.

The development of the FFT algorithm has given impetus to filtering in the frequency domain. There are several advantages to this approach. The foremost benefit is that convolution in the spatial domain corresponds to multiplication in the frequency domain. As a result, when a convolution kernel is sufficiently large, it becomes more cost-effective to transform the image and the kernel into the frequency domain, multiply them, and then transform the product back into the spatial domain. A second benefit is that important questions relating to sampling, interpolation, and aliasing can be answered rigorously. These topics are addressed in subsequent chapters.

2.2. IMAGE ACQUISITION

Before a digital computer can begin to process an image, that image must first be available in digital form. This is made possible by a *digital image acquisition system*, a device that scans the scene and generates an array of numbers representing the light intensities at a discrete set of points. Also known as a *digitizer*, this device serves as the front-end to any image processing system, as depicted in Fig. 2.7.

Digital image acquistion systems consist of three basic components: an imaging sensor to measure light, scanning hardware to collect measurements across the entire scene, and an analog-to-digital converter to discretize the continuous values into finite-precision numbers suitable for computer processing. The remainder of this chapter is devoted to describing these components. However, since a full description that does justice to this topic falls outside the scope of this book, our discussion will be brief and incomplete. Readers can find this material in most image processing textbooks. Useful reviews can also be found in [Nagy 83] and [Schreiber 86].

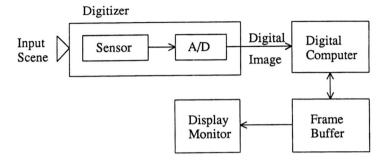

Figure 2.7: Elements of an image processing system.

Consider the image acquisition system shown in Fig. 2.8. The entire imaging process can be viewed as a cascade of filters applied to the input image. The scene radiance $f(x,y)$ is a continuous two-dimensional image. It passes through an imaging subsystem, which acts as the first stage of data acquisition. Section 2.3 describes the operation of several common imaging systems. Due to the point spread function of the image sensor, $h(x,y)$, the output $g(x,y)$ is a degraded version of $f(x,y)$.

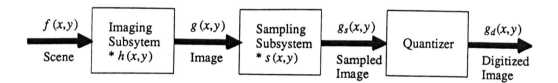

Figure 2.8: Image acquisition system.

By definition,

$$g(x,y) = f(x,y) * h(x,y) \qquad (2.2.1)$$

where * denotes convolution. If the PSF profile is identical in all orientations, the PSF is said to be *rotationally-symmetric*. Furthermore, if the PSF retains the same shape throughout the image, it is said to be *spatially-invariant*. Also, if the two-dimensional PSF can be decomposed into two one-dimensional filters, e.g., $h(x,y) = h_x(x,y) h_y(x,y)$, it is said to be *separable*. In practice, though, point spread functions are usually not rotationally-symmetric, spatially-invariant, or separable. As a result, most imaging devices induce geometric distortion in addition to blurring.

The continuous image $g(x,y)$ then enters a sampling subsystem, generating the discrete-continuous image $g_s(x,y)$. The sampled image $g_s(x,y)$ is given by

$$g_s(x,y) = g(x,y)s(x,y) \tag{2.2.2}$$

where

$$s(x,y) = \sum_{m=-\infty}^{\infty} \sum_{n=-\infty}^{\infty} \delta(x-m, y-n) \tag{2.2.3}$$

is the two-dimensional *comb function*, depicted in Fig. 2.9, and $\delta(x,y)$ is the impulse function. The comb function comprises our *sampling grid* which is conveniently nonzero only at integral (x,y) coordinates. Therefore, $g_s(x,y)$ is now a discrete-continuous image with intensity values defined only over integral indices of x and y.

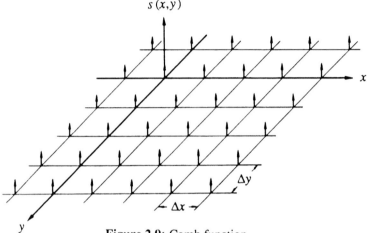

Figure 2.9: Comb function.

Even after sampling, the intensity values continue to retain infinite precision. Since computers have finite memory, each sampled point must be quantized. Quantization is a point process that satisfies a nonlinear function of the form shown in Fig. 2.10. It reflects the fact that accuracy is limited by the system's resolution.

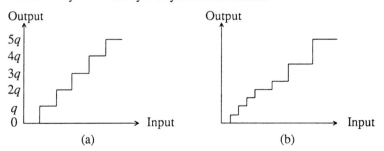

Figure 2.10: Quantization function. (a) Uniform; (b) Nonuniform.

The horizontal plateaus in Fig. 2.10a are due to the fact that the continuous input is truncated to a fixed number of bits, e.g., N bits. Consequently, all input ranges that share the first N bits become indistinguishable and are assigned the same output value. This form of quantization is known as *uniform quantization*. The difference q between successive output values is inversely proportional to N. That is, as the precision rises, the increments between successive numbers grows smaller. In practice, quantization is intimately coupled with the precision of the image pickup device in the imaging system.

Quantization is not restricted to be uniform. Figure 2.10b depicts *nonuniform quantization* for functions that do not require equispaced plateau intervals. This permits us to incorporate properties of the imaged scene and the imaging sensor when assigning discrete values to the input. For instance, it is generally known that the human visual system has greater acuity for low intensities. In that case, it is reasonable to assign more quantization levels in the low intensity range at the expense of accuracy in the high intensity range where the visual system is less sensitive anyway. Such a *nonuniform quantization* scheme is depicted in Fig. 2.10b. Notice that the nonuniformity appears in both the increments between successive levels, as well as the extent of these intervals. This is equivalent to performing a nonlinear point transformation prior to performing uniform quantization.

Returning to Fig. 2.8, we see that $g_s(x,y)$ passes through a quantizer to yield the discrete-discrete (digital) image $g_d(x,y)$. The actual quantization is achieved through the use of an *analog-to-digital* converter. Together, sampling and quantization comprise the process known as *digitization*. Note that sampling actually refers to spatial quantization (e.g., only a discrete set of spatial positions are defined) while the term quantization is typically left to refer to the discretization of image values.

A digital image is an approximation to a continuous image $f(x,y)$. It is usually stored in a computer as an $N \times M$ array of equally spaced discrete samples:

$$f(x,y) \approx \begin{bmatrix} f(0,0) & f(0,1) & .... & f(0,M-1) \\ f(1,0) & f(1,1) & .... & f(1,M-1) \\ & \cdot \cdot \cdot & & \\ & \cdot \cdot \cdot & & \\ f(N-1,0) & f(N-1,1) & .... & f(N-1,M-1) \end{bmatrix} \tag{2.2.4}$$

Each sample is referred to as an *image element, picture element, pixel*, or *pel*, with the last two names being commonly used abbreviations of "picture elements." Collectively, they comprise the 2-D array of pixels that serve as input to subsequent computer processing. Each pixel can be thought of as a finite-sized rectangular region on the screen, much like a tile in a mosaic. Many applications typically select $N = M = 512$ with 8-bits per pixel (per channel). In digital image processing, it is common practice to let the number of samples and quantization levels be integer powers of two. These standards are derived from hardware and software considerations. For example, even if only 6-bit pixels are required, an entire 8-bit byte is devoted to it because packing 6-bit quantities in multiples of 8-bit memory locations is impractical.

Digital images are the product of both spatial sampling and intensity quantization. As stated earlier, sampling can actually be considered to be a form of spatial quantization, although it is normally treated as the product of the continuous input image with a sampling grid. Intensity quantization is the result of discretizing pixel values to a finite number of bits. Note that these two forms of quantization apply to the image indices and values, respectively. A tradeoff exists between sampling rate and quantization levels. An interesting review of work in this area, as well as related work in image coding, is described in [Netravali 80, 88]. Finally, a recent analysis on the tradeoff between sampling and quantization can be found in [Lee 87].

2.3. IMAGING SYSTEMS

A continuous image is generally presented to a digitization system in the form of analog voltage or current. This is usually the output of a transducer that transforms light into an electrical signal that represents brightness. This electrical signal is then *digitized* by an analog-to-digital (A/D) converter to produce a discrete representation that is suitable for computer processing. In this section, we shall examine several imaging systems that produce an analog signal from scene radiance.

There are three broad categories of imaging systems: electronic, solid-state, and mechanical. They comprise some of the most commonly used input devices, including vidicon cameras, CCD cameras, film scanners, flat-bed scanners, microdensitometers, and image dissectors. The imaging sensors in these devices are essentially transducers that convert optical signals into electrical voltages.

The primary distinction between these systems is the imaging and scanning mechanisms. Electronic scanners use an electron beam to measure light falling on a photosensitive surface. Solid-state imaging systems use arrays of photosensitive cells to sense incident light. In these two classes, the scanned material and sensors are stationary. Mechanical scanners are characterized by a moving assembly that transports the scanned material and sensors past one another. Note that either electronic or solid-state sensors can be used here. We now describe each of these three categories of digital image acquisition systems in more detail.

2.3.1. Electronic Scanners

The name *flying spot scanner* is given to a class of electronic scanners that operate on the principle of focusing an electron beam on a photodetector. The photodetector is a surface coated with photosensitive material that responds to incident light projected from an image. In this assembly, the image and photodetector remain stationary. Scanning is accomplished with a "flying spot," which is a moving point of light on the face of a cathode-ray tube (CRT), or a laser beam directed by mirrors. The motion of the point is controlled electronically, usually through deflections induced by electromagnets or electrostatics. This permits high scanning speeds and flexible control of the scanning pattern.

2.3.1.1. Vidicon Systems

One of the most frequently utilized imaging devices that fall into this class are vidicon systems, shown in Fig. 2.11. These devices have traditionally been used in TV cameras to generate analog video signals. The main component is a glass vidicon tube containing a scanning electron beam mechanism at one end and a photosensitive surface at the other. An image is focused on the front (outer) side of the photosensitive surface, producing a charge depletion on the back (inner) side that is proportional to the incident light. This yields a charge distribution with a high density of electrons in the dark image regions and a low electron density in the lighter regions. This is an electrical analog to the photographic process that produces a negative image.

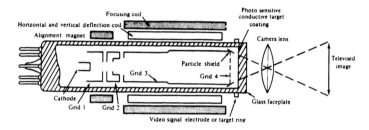

Figure 2.11: Vidicon tube [Ballard 82].

The charge distribution is "read" through the use of a scanning electron beam. The beam, emanating from the cathode at the rear of the tube, is made to scan the charge distribution in raster order, i.e., row by row. Upon contact with the photosensitive surface, it replaces the electron charge in the regions where the charge was depleted by exposure to the light. This charge neutralization process generates fluctuations in the electron beam current, generating the analog video signal. In this manner, the intensity values across an image are encoded as analog currents or voltages with fluctuations that are proportional to the incident light. Once a physical image has been converted to an analog signal, it is sampled and digitized to produce a 2-D array of integers that becomes available for computer processing.

The spatial resolution of the acquired image is determined by the spatial scanning frequency and the sampling rate: higher rates produce more samples. Sampling rates also have an impact on the choice of photosensitive material used. Slower scan rates require photosensitive material that decays slowly. This can introduce several artifacts. First, high retention capabilities may cause incomplete readout of the charge distribution due to the sluggish response. Second, slowly decaying charge gives rise to temporal blurring in time-varying images whereby charge distributions of several images may get merged together. This problem can be alleviated by saturating the surface with electrical charge between exposures in order to reduce any residual images.

Vidicon systems often suffer from geometric distortions. This is caused by several factors. First, the scanning electron beam often does not precisely retain positional linearity across the full face of the surface. Second, the electron beam can be deflected off course by high contrast charge (image) boundaries. This is particularly troublesome because it is an image-dependent artifact. Third, the photosensitive material may be defective with uneven charge retention due to nonuniform coatings. Several related systems offer more stable performance, including those using image orthicon, plumbicon, and saticon tubes. Orthicon tubes have the additional advantage of accommodating flexible scan patterns.

2.3.1.2. Image Dissectors

Video signals can also be generated by using image dissectors. As with vidicon cameras, an image is focused directly onto a cathode coated with a photosensitive layer. This time, however, the cathode *emits* electrons in proportion to the incident light. This produces an electron beam whose cross section is roughly the same as the geometry of the tube surface. The beam is accelerated toward a target by the anode. The target is an electron multiplier covered by a small aperture, or pinhole, which allows only a small part of the electron beam emitted by the cathode to reach the target. Focusing coils focus the beam, and deflection coils then scan it past the target aperture, where the electron multiplier produces a varying voltage representing the video signal. The name "dissector" is derived from the manner in which the image is scanned past the target. Figure 2.12 shows a schematic diagram.

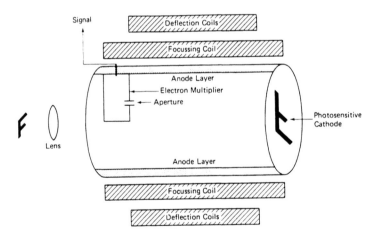

Figure 2.12: Image dissector [Ballard 82].

Image dissectors differ from vidicon systems in that dissectors are based on the principle of photoemission, whereas vidicon tubes are based on the principle of photoconductivity. This manifests itself in the manner in which these devices sense the image. In vidicon tubes, a narrow beam emanates from the cathode and is deflected across the photosensitive surface to sense each point. In image dissectors, a wide electron beam is produced by the photosensitive cathode, and each point is sensed by deflecting the entire beam past a pinhole onto some pickup device. This method facilitates noise reduction by integrating the emission of each input point over a specified time interval. Although the slow response of photoemissive materials limits the speed of image dissectors, the integration capability makes image dissectors attractive in applications requiring high signal-to-noise ratios for stationary images.

2.3.2. Solid-State Sensors

The most recent developments in image acquisition have come from solid-state imaging sensors, known as *charge transfer devices* (CTD). There are two main classes of CTDs: *charge-coupled devices* (CCDs) and *charge-injection devices* (CIDs). They differ primarily in the way in which information is read out.

2.3.2.1. CCD Cameras

A CCD is a monolithic array of closely spaced MOS (metal-oxide semiconductor) capacitors on a small rectangular solid-state surface. Each capacitor is often referred to as a photosite, or potential well, storing charge in response to the incident light intensity. An image is acquired by exposing the array to the desired scene. The exposure creates a distribution of electric potential throughout all the capacitors. The sampled analog, or discrete-continuous, video signal is generated by reading each well sequentially. This signal is then digitized to produce a digital image.

The electric potential is read from the CCD in a process known as *bucket brigade* due to its resemblance to shift registers in computer logic circuits. The first potential well on each line is read out. Then, the electric potential along each line is shifted by one position. Note that connections between capacitors along a line permit charge to shift from element to element along a row. The read-shift cycle is then repeated until all the potential wells have been shifted out of the monolithic array. This process is depicted in Fig. 2.13.

CCD arrays are packaged as either line sensors or area sensors. Line sensors consist of a scanline of photosites and produce a 2-D image by relative motion with the scene. This is usually integrated as part of a mechanical scanner (more on this later) whereby some mechanical assembly moves the line sensor across the entire physical image. Area sensors are composed of a 2-D matrix of photosites.

CCDs have several advantages over vidicon systems. The chief benefits are derived from the extremely linear radiometric (intensity) response and increased sensitivity. Unlike vidicon systems that can yield no more than 8 bits of precision because of analog noise, a CCD can easily provide 12 bits of precision. Furthermore, the fixed position of each photosite yields high geometric precision. The devices are small, portable, reliable, cheap, operate at low voltage, consume little power, are not damaged by intense light, and can provide images of up to 2000×2000 samples. As a result, they have made their way into virtually all modern TV cameras and camcorders. CCD cameras also offer superior performance in low lighting and low temperature conditions. As a result, they

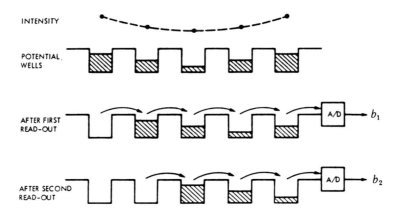

Figure 2.13: CCD readout mechanism [Green 89].

are even utilized in the NASA Space Telescope project and are found aboard the Galileo spacecraft that is due to orbit Jupiter in the early 1990s. Interested readers are referred to [Janesick 87] for a thorough treatment of CCD technology.

2.3.2.2. CID Cameras

Charge-injection devices resemble charge-coupled devices except that the readout, or sensing, process is different. Instead of behaving like a shift register during sensing, the charges are confined to the photosites where they were generated. They are read by using a row-column addressing technique similar to that used in conventional computer memories. Basically, the stored charge is "injected" into the substrate and the resulting displacement current is detected to create the video signal. CIDs are better than CCDs in the following respects: they offer wider spectral and dynamic range, increased tolerance to processing defects, simple mechanization, avoidance of charge transfer losses, and minimized blooming. They are, however, not superior to CCD cameras in low light or low temperature settings.

2.3.3. Mechanical Scanners

A *mechanical scanner* is an imaging device that operates by mechanically passing the photosensors and images past one another. This is in contrast to electronic and solid-state scanners in which the image and photodetector both remain stationary. However, it is important to note that either of these two classes of systems can be used in a mechanical scanner.

There are three primary types of mechanical scanners: *flat-bed*, *drum*, and *scanning cameras*. In flat-bed scanners, a film or photograph is laid on a flat surface over which the light source and the sensor are transported in a raster fashion. In a drum digitizer, the image is mounted on a rotating drum, while the light beam moves along the drum parallel to its axis of rotation. Finally, scanning cameras embed a scanning mechanism directly in the camera. In one manifestation, they use stationary line sensors with a mirror to deflect the light from successive image rows onto the sensor. In a second manifestation, the actual line sensor is physically moved *inside* the camera. These techniques basically address the manner in which the image is presented to the photosensors. The actual choice of sensors, however, can be taken from electronic scanners or solid-state imaging devices. Futhermore, the light sources can be generated by a CRT, laser beam, lamp, or light-emitting diodes (LEDs).

Microdensitometers are film scanners used for digitizing film transparencies or photographs at spot sizes ranging down to one micron. These devices are usually flat-bed scanners, requiring the scanned material to be mounted on a flat surface which is translated in relation to a light beam. The light beam passes through the transparency, or it is reflected from the surface of the photograph. In either case, a photodetector senses the transmitted light intensity. Since microdensitometers are mechanically controlled, they are slow image acquisition devices, but offer high geometric precision.

2.4. VIDEO DIGITIZERS

Many image acquisition systems generate television signals. These are analog video signals that are acquired in a fixed format, according to one of the three color television standards: National Television Systems Committee (NTSC), Sequential Couleur Avec Memoire (SECAM, or sequential chrominance signal with memory), and Phase Alternating Line (PAL). These systems establish format conventions and standards for broadcast video transmission in different parts of the world. NTSC is used in North America and Japan; SECAM is prevalent in France, Eastern Europe, the Soviet Union, and the Middle East; and PAL is used in most of Western Europe, including West Germany and the United Kingdom, as well as South America, Asia, and Africa.

The NTSC system requires the video signal to consist of a sequence of *frames*, with 525 lines per frame, and 30 frames per second. Each frame is a complete scan of the target. In order to reduce transmission bandwidth, a frame is composed of two interlaced fields, each consisting of 262.5 lines. The first field contains all the odd lines and the second field contains the even lines. To reduce flicker, alternate fields are sent at a rate of 60 fields per second.

The NTSC system further reduces transmission bandwidth by compressing chrominance information. Colors are represented in the YIQ color space, a linear transformation of RGB. The term *Y* refers to the monochrome intensity. This is the only signal that is used in black-and-white televisions. Color televisions have receivers that make use of *I* and *Q*, the in-phase and quadrature chrominance components, respectively. The conversion between the RGB and YIQ color spaces is given in [Foley 90]. Somewhat better quality is achieved with the SECAM and PAL systems. Although they also bandlimit chrominance, they both use 625 lines per frame, 25 frames per second, and 2:1 line interlacing.

In recent years, many devices have been designed to digitize video signals. The basic idea of video digitizers involves freezing a video frame and then digitizing it. Each NTSC frame contains 482 visible lines with 640 samples per line. This is in accord with the standard 4:3 aspect ratio of the screen. At 8 bits/pixel, this equates to roughly one quarter of a megabyte for a monochrome image. Color images require three times this amount. Even more memory is needed for high-definition television (HDTV) images. Although no HDTV standard has yet been formally established, HDTV color images with a resolution of, say, 1050×1024 requires approximately 3 Mbytes of data! Most general-purpose computers cannot handle the bandwidth necessary to transfer and process this much information, especially at a rate of 30 frames per second. As a result, some form of rate buffering is required.

Rate buffering is a process through which high rate data are stored in an intermediate storage device as they are acquired at a high rate and then read out from the intermediate storage at a lower rate. The intermediate memory is known as a *frame buffer* or *frame store*. Its single most distinguishing characteristic is that its contents can be written or read at TV rates. In addition, it is sometimes enhanced with many memory-addressing modes, including real-time zoom (pixel replication), scroll (vertical shifts), and pan (horizontal shifts). Such video digitizers operate at frame rates, and are also known as *frame grabbers*. Frame grabbers attached to CCD or vidicon cameras have become popular digital image acquisition systems due to their low price, general-purpose use, and accuracy.

2.5. DIGITIZED IMAGERY

Several images will be used repeatedly throughout this book to demonstrate algorithms. They are shown in Fig. 2.14 in order to avoid duplicating them in later examples. We shall refer to them as the Checkerboard, Madonna,[†] Mandrill, and Star images, respectively.

All four images are stored as arrays of 512×512 24-bit color pixels. They each have particular properties that make them interesting examples. The Checkerboard image is useful in that it has a regular grid structure that is readily perceived under any geometric transformation. In order to enhance this effect a green color ramp, rising from top to bottom, has been added to the underlying red-blue checkerboard pattern. This enables readers to easily track the checkerboard tiling in a warped output image.

The Madonna image is a digitized frame from one of her earlier music videos. It is an example of a natural image that has both highly textured regions (hair) and smoothly varying areas (face). This helps the reader assess the quality of filtering among disparate image characteristics. The Mandrill image is used for similar reasons.

Perhaps no image pattern is more troubling to a digital image warping algorithm than the Star image taken from the IEEE Facsimile Chart. It contains a wide range of spatial frequencies that steadily increase towards the center. This serves to push

---

† Madonna is reprinted with permission of Warner Bros. Records.

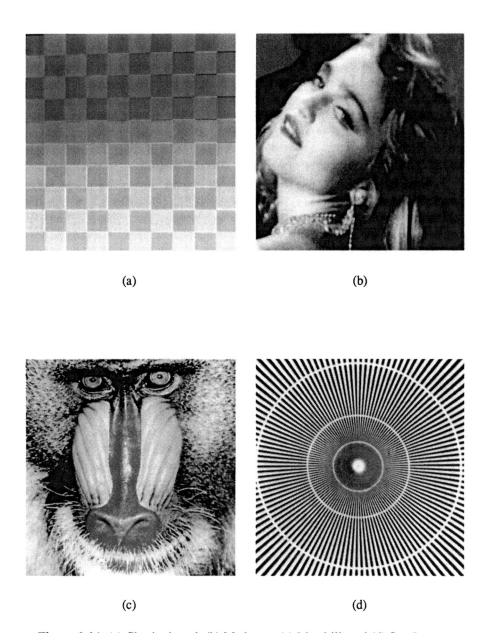

(a)   (b)

(c)   (d)

Figure 2.14: (a) Checkerboard; (b) Madonna; (c) Mandrill; and (d) Star Images.

algorithmic approximations to the limit. As a result, this image is a useful benchmark for evaluating the filtering quality of a warping algorithm.

2.6. SUMMARY

Input imagery appears in many different media, including photographs, film, and surface radiance. The purpose of digital image acquisition systems is to convert these input sources into digital form, thereby meeting the most basic requirement for computer processing of images. This is a two stage process. First, imaging systems are used to generate analog signals in response to incident light. These signals, however, cannot be directly manipulated by digital computers. Consequently, an analog-to-digital converter is used to discretize the input. This involves sampling and quantizing the analog signal. The result is a digital image, an array of integer intensity values.

The material contained in this chapter is found in most introductory image processing texts. Readers are referred to [Pratt 78], [Pavlidis 82], [Gonzalez 87], and [Jain 89] for a thorough treatment of basic image processing concepts. [Schreiber 86] is an excellent monograph on the fundamentals of electronic imaging systems. A fine overview of optical scanners is found in [Nagy 83]. Remote sensing applications for the topics discussed in this chapter can be found in [Green 89] and [Schowengerdt 83].

3

SPATIAL TRANSFORMATIONS

This chapter describes common spatial transformations derived for digital image warping applications in remote sensing, medical imaging, computer vision, and computer graphics. A spatial transformation is a mapping function that establishes a spatial correspondence between all points in an image and its warped counterpart. Due to the inherently wide scope of this subject, our discussion is by no means a complete review. Instead, we concentrate on widely used formulations, putting emphasis on an intuitive understanding of the mathematics that underly their usage. In this manner, we attempt to capture the essential methods from which peripheral techniques may be easily extrapolated.

The most elementary formulations we shall consider are those that stem from a general homogeneous transformation matrix. They span two classes of simple planar mappings: affine and perspective transformations. More general nonplanar results are possible with bilinear transformations. We discuss the geometric properties of these three classes of transformations and review the mathematics necessary to invert and infer these mappings.

In many fields, warps are often specified by polynomial transformations. This is common practice in geometric correction applications, where spatial distortions are adequately modeled by low-order polynomials. It becomes critically important in these cases to accurately estimate (infer) the unknown polynomial coefficients. We draw upon several techniques from numerical analysis to solve for these coefficients. For those instances where local distortions are present, we describe piecewise polynomial transformations which permit the coefficients to vary from region to region.

A more general framework, expressed in terms of surface interpolation, yields greater insight into this problem (and its solution). This broader outlook stems from the realization that a mapping function can be represented as two surfaces, each relating the point-to-point correspondences of 2-D points in the original and warped images. This approach facilitates the use of mapping functions more sophisticated than polynomials. We discuss this reformulation of the problem, and review various surface interpolation algorithms.

3.1. DEFINITIONS

A *spatial transformation* defines a geometric relationship between each point in the input and output images. An input image consists entirely of reference points whose coordinate values are known precisely. The output image is comprised of the observed (warped) data. The general mapping function can be given in two forms: either relating the output coordinate system to that of the input, or vice versa. Respectively, they can be expressed as

$$[x, y] = [X(u,v), Y(u,v)] \qquad (3.1.1)$$

or

$$[u, v] = [U(x,y), V(x,y)] \qquad (3.1.2)$$

where $[u,v]$ refers to the input image coordinates corresponding to output pixel $[x,y]$, and X, Y, U, and V are arbitrary mapping functions that uniquely specify the spatial transformation. Since X and Y map the input onto the output, they are referred to as the forward mapping. Similarly, the U and V functions are known as the inverse mapping since they map the output onto the input.

3.1.1. Forward Mapping

The *forward mapping* consists of copying each input pixel onto the output image at positions determined by the X and Y mapping functions. Figure 3.1 illustrates the forward mapping for the 1-D case. The discrete input and output are each depicted as a string of pixels lying on an integer grid (dots). Each input pixel is passed through the spatial transformation where it is assigned new output coordinate values. Notice that the input pixels are mapped from the set of integers to the set of real numbers. In the figure, this corresponds to the regularly spaced input samples and the irregular output distribution.

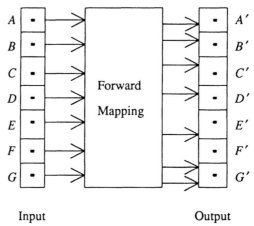

Figure 3.1: Forward mapping.

The real-valued output positions assigned by X and Y present complications at the discrete output. In the continuous domain, where pixels may be viewed as points, the mapping is straightforward. However, in the discrete domain pixels are now taken to be finite elements defined to lie on a (discrete) integer lattice. It is therefore inappropriate to implement the spatial transformation as a point-to-point mapping. Doing so can give rise to two types of problems: holes and overlaps. Holes, or patches of undefined pixels, occur when mapping contiguous input samples to sparse positions on the output grid. In Fig. 3.1, F' is a hole since it is bypassed in the input-output mapping. In contrast, overlaps occur when consecutive input samples collapse into one output pixel, as depicted in Fig. 3.1 by output pixel G'.

The shortcomings of a point-to-point mapping are avoided by using a *four-corner mapping* paradigm. This considers input pixels as square patches that may be transformed into arbitrary quadrilaterals in the output image. This has the effect of allowing the input to remain contiguous after the mapping.

Due to the fact that the projected input is free to lie anywhere in the output image, input pixels often straddle several output pixels or lie embedded in one. These two instances are illustrated in Fig. 3.2. An *accumulator array* is required to properly integrate the input contributions at each output pixel. It does so by determining which fragments contribute to each output pixel and then integrating over all contributing fragments. The partial contributions are handled by scaling the input intensity in proportion to the fractional part of the pixel that it covers. Intersection tests must be performed to compute the coverage. Thus, each position in the accumulator array evaluates $\sum_{i=0}^{N} w_i f_i$, where f_i is the input value, w_i is the weight reflecting its coverage of the output pixel, and N is the total number of deposits into the cell. Note that N is free to vary among pixels and is determined only by the mapping function and the output discretization.

Formulating the transformation as a four-corner mapping problem allows us to avoid holes in the output image. Nevertheless, this paradigm introduces two problems in the forward mapping process. First, costly intersection tests are needed to derive the weights. Second, magnification may cause the same input value to be applied onto many output pixels unless additional filtering is employed.

Both problems can be resolved by adaptively sampling the input based on the size of the projected quadrilateral. In other words, if the input pixel is mapped onto a large area in the output image, then it is best to repeatedly subdivide the input pixel until the projected area reaches some acceptably low limit, i.e., one pixel size. As the sampling rate rises, the weights converge to a single value, the input is resampled more densely, and the resulting computation is performed at higher precision.

It is important to note that uniformly sampling the input image does not guarantee uniform sampling in the output image unless X and Y are affine (linear) mappings. Thus, for nonaffine mappings (e.g., perspective or bilinear) the input image must be adaptively sampled at rates that are spatially varying. For example, the oblique surface shown in Fig. 3.3 must be sampled more densely near the horizon to account for the foreshortening

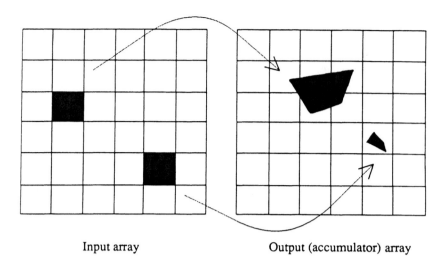

Input array   Output (accumulator) array

Figure 3.2: Accumulator array.

due to the bilinear mapping. In general, forward mapping is useful when the input image must be read sequentially or when it does not reside entirely in memory. It is particularly useful for separable algorithms that operate in scanline order (see Chapter 7).

Figure 3.3: An oblique surface requiring adaptive sampling.

3.1.2. Inverse Mapping

The *inverse mapping* operates in screen order, projecting each output coordinate into the input image via U and V. The value of the data sample at that point is copied onto the output pixel. Again, filtering is necessary to combat the aliasing artifacts described in more detail later. This is the most common method since no accumulator array is necessary and since output pixels that lie outside a clipping window need not be evaluated. This method is useful when the screen is to be written sequentially, U and V are readily available, and the input image can be stored entirely in memory.

Figure 3.4 depicts the inverse mapping, with each output pixel mapped back onto the input via the spatial transformation (inverse) mapping function. Notice that the output pixels are centered on integer coordinate values. They are projected onto the input at real-valued positions. As we will see later, an interpolation stage must be introduced in

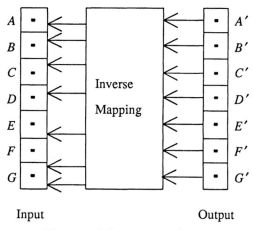

Figure 3.4: Inverse mapping.

order to retrieve input values at undefined (nonintegral) input positions.

Unlike the point-to-point forward mapping scheme, the inverse mapping guarantees that all output pixels are computed. However, the analogous problem remains to determine whether large holes are left when sampling the input. If this is the case, large amounts of input data may have been discarded while evaluating the output, thereby giving rise to artifacts described in Chapter 6. Thus, filtering is necessary to integrate the area projected onto the input. In general, though, this arrangement has the advantage of allowing interpolation to occur in the input space instead of the output space. This proves to be a much more convenient approach than forward mapping. Graphically, this is equivalent to the dual of Fig. 3.2, where the input and output captions are interchanged.

In their most unconstrained form, U and V can serve to scramble the image by defining a discontinuous function. The image remains coherent only if U and V are piecewise continuous. Although there exists an infinite number of possible mapping functions, several common forms of U and V have been isolated for geometric correction and geometric distortion. The remainder of this chapter addresses these formulations.

3.2. GENERAL TRANSFORMATION MATRIX

Many simple spatial transformations can be expressed in terms of the general 3×3 transformation matrix T_1 shown in Eq. (3.2.1). It handles scaling, shearing, rotation, reflection, translation, and perspective in 2-D. Without loss of generality, we shall ignore the component in the third dimension since we are only interested in 2-D image projections (e.g., mappings between the uv- and xy-coordinate systems).

$$[x', y', w'] = [u, v, w]T_1 \tag{3.2.1}$$

where

$$T_1 = \begin{bmatrix} a_{11} & a_{12} & a_{13} \\ a_{21} & a_{22} & a_{23} \\ a_{31} & a_{32} & a_{33} \end{bmatrix}$$

The 3×3 transformation matrix can be best understood by partitioning it into four separate sections. The 2×2 submatrix

$$T_2 = \begin{bmatrix} a_{11} & a_{12} \\ a_{21} & a_{22} \end{bmatrix}$$

specifies a linear transformation for scaling, shearing, and rotation. The 1×2 matrix $[a_{31} \ a_{32}]$ produces translation. The 2×1 matrix $[a_{13} \ a_{23}]^T$ produces perspective transformation. Note that the superscript T denotes matrix transposition, whereby rows and columns are interchanged. The final element a_{33} is responsible for overall scaling.

For consistency, the transformations that follow are cast in terms of forward mapping functions X and Y that transform source images in the uv-coordinate system onto target images in the xy-coordinate system. Similar derivations apply for inverse mapping functions U and V. We note that the transformations are written in postmultiplication form. That is, the transformation matrix is written *after* the position row vector. This is equivalent to the premultiplication form where the transformation matrix precedes the position column vector. The latter form is more common in the remote sensing, computer vision, and robotics literature.

3.2.1. Homogeneous Coordinates

The general 3×3 matrix used to specify 2-D coordinate transformations operates in the *homogeneous coordinate system*. The use of homogeneous coordinates was introduced into computer graphics by Roberts to provide a consistent representation for affine and perspective transformations [Roberts 66]. In the discussion that follows, we briefly motivate and outline the homogeneous notation.

Elementary 2-D mapping functions can be specified with the general 2×2 transformation matrix T_2. Applying T_2 to a 2-D position vector $[u,v]$ yields the following linear mapping functions for X and Y.

$$x = a_{11}u + a_{21}v \qquad (3.2.2a)$$

$$y = a_{12}u + a_{22}v \qquad (3.2.2b)$$

Equations (3.2.2a) and (3.2.2b) are said to be linear because they satisfy the following two conditions necessary for any linear function $L(x)$: $L(x+y) = L(x) + L(y)$ and $L(cx) = cL(x)$ for any scalar c, and position vectors x and y. Unfortunately, linear transformations do not account for translations since there is no facility for adding constants. Therefore, we define $A(x)$ to be an affine transformation if and only if there exists a constant t and a linear transformation $L(x)$ such that $A(x) = L(x) + t$ for all x. Clearly linear transformations are a subset of affine transformations.

In order to accommodate affine mappings, the position vectors are augmented with an additional component, turning $[x, y]$ into $[x, y, 1]$. In addition, the translation parameters are appended to T_2 yielding

$$T_3 = \begin{bmatrix} a_{11} & a_{12} \\ a_{21} & a_{22} \\ a_{31} & a_{32} \end{bmatrix}$$

The affine mapping is given as $[x, y] = [u, v, 1] T_3$. Note that the added component to $[u, v]$ has no physical significance. It simply allows us to incorporate translations into the general transformation scheme.

The 3×2 matrix T_3 used to specify an affine transformation is not square and thus does not have an inverse. Since inverses are necessary to relate the two coordinate systems (before and after a transformation), the coefficients are embedded into a 3×3 transformation matrix in order to make it invertible. Thus, the additional row introduced to T_2 by translation is balanced by appending an additional column to T_3. This serves to introduce a third component w' to the transformed 2-D position vector (Eq. 3.2.1). The use of homogeneous coordinates to represent affine transformations is derived from this need to retain an inverse for T_3.

All 2-D position vectors are now represented with three components in a representation known as *homogeneous notation*. In general, n-dimensional position vectors now consist of $n+1$ elements. This formulation forces the homogeneous coordinate w' to take on physical significance: it refers to the plane upon which the transformation operates. That is, a 2-D position vector $[u, v]$ lying on the $w = 1$ plane becomes a 3-D homogeneous vector $[u, v, 1]$. For convenience, all input points lie on the $w = 1$ plane to trivially facilitate translation by $[a_{31} \ a_{32}]$.

Since only 2-D transformations are of interest to us, the results of the transformation must lie on the same plane, e.g., $w' = w = 1$. However, since w' is free to take on any value in the general case, the homogeneous coordinates must be divided by w' in order to be left with results in the plane $w' = w = 1$. This leads us to an important property of the homogeneous notation: *the representation of a point is no longer unique.*

Consider the implicit equation of a line in two dimensions, $ax + by + c = 0$. The coefficients a, b, and c are not unique. Instead, it is the *ratio* among coefficients that is important. Not surprisingly, equations of the form $f(x) = 0$ are said to be homogeneous equations because equality is preserved after scalar multiplication. Similarly, scalar multiples of a 2-D position vector represent the same point in a homogeneous coordinate system.

Any 2-D position vector $p = [x, y]$ is represented by the homogeneous vector $p_h = [x', y', w'] = [xw', yw', w']$ where $w' \neq 0$. To recover p from p_h, we simply divide by the homogeneous coordinate w' so that $[x, y] = [x'/w', y'/w']$. Consequently, vectors of the form $[xw', yw', w']$ form an equivalence class of homogeneous representations for the vector p. The division that cancels the effect of multiplication with w' corresponds to a projection onto the $w' = 1$ plane using rays passing through the origin. Interested readers are referred to [Rogers 76, Pavlidis 82, Penna 86, Foley 90] for a thorough treatment of homogeneous coordinates.

3.3. AFFINE TRANSFORMATIONS

The general representation of an *affine transformation* is

$$[x, y, 1] = [u, v, 1] \begin{bmatrix} a_{11} & a_{12} & 0 \\ a_{21} & a_{22} & 0 \\ a_{31} & a_{32} & 1 \end{bmatrix} \qquad (3.3.1)$$

Division by the homogeneous coordinate w' is avoided by selecting $w = w' = 1$. Consequently, an affine mapping is characterized by a transformation matrix whose last column is equal to $[\,0\;0\;1\,]^T$. This corresponds to an *orthographic* or *parallel plane projection* from the source uv-plane onto the target xy-plane. As a result, affine mappings preserve parallel lines, allowing us to avoid foreshortened axes when performing 2-D projections. Furthermore, equispaced points are preserved (although the actual spacing in the two coordinate systems may differ). As we shall see later, affine transformations accommodate planar mappings. For instance, they can map triangles to triangles. They are, however, not general enough to map quadrilaterals to quadrilaterals. That is reserved for perspective transformations (see Section 3.4). Examples of three affine warps applied to the Checkerboard image are shown in Fig. 3.5.

Figure 3.5: Affine warps.

For affine transformations, the forward mapping functions are

$$x = a_{11}u + a_{21}v + a_{31} \tag{3.3.2a}$$

$$y = a_{12}u + a_{22}v + a_{32} \tag{3.3.2b}$$

This accommodates translations, rotations, scale, and shear. Since the product of affine transformations is also affine, they can be used to perform a general orientation of a set of points relative to an arbitrary coordinate system while still maintaining a unity value for the homogeneous coordinate. This is necessary for generating composite transformations. We now consider special cases of the affine transformation and its properties.

3.3.1. Translation

All points are translated to new positions by adding offsets T_u and T_v to u and v, respectively. The translate transform is

$$[x, y, 1] = [u, v, 1] \begin{bmatrix} 1 & 0 & 0 \\ 0 & 1 & 0 \\ T_u & T_v & 1 \end{bmatrix} \tag{3.3.3}$$

3.3.2. Rotation

All points in the *uv*-plane are rotated about the origin through the counterclockwise angle θ.

$$[x, y, 1] = [u, v, 1] \begin{bmatrix} \cos\theta & \sin\theta & 0 \\ -\sin\theta & \cos\theta & 0 \\ 0 & 0 & 1 \end{bmatrix} \qquad (3.3.4)$$

3.3.3. Scale

All points are scaled by applying the scale factors S_u and S_v to the *u* and *v* coordinates, respectively. Enlargements (reductions) are specified with positive scale factors that are larger (smaller) than unity. Negative scale factors cause the image to be reflected, yielding a mirrored image. Finally, if the scale factors are not identical, then the image proportions are altered resulting in a differentially scaled image.

$$[x, y, 1] = [u, v, 1] \begin{bmatrix} S_u & 0 & 0 \\ 0 & S_v & 0 \\ 0 & 0 & 1 \end{bmatrix} \qquad (3.3.5)$$

3.3.4. Shear

The coordinate scaling described above involves only the diagonal terms a_{11} and a_{22}. We now consider the case where $a_{11} = a_{22} = 1$, and $a_{12} = 0$. By allowing a_{21} to be nonzero, *x* is made linearly dependent on both *u* and *v*, while *y* remains identical to *v*. A similar operation can be applied along the *v*-axis to compute new values for *y* while *x* remains unaffected. This effect, called shear, is therefore produced by using the off-diagonal terms. The shear transform along the *u*-axis is

$$[x, y, 1] = [u, v, 1] \begin{bmatrix} 1 & 0 & 0 \\ H_v & 1 & 0 \\ 0 & 0 & 1 \end{bmatrix} \qquad (3.3.6a)$$

where H_v is used to make *x* linearly dependent on *v* as well as *u*. Similarly, the shear transform along the *v*-axis is

$$[x, y, 1] = [u, v, 1] \begin{bmatrix} 1 & H_u & 0 \\ 0 & 1 & 0 \\ 0 & 0 & 1 \end{bmatrix} \qquad (3.3.6b)$$

3.3.5. Composite Transformations

Multiple transforms can be collapsed into a single composite transformation. The transforms are combined by taking the product of the 3×3 matrices. This is generally not a commutative operation. An example of a composite transformation representing a translation followed by a rotation and a scale change is given below.

$$[x, y, 1] = [u, v, 1]\, \mathbf{M}_{comp} \tag{3.3.7}$$

where

$$
\mathbf{M}_{comp} = \begin{bmatrix} 1 & 0 & 0 \\ 0 & 1 & 0 \\ T_u & T_v & 1 \end{bmatrix} \begin{bmatrix} \cos\theta & \sin\theta & 0 \\ -\sin\theta & \cos\theta & 0 \\ 0 & 0 & 1 \end{bmatrix} \begin{bmatrix} S_u & 0 & 0 \\ 0 & S_v & 0 \\ 0 & 0 & 1 \end{bmatrix}
$$

$$
= \begin{bmatrix} S_u\cos\theta & S_v\sin\theta & 0 \\ -S_u\sin\theta & S_v\cos\theta & 0 \\ S_u(T_u\cos\theta - T_v\sin\theta) & S_v(T_u\sin\theta + T_v\cos\theta) & 1 \end{bmatrix}
$$

3.3.6. Inverse

The inverse of an affine transformation is itself affine. It can be readily computed from the adjoint $adj(T_1)$ and determinant $det(T_1)$ of transformation matrix T_1. From linear algebra, we know that $T_1^{-1} = adj(T_1) / det(T_1)$ where the adjoint of a matrix is simply the transpose of the matrix of cofactors [Strang 80]. This yields

$$[u, v, 1] = [x, y, 1] \begin{bmatrix} A_{11} & A_{12} & 0 \\ A_{21} & A_{22} & 0 \\ A_{31} & A_{32} & 1 \end{bmatrix} \tag{3.3.8}$$

$$= [x, y, 1]\, \frac{1}{a_{11}a_{22} - a_{21}a_{12}} \begin{bmatrix} a_{22} & -a_{12} & 0 \\ -a_{21} & a_{11} & 0 \\ a_{21}a_{32} - a_{31}a_{22} & a_{31}a_{12} - a_{11}a_{32} & a_{11}a_{22} - a_{21}a_{12} \end{bmatrix}$$

3.3.7. Inferring Affine Transformations

An affine transformation has six degrees of freedom, relating directly to coefficients $a_{11}, a_{21}, a_{31}, a_{12}, a_{22}$, and a_{32}. In computer graphics, these coefficients are known by virtue of the applied coordinate transformation. In areas such as remote sensing, however, it is usually of interest to infer the mapping given only a reference image and an observed image. If an affine mapping is deemed adequate to describe the transformation, the six coefficients may be derived by specifying the coordinate correspondence of three noncollinear points in both images. Let (u_k, v_k) and (x_k, y_k) for $k = 0, 1, 2$ be these three points in the reference and observed images, respectively. Equation (3.3.9) expresses their relationship in the form of a matrix equation. The six unknown coefficients of the affine mapping are determined by solving the system of six linear equations contained in Eq. (3.3.9).

$$\begin{bmatrix} x_0 & y_0 & 1 \\ x_1 & y_1 & 1 \\ x_2 & y_2 & 1 \end{bmatrix} = \begin{bmatrix} u_0 & v_0 & 1 \\ u_1 & v_1 & 1 \\ u_2 & v_2 & 1 \end{bmatrix} \begin{bmatrix} a_{11} & a_{12} & 0 \\ a_{21} & a_{22} & 0 \\ a_{31} & a_{32} & 1 \end{bmatrix} \tag{3.3.9}$$

Let the system of equations given above be denoted as $X = UA$. In order to determine the coefficients, we isolate A by multiplying both sides with U^{-1}, the inverse of the matrix containing points (u_k, v_k). As before, $U^{-1} = adj(U) / det(U)$ where $adj(U)$ is the adjoint of U and $det(U)$ is the determinant. Although the adjoint is always computable, an inverse will not exist unless the determinant is nonzero. Fortunately, the constraint on U to consist of *noncollinear* points serves to ensure that U is nonsingular, i.e., $det(U) \neq 0$. Consequently, the inverse U^{-1} is guaranteed to exist. Solving for the coefficients in terms of the known (u_k, v_k) and (x_k, y_k) pairs, we have

$$A = U^{-1}X \tag{3.3.10}$$

or equivalently,

$$\begin{bmatrix} a_{11} & a_{12} & 0 \\ a_{21} & a_{22} & 0 \\ a_{31} & a_{32} & 1 \end{bmatrix} = \frac{1}{det(U)} \begin{bmatrix} v_1 - v_2 & v_2 - v_0 & v_0 - v_1 \\ u_2 - u_1 & u_0 - u_2 & u_1 - u_0 \\ u_1 v_2 - u_2 v_1 & u_2 v_0 - u_0 v_2 & u_2 v_1 - u_1 v_0 \end{bmatrix} \begin{bmatrix} x_0 & y_0 & 1 \\ x_1 & y_1 & 1 \\ x_2 & y_2 & 1 \end{bmatrix}$$

where

$$det(U) = u_0(v_1 - v_2) - v_0(u_1 - u_2) + (u_1 v_2 - u_2 v_1)$$

When more than three correspondence points are available, and when these points are known to contain errors, it is common practice to approximate the coefficients by solving an overdetermined system of equations. In that case, matrix U is no longer a square 3×3 matrix and it must be inverted using any technique that solves a least-squares linear system problem.

Since only three points are needed to infer an affine mapping, it is clear that affine transformations realize a limited set of planar mappings. Essentially, they can map an input triangle into an arbitrary triangle at the output. An input rectangle can be mapped into a parallelogram at the output. More general distortions, however, cannot be handled by affine transformations. For example, to map a rectangle into an arbitrary quadrilateral requires a perspective, bilinear, or more complex transformation. Fast incremental methods for computing affine mappings are discussed in Chapter 7.

3.4. PERSPECTIVE TRANSFORMATIONS

The general representation of a *perspective transformation* is

$$[x', y', w'] = [u, v, w] \begin{bmatrix} a_{11} & a_{12} & a_{13} \\ a_{21} & a_{22} & a_{23} \\ a_{31} & a_{32} & a_{33} \end{bmatrix} \qquad (3.4.1)$$

where $x = x'/w'$ and $y = y'/w'$.

A *perspective transformation*, or *projective mapping*, is produced when $[a_{13} \, a_{23}]^T$ is nonzero. It is used in conjunction with a projection onto a viewing plane in what is known as a *perspective* or *central projection*. Perspective transformations preserve parallel lines only when they are parallel to the projection plane. Otherwise, lines converge to a vanishing point. This has the property of foreshortening distant lines, a useful technique for rendering realistic images. For perspective transformations, the forward mapping functions are

$$x = \frac{x'}{w'} = \frac{a_{11}u + a_{21}v + a_{31}}{a_{13}u + a_{23}v + a_{33}} \qquad (3.4.2a)$$

$$y = \frac{y'}{w'} = \frac{a_{12}u + a_{22}v + a_{32}}{a_{13}u + a_{23}v + a_{33}} \qquad (3.4.2b)$$

They take advantage of the fact that w' is allowed to vary at each point and division by w' is equivalent to a projection using rays passing through the origin. Note that affine transformations are a special case of perspective transformations where w' is constant over the entire image, i.e., $a_{13} = a_{23} = 0$.

Perspective transformations share several important properties with affine transformations. They are planar mappings, and thus their forward and inverse transforms are single-valued. They preserve lines in all orientations. That is, lines map onto lines (although not of the same orientation). As we shall see, this desirable property is lacking in more general mappings. Furthermore, the nine degrees of freedom in Eq. (3.4.1) is sufficient to permit planar quadrilateral-to-quadrilateral mappings. In contrast, affine transformations offer only six degrees of freedom (Eq. 3.3.1) and thereby facilitates only triangle-to-triangle mappings.

Examples of projective warps are shown in Fig. 3.6. Note that the intersections along the edges are no longer equispaced. Also, in the rightmost image the horizontal lines remain parallel because they lie parallel to the projection plane.

3.4.1. Inverse

The inverse of a projective mapping can be easily computed in terms of the adjoint of the transformation matrix T_1. Thus, $T_1^{-1} = adj(T_1)/det(T_1)$ where $adj(T_1)$ is the adjoint of T_1 and $det(T_1)$ is the determinant. Since two matrices which are (nonzero) scalar multiples of each other are equivalent in the homogeneous coordinate system,

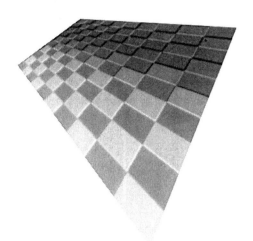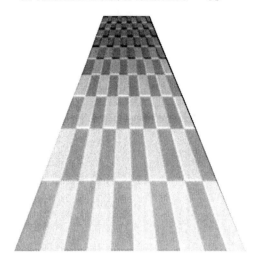

Figure 3.6: Perspective warps.

there is no need to divide by the determinant (a scalar). Consequently, the adjoint matrix can be used in place of the inverse matrix. This proves to be a very useful result, especially since the adjoint will be well-behaved even if the determinant is very small when the matrix is nearly singular. Note that if the matrix is singular, the inverse is undefined and therefore the adjoint cannot be a scalar multiple of it. Due to these results from linear algebra, the inverse is expressed below in terms of the elements in T_1 that are used to realize the forward mapping.

$$[u, v, w] = [x', y', w'] \begin{bmatrix} A_{11} & A_{12} & A_{13} \\ A_{21} & A_{22} & A_{23} \\ A_{31} & A_{32} & A_{33} \end{bmatrix} \qquad (3.4.3)$$

$$= [x', y', w'] \begin{bmatrix} a_{22}a_{33} - a_{23}a_{32} & a_{13}a_{32} - a_{12}a_{33} & a_{12}a_{23} - a_{13}a_{22} \\ a_{23}a_{31} - a_{21}a_{33} & a_{11}a_{33} - a_{13}a_{31} & a_{13}a_{21} - a_{11}a_{23} \\ a_{21}a_{32} - a_{22}a_{31} & a_{12}a_{31} - a_{11}a_{32} & a_{11}a_{22} - a_{12}a_{21} \end{bmatrix}$$

3.4.2. Inferring Perspective Transformations

A perspective transformation is expressed in terms of the nine coefficients in the general 3×3 matrix T_1. Without loss of generality, T_1 can be normalized so that $a_{33} = 1$. This leaves eight degrees of freedom for a projective mapping. The eight coefficients can be determined by establishing correspondence between four points in the reference and observed images. Let (u_k, v_k) and (x_k, y_k) for $k = 0, 1, 2, 3$ be these four points in the reference and observed images, respectively. Assuming $a_{33} = 1$, Eqs. (3.4.2a) and (3.4.2b) can be rewritten as

$$x = a_{11}u + a_{21}v + a_{31} - a_{13}ux - a_{23}vx \qquad (3.4.4a)$$

$$y = a_{12}u + a_{22}v + a_{32} - a_{13}uy - a_{23}vy \qquad (3.4.4b)$$

Applying Eqs. (3.4.4a) and (3.4.4b) to the four pairs of correspondence points yields the 8×8 system of equations shown in Eq. (3.4.5).

$$\begin{bmatrix} u_0 & v_0 & 1 & 0 & 0 & 0 & -u_0x_0 & -v_0x_0 \\ u_1 & v_1 & 1 & 0 & 0 & 0 & -u_1x_1 & -v_1x_1 \\ u_2 & v_2 & 1 & 0 & 0 & 0 & -u_2x_2 & -v_2x_2 \\ u_3 & v_3 & 1 & 0 & 0 & 0 & -u_3x_3 & -v_3x_3 \\ 0 & 0 & 0 & u_0 & v_0 & 1 & -u_0y_0 & -v_0y_0 \\ 0 & 0 & 0 & u_1 & v_1 & 1 & -u_1y_1 & -v_1y_1 \\ 0 & 0 & 0 & u_2 & v_2 & 1 & -u_2y_2 & -v_2y_2 \\ 0 & 0 & 0 & u_3 & v_3 & 1 & -u_3y_3 & -v_3y_3 \end{bmatrix} A = X \qquad (3.4.5)$$

The coefficients are determined by solving the linear system. This yields a solution to the general (planar) quadrilateral-to-quadrilateral problem. Speedups are possible when considering several special cases: square-to-quadrilateral, quadrilateral-to-square, and quadrilateral-to-quadrilateral using the results of the last two cases. We now consider each case individually. A detailed exposition is found in [Heckbert 89].

3.4.2.1. Case 1: Square-to-Quadrilateral

Consider mapping a unit square onto an arbitrary quadrilateral. The following four-point correspondences are established from the uv-plane onto the xy-plane.

$$(0,0) \rightarrow (x_0,y_0)$$
$$(1,0) \rightarrow (x_1,y_1)$$
$$(0,1) \rightarrow (x_2,y_2)$$
$$(1,1) \rightarrow (x_3,y_3)$$

In this case, the eight equations become

$$a_{31} = x_0$$

$$a_{11} + a_{31} - a_{13}x_1 = x_1$$

$$a_{11} + a_{21} + a_{31} - a_{13}x_2 + a_{23}x_2 = x_2$$

$$a_{12} + a_{13} - a_{23}x_3 = x_3$$

$$a_{32} = y_0$$

$$a_{12} + a_{32} - a_{13}y_1 = y_1$$

$$a_{12} + a_{22} + a_{32} - a_{13}y_2 + a_{23}y_2 = y_2$$

$$a_{22} + a_{32} - a_{23}y_3 = y_3$$

The solution can take two forms, depending on whether the mapping is affine or perspective. We define the following terms for our discussion.

$$\Delta x_1 = x_1 - x_2 \qquad \Delta x_2 = x_3 - x_2 \qquad \Delta x_3 = x_0 - x_1 + x_2 - x_3$$

$$\Delta y_1 = y_1 - y_2 \qquad \Delta y_2 = y_3 - y_2 \qquad \Delta y_3 = y_0 - y_1 + y_2 - y_3$$

If $\Delta x_3 = 0$ and $\Delta y_3 = 0$, then the xy quadrilateral is a parallelogram. This implies that the mapping is affine. As a result, we obtain the following coefficients.

$$a_{11} = x_1 - x_0$$

$$a_{21} = x_2 - x_1$$

$$a_{31} = x_0$$

$$a_{12} = y_1 - y_0$$

$$a_{22} = y_2 - y_1$$

$$a_{32} = y_0$$

$$a_{13} = 0$$

$$a_{23} = 0$$

If, however, $\Delta x_3 \neq 0$ or $\Delta y_3 \neq 0$, then the mapping is projective. The coefficients of the perspective transformation are

$$a_{13} = \begin{vmatrix} \Delta x_3 & \Delta x_2 \\ \Delta y_3 & \Delta y_2 \end{vmatrix} \Big/ \begin{vmatrix} \Delta x_1 & \Delta x_2 \\ \Delta y_1 & \Delta y_2 \end{vmatrix}$$

$$a_{23} = \begin{vmatrix} \Delta x_1 & \Delta x_3 \\ \Delta y_1 & \Delta y_3 \end{vmatrix} \Big/ \begin{vmatrix} \Delta x_1 & \Delta x_2 \\ \Delta y_1 & \Delta y_2 \end{vmatrix}$$

$$a_{11} = x_1 - x_0 + a_{13}x_1$$
$$a_{21} = x_3 - x_0 + a_{23}x_3$$
$$a_{31} = x_0$$
$$a_{12} = y_1 - y_0 + a_{13}y_1$$
$$a_{22} = y_3 - y_0 + a_{23}y_3$$
$$a_{32} = y_0$$

This proves to be faster than the direct solution with a linear system solver. The computation may be generalized to map arbitrary rectangles onto quadrilaterals by premultiplying with a scale and translation matrix.

3.4.2.2. Case 2: Quadrilateral-to-Square

This case is the inverse of the mapping already considered. As discussed earlier, the adjoint of a projective mapping can be used in place of the inverse. Thus, the simplest solution is to compute the square-to-quadrilateral mapping coefficients described above to find the inverse of the desired mapping, and then take its adjoint to compute the quadrilateral-to-square mapping.

3.4.2.3. Case 3: Quadrilateral-to-Quadrilateral

The results of the last two cases may be cascaded to yield a fast solution to the general quadrilateral-to-quadrilateral mapping problem. Figure 3.7 depicts this formulation.

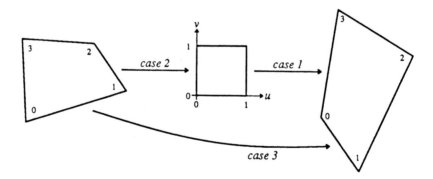

Figure 3.7: Quadrilateral-to-quadrilateral mapping [Heckbert 89].

The general quadrilateral-to-quadrilateral problem is also known as *four-corner mapping*. Perspective transformations offer a planar solution to this problem. When the quadrilaterals become nonplanar, however, more general solutions are necessary. Bilinear transformations are an example of the simplest mapping functions that address four-corner mappings for nonplanar quadrilaterals.

3.5. BILINEAR TRANSFORMATIONS

The general representation of a *bilinear transformation* is

$$[x, y] = [uv, u, v, 1] \begin{bmatrix} a_3 & b_3 \\ a_2 & b_2 \\ a_1 & b_1 \\ a_0 & b_0 \end{bmatrix} \qquad (3.5.1)$$

A bilinear transformation, or *bilinear mapping*, handles the four-corner mapping problem for nonplanar quadrilaterals. It is most commonly used in the forward mapping formulation where rectangles are mapped onto nonplanar quadrilaterals. It is pervasive in remote sensing and medical imaging where a grid of markings on the sensor are imaged and registered with their known positions for calibration purposes. It is also common in computer graphics where it plays a central role in forward mapping algorithms for texture mapping.

Bilinear mappings preserve lines that are horizontal or vertical in the source image. This follows from the bilinear interpolation used to realize the transformation. Thus, points along horizontal and vertical lines in the source image (including borders) remain equispaced. This is a property shared with affine transformations. However, lines not oriented along these two directions (e.g., diagonals) are not preserved as lines. Instead, diagonal lines map onto quadratic curves at the output. Examples of bilinear warps are shown in Fig. 3.8.

Figure 3.8: Bilinear warps.

Bilinear mappings are defined through piecewise functions that must interpolate the coordinate assignments specified at the vertices. This scheme is based on bilinear interpolation to evaluate the X and Y mapping functions. We illustrate this method below for computing $X(u,v)$. An identical procedure is performed to compute $Y(u,v)$.

3.5.1. Bilinear Interpolation

Bilinear interpolation utilizes a linear combination of the four "closest" pixel values to produce a new, interpolated value. Given four points, (u_0, v_0), (u_1, v_1), (u_2, v_2), and (u_3, v_3), and their respective function values x_0, x_1, x_2, and x_3, any intermediate coordinate $X(u, v)$ may be computed by the expression

$$X(u,v) = a_0 + a_1 u + a_2 v + a_3 uv \tag{3.5.2}$$

where the a_i coefficients are obtained by solving

$$\begin{bmatrix} x_0 \\ x_1 \\ x_2 \\ x_3 \end{bmatrix} = \begin{bmatrix} 1 & u_0 & v_0 & u_0 v_0 \\ 1 & u_1 & v_1 & u_1 v_1 \\ 1 & u_2 & v_2 & u_2 v_2 \\ 1 & u_3 & v_3 & u_3 v_3 \end{bmatrix} \begin{bmatrix} a_0 \\ a_1 \\ a_2 \\ a_3 \end{bmatrix} \tag{3.5.3}$$

Since the four points are assumed to lie on a rectangular grid, we rewrite them in the above matrix in terms of u_0, u_1, v_0, and v_2. Namely, the points are (u_0, v_0), (u_1, v_0), (u_0, v_2), and (u_1, v_2), respectively. Solving for a_i and substituting into Eq. (3.5.2) yields

$$X(u', v') = x_0 + (x_1 - x_0) u' + (x_2 - x_0) v' + (x_3 - x_2 - x_1 + x_0) u' v' \tag{3.5.4}$$

where u' and $v' \in (0,1)$ are normalized coordinates that span the rectangle, and

$$u = u_0 + (u_1 - u_0) u'$$
$$v = v_0 + (v_1 - v_0) v'$$

Therefore, given coordinates (u, v) and function values (x_0, x_1, x_2, x_3), the normalized coordinates (u', v') are computed and the point correspondence (x, y) in the arbitrary quadrilateral is determined by Eq. (3.5.4). Figure 3.9 depicts this bilinear interpolation for the X mapping function.

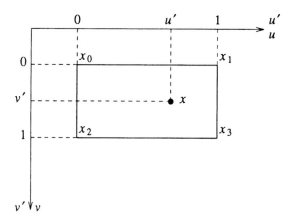

Figure 3.9: Bilinear interpolation.

3.5.2. Separability

The bilinear mapping is a separable transformation: it is the product of two 1-D mappings, each operating along orthogonal axes. This property enables us to easily extend 1-D linear interpolation into two dimensions, resulting in a computationally efficient algorithm. The algorithm requires two passes, with the first pass applying 1-D linear interpolation along the horizontal direction, and the second pass interpolating along the vertical direction. For example, consider the rectangle shown in Fig. 3.10. Points x_{01} and x_{23} are interpolated in the first pass. These results are then used in the second pass to compute the final value x.

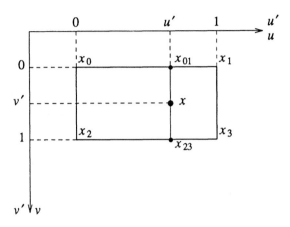

Figure 3.10: Separable bilinear interpolation.

Up to numerical inaccuracies, the separable results can be shown to be identical with the solution given in Eq. (3.5.4). In the first (horizontal) pass, we compute

$$x_{01} = x_0 + (x_1 - x_0) u'$$ (3.5.5)

$$x_{23} = x_2 + (x_3 - x_2) u'$$

These two intermediate results are then combined in the second (vertical) pass to yield the final value

$$x = x_{01} + (x_{23} - x_{01}) v'$$ (3.5.6)

$$= x_0 + (x_1 - x_0) u' + [(x_2 - x_0) + (x_3 - x_2 - x_1 + x_0) u'] v'$$

$$= x_0 + (x_1 - x_0) u' + (x_2 - x_0) v' + (x_3 - x_2 - x_1 + x_0) u' v'$$

Notice that this result is identical with the classic solution derived in Eq. (3.5.4).

3.5.3. Inverse

In remote sensing, the opposite problem is posed: given a normalized coordinate (x',y') in an arbitrary (distorted) quadrilateral, find its position in the rectangle. Two solutions are presented below.

By inverting Eq. (3.5.2), we can determine the normalized coordinate (u',v') corresponding to the given coordinate (x,y). The derivation is given below. First, we rewrite the expressions for x and y in terms of u and v, as given in Eq. (3.5.2).

$$x = a_0 + a_1 u + a_2 v + a_3 uv \qquad (3.5.7a)$$

$$y = b_0 + b_1 u + b_2 v + b_3 uv \qquad (3.5.7b)$$

Isolating u in Eq. (3.5.7a) gives us

$$u = \frac{x - a_0 - a_2 v}{a_1 + a_3 v} \qquad (3.5.8)$$

In order to solve this, we must determine v. This can be done by substituting Eq. (3.5.8) into Eq. (3.5.7b). Multiplying both sides by $(a_1 + a_3 v)$ yields

$$y(a_1 + a_3 v) = b_0(a_1 + a_3 v) + b_1(x - a_0 - a_2 v) + b_2 v(a_1 + a_3 v) + b_3 v(x - a_0 - a_2 v) \quad (3.5.9)$$

This can be rewritten as

$$c_2 v^2 + c_1 v + c_0 = 0 \qquad (3.5.10)$$

where

$$c_0 = a_1 (b_0 - y) + b_1 (x - a_0)$$

$$c_1 = a_3 (b_0 - y) + b_3 (x - a_0) + a_1 b_2 - a_2 b_1$$

$$c_2 = a_3 b_2 - a_2 b_3$$

The inverse mapping for v thus requires the solution of a quadratic equation. Once v is determined, it is plugged into Eq. (3.5.8) to compute u. Clearly, the inverse transform is multi-valued and is more difficult to compute than the forward transform.

3.5.4. Interpolation Grid

Mapping from an arbitrary grid to a rectangular grid is an important step in performing any 2-D interpolation within an arbitrary quadrilateral. The procedure is given as follows.

1. To any point (x,y) inside an interpolation region defined by four arbitrary points, a normalized coordinate (u',v') is associated in a rectangular region. This makes use of the results derived above. A geometric interpretation is given in Fig. 3.11, where the normalized coordinates can be found by determining the grid lines that intersect at (x,y) (point P). Given the positions labeled at the vertices, the normalized coordinates (u',v') are given as

$$u' = \frac{P_{01}P_0}{P_1P_0} = \frac{P_{23}P_2}{P_3P_2}$$

$$v' = \frac{P_{02}P_0}{P_2P_0} = \frac{P_{13}P_1}{P_3P_1}$$

(3.5.11)

2.   The function values at the four quadrilateral vertices are assigned to the rectangle vertices.

3.   A rectangular grid interpolation is then performed, using the normalized coordinates to index the interpolation function.

4.   The result is then assigned to point (x,y) in the distorted plane.

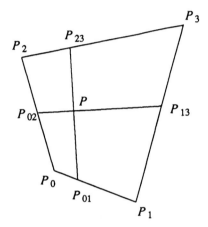

Figure 3.11: Geometric interpretation of arbitrary grid interpolation.

It is important to note that the primary benefit of this procedure is that higher-order interpolation methods (e.g., spline interpolation) that are commonly defined to operate on rectangular lattices can now be extended into the domain of non-rectangular grids. This thereby allows the generation of a continuous interpolation function for any arbitrary grid [Bizais 83]. More will be said about this in Chapter 7, when we discuss separable mesh warping.

3.6. POLYNOMIAL TRANSFORMATIONS

Geometric correction requires a spatial transformation to invert an unknown distortion function. The mapping functions, U and V, have been almost universally chosen to be global bivariate *polynomial transformations* of the form

$$u = \sum_{i=0}^{N} \sum_{j=0}^{N-i} a_{ij}x^iy^j$$

(3.6.1)

$$v = \sum_{i=0}^{N} \sum_{j=0}^{N-i} b_{ij}x^iy^j$$

where a_{ij} and b_{ij} are the constant polynomial coefficients. Since this formulation for geometric correction originated in remote sensing [Markarian 71], the discussion below will center on its use in that field. All the examples, though, have direct analogs in other related areas such as medical imaging [Singh 79] and computer vision [Rosenfeld 82].

The polynomial transformations given above are low-order global mapping functions operating on the entire image. They are intended to account for sensor-related spatial distortions such as centering, scale, skew, and pincushion effects, as well as errors due to earth curvature, viewing geometry, and camera attitude and altitude deviations. Due to dynamic operating conditions, these errors are comprised of internal and external components. The internal errors are sensor-related distortions. External errors are due to platform perturbations and scene characteristics. The effects of these errors have been categorized in [Bernstein 71] and are shown in Fig. 3.12.

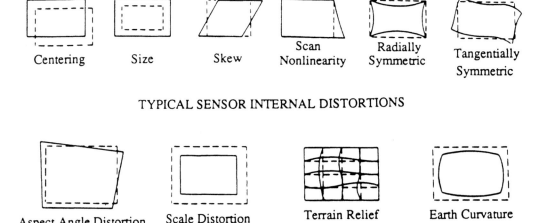

Centering   Size   Skew   Scan Nonlinearity   Radially Symmetric   Tangentially Symmetric

TYPICAL SENSOR INTERNAL DISTORTIONS

Aspect Angle Distortion (Attitude Effects)   Scale Distortion (Altitude Effect)   Terrain Relief   Earth Curvature

TYPICAL EXTERNAL IMAGE DISTORTIONS

Figure 3.12: Common geometric image distortions.

These errors are characterized as low-frequency (smoothly varying) distortions. The global effects of the polynomial mapping will not account for high-frequency deformations that are local in nature. Since most sensor-related errors tend to be low-frequency, modeling the spatial transformation with low-order polynomials appears justified. Common values of N that have been used in the polynomials of Eq. (3.6.1) include $N = 1$ [Steiner 77], $N = 2$ [Nack 77], $N = 3$ [Van Wie 77], and $N = 4$ [Leckie 80]. For many practical problems, a second-degree ($N = 2$) approximation has been shown to be adequate [Lillestrand 72].

Note that a first-degree ($N = 1$) bivariate polynomial defines those mapping functions that are exactly given by a general 3×3 affine transformation matrix. As discussed in the previous section, these polynomials characterize common physical distortions, i.e., affine transformations. When the viewing geometry is known in advance, the selection of the polynomial coefficients is determined directly from the scale, translation, rotation, and skew specifications. This is an example typical of computer graphics. For example, given a mathematical model of the world, including objects and the viewing plane, it is relatively straightforward to cascade transformation matrices such that a series of projections onto the viewing plane can be realized.

In the fields of remote sensing, medical imaging, and computer vision, however, the task of computing the spatial transformation is not so straightforward. In the vast majority of applications, the polynomial coefficients are not given directly. Instead, spatial information is supplied by means of *tiepoints* or *control points*, corresponding positions in the input and output images whose coordinates can be defined precisely. In these cases, the central task of the spatial transformation stage is to infer the coefficients of the polynomial that models the unknown distortion. Once these coefficients are known, Eq. (3.6.1) is fully specified and it is used to map the observed (x,y) points onto the reference (u,v) coordinate system. The process of using tiepoints to infer the polynomial coefficients necessary for defining the spatial transformation is known as *spatial interpolation* [Green 89].

Rather than apply the mapping functions over the entire set of points, an *interpolation grid* is often introduced to reduce the computational complexity. This method evaluates the mapping function at a relatively sparse set of grid, or mesh, points. The spatial correspondence of points internal to the mesh is computed by bilinear interpolation from the corner points [Bernstein 76] or by fitting cubic surface patches to the mesh [Goshtasby 89].

3.6.1. Inferring Polynomial Coefficients

Auxiliary information is needed to determine the polynomial coefficients. This information includes reseau marks, platform attitude and altitude data, and ground control points. Reseau marks are small cruciform markings inscribed on the faceplate of the sensor. Since the locations of the reseau marks can be accurately calibrated, the measured differences between their true locations and imaged (distorted) locations yields a sparse sensor distortion mapping. This accounts for the internal errors.

External errors can be directly characterized from platform attitude, altitude, and ephemerides data. However, this data is not generally precisely known. Consequently, ground control points are used to determine the external error. A *ground control point* (GCP) is an identifiable natural landmark detectable in a scene, whose location and elevation are known precisely. This establishes a correspondence between image coordinates (measured in rows and columns) and map coordinates (measured in latitude/longitude angles, feet, or meters). Typical GCPs include airports, highway intersections, land-water interfaces, and geological patterns [Bernstein 71, 76].

A number of these points are located and differences between their observed and actual locations are used to characterize the external error component. Together with the internal distortion function, this serves to fully define the spatial transformation that inverts the distortions present in the input image, yielding a corrected output image. Since there are more ground control points than undetermined polynomial coefficients, a least-squared-error fit is used. In the discussion that follows, we describe several techniques to solve for the unknown polynomial coefficients. They include the pseudoinverse solution, least-squares with ordinary and orthogonal polynomials, and weighted least-squares with orthogonal polynomials.

3.6.2. Pseudoinverse Solution

Let a correspondence be established between M points in the observed and reference images. The spatial transformation that approximates this correspondence is chosen to be a polynomial of degree N. In two variables (e.g., x and y), such a polynomial has K coefficients where

$$K = \sum_{i=0}^{N} \sum_{j=0}^{N-i} 1 = \frac{(N+1)(N+2)}{2}$$

For example, a second-degree approximation requires only six coefficients to be solved. In this case, $N=2$ and $K=6$. We thus have

$$
\begin{bmatrix} u_1 \\ u_2 \\ u_3 \\ \ldots \\ \ldots \\ u_M \end{bmatrix}
=
\begin{bmatrix}
1 & x_1 & y_1 & x_1 y_1 & x_1^2 & y_1^2 \\
1 & x_2 & y_2 & x_2 y_2 & x_2^2 & y_2^2 \\
1 & x_3 & y_3 & x_3 y_3 & x_3^2 & y_3^2 \\
 & & & \ldots & & \\
 & & & \ldots & & \\
1 & x_M & y_M & x_M y_M & x_M^2 & y_M^2
\end{bmatrix}
\begin{bmatrix} a_{00} \\ a_{10} \\ a_{01} \\ a_{11} \\ a_{20} \\ a_{02} \end{bmatrix}
\tag{3.6.2}
$$

where $M \geq 6$. A similar equation holds for v and b_{ij}. Both of these expressions may be written in matrix notation as

$$U = WA \tag{3.6.3}$$

$$V = WB$$

In order to solve for A and B, we must compute the inverse of W. However, since W has dimensions $M \times K$, it is not a square matrix and thus it has no inverse. Instead, we first multiply both sides by W^T before isolating the desired A and B vectors. This serves to cast W into a $K \times K$ square matrix that may be readily inverted [Wong 77]. This gives us

$$W^T U = W^T W A \tag{3.6.4}$$

Expanding the matrix notation, we obtain Eq. (3.6.5), the following system of linear equations. For notational convenience, we omit the limits of the summations and the associated subscripts. Note that all summations run from $k=1$ to M, operating on data x_k, y_k, and u_k.

$$
\begin{bmatrix} \sum u \\ \sum xu \\ \sum yu \\ \sum xyu \\ \sum x^2u \\ \sum y^2u \end{bmatrix} = \begin{bmatrix} M & \sum x & \sum y & \sum xy & \sum x^2 & \sum y^2 \\ \sum x & \sum x^2 & \sum xy & \sum x^2y & \sum x^3 & \sum xy^2 \\ \sum y & \sum xy & \sum y^2 & \sum xy^2 & \sum x^2y & \sum y^3 \\ \sum xy & \sum x^2y & \sum xy^2 & \sum x^2y^2 & \sum x^3y & \sum xy^3 \\ \sum x^2 & \sum x^3 & \sum x^2y & \sum x^3y & \sum x^4 & \sum x^2y^2 \\ \sum y^2 & \sum xy^2 & \sum y^3 & \sum xy^3 & \sum x^2y^2 & \sum y^4 \end{bmatrix} \begin{bmatrix} a_{00} \\ a_{10} \\ a_{01} \\ a_{11} \\ a_{20} \\ a_{02} \end{bmatrix}
$$

A similar procedure is performed for v and b_{ij}. Solving for A and B, we have

$$
A = (W^TW)^{-1}W^TU \tag{3.6.6}
$$

$$
B = (W^TW)^{-1}W^TV
$$

This technique is known as the *pseudoinverse solution* to the linear least-squares problem. It leaves us with K-element vectors A and B, the polynomial coefficients for the U and V mapping functions, respectively.

Unfortunately, this method suffers from several problems. The primary difficulty lies in multiplying Eq. (3.6.3) with W^T. This squares the condition number, thereby reducing the precision of the coefficients by one half. As a result, alternate solutions are recommended. For example, it is preferable to compute a decomposition of W rather than solve for its inverse directly. The reader is referred to the linear algebra literature for a discussion of singular value decomposition, and LU and QR decomposition techniques [Strang 80]. An exposition can be found in [Press 88], where emphasis is given to an intuitive understanding of the benefits and drawbacks of these techniques. The text also provides useful source code written in the C programming language. (Versions of the book with Pascal and Fortran programs are available as well).

3.6.3. Least-Squares With Ordinary Polynomials

The pseudoinverse solution proves to be identical to that of the classic least-squares formulation with ordinary polynomials. Although both approaches share some of the same problems, the least-squares method is discussed here due to its prominence in the solution of overdetermined systems of linear equations. Furthermore, it can be altered to yield a stable closed-form solution for the unknown coefficients, as described in the next section.

Referring back to Eq. (3.6.1) with $N=2$, coefficients a_{ij} are determined by minimizing

$$E = \sum_{k=1}^{M} \delta_k^2$$

$$= \sum_{k=1}^{M} [\, U(x_k, y_k) - u_k \,]^2 \qquad (3.6.7)$$

$$= \sum_{k=1}^{M} [\, a_{00} + a_{10}x_k + a_{01}y_k + a_{11}x_ky_k + a_{20}x_k^2 + a_{02}y_k^2 - u_k \,]^2$$

This is achieved by determining the partial derivatives of E with respect to coefficients a_{ij}, and equating them to zero. For each coefficient a_{ij}, we have

$$\frac{\partial E}{\partial a_{ij}} = 2 \sum_{k=1}^{M} \delta_k \frac{\partial \delta_k}{\partial a_{ij}} = 0 \qquad (3.6.8)$$

By considering the partial derivative of E with respect to all six coefficients, we obtain Eq. (3.6.9), the following system of linear equations. For notational convenience, we have omitted the limits of summation and subscripts.

$$\begin{aligned}
\sum u &= a_{00}M &&+ a_{10}\sum x &&+ a_{01}\sum y &&+ a_{11}\sum xy &&+ a_{20}\sum x^2 &&+ a_{02}\sum y^2 \\
\sum xu &= a_{00}\sum x &&+ a_{10}\sum x^2 &&+ a_{01}\sum xy &&+ a_{11}\sum x^2y &&+ a_{20}\sum x^3 &&+ a_{02}\sum xy^2 \\
\sum yu &= a_{00}\sum y &&+ a_{10}\sum xy &&+ a_{01}\sum y^2 &&+ a_{11}\sum xy^2 &&+ a_{20}\sum x^2y &&+ a_{02}\sum y^3 \\
\sum xyu &= a_{00}\sum xy &&+ a_{10}\sum x^2y &&+ a_{01}\sum xy^2 &&+ a_{11}\sum x^2y^2 &&+ a_{20}\sum x^3y &&+ a_{02}\sum xy^3 \\
\sum x^2u &= a_{00}\sum x^2 &&+ a_{10}\sum x^3 &&+ a_{01}\sum x^2y &&+ a_{11}\sum x^3y &&+ a_{20}\sum x^4 &&+ a_{02}\sum x^2y^2 \\
\sum y^2u &= a_{00}\sum y^2 &&+ a_{10}\sum xy^2 &&+ a_{01}\sum y^3 &&+ a_{11}\sum xy^3 &&+ a_{20}\sum x^2y^2 &&+ a_{02}\sum y^4
\end{aligned}$$

This is a symmetric 6×6 system of linear equations, whose coefficients are all summations from $k = 1$ to M which are evaluated from the original data. By inspection, this result is equivalent to Eq. (3.6.5), the system of equations derived earlier in the pseudoinverse solution. Known as the *normal equations*, this system of linear equations can be compactly expressed in the following notation.

$$\sum_{i=0}^{N} \sum_{j=0}^{N-i} a_{ij} \left[\sum_{k=1}^{M} x_k^i y_k^j x_k^l y_k^m \right] = \sum_{k=1}^{M} u_k x_k^l y_k^m \qquad (3.6.10)$$

for $l = 0, ..., N$ and $m = 0, ..., N - l$. Note that (i, j) are running indices along rows, where each row is associated with a constant (l, m) pair.

The least-squares procedure operates on an overdetermined system of linear equations, (i.e., M data points are used to determine K coefficients, where $M > K$). As a result, we have only an *approximating* mapping function. If we substitute the (x, y) control point coordinates back into the polynomial, the computed results will lie near, but will generally not coincide exactly with their counterpart (u, v) coordinates. Stated intuitively, the polynomial of order K must yield the best compromise among all M control points. As we have seen, the best fit is determined by minimizing the sum of the squared-error $[U(x_k, y_k) - u_k]^2$ for $k = 1, ..., M$. Of course, if $M = K$ then no compromise is necessary and the polynomial mapping function will *interpolate* the points, actually passing through them.

The results of the least-squares approach and the pseudoinverse solution will yield those coefficients that best approximate the true mapping function as given by the M control points. We can refine the approximation by considering the error at each data point and throwing away that point which has maximum deviation from the model. The coefficients are then recomputed with $M - 1$ points. This process can be repeated until we have K points remaining (matching the number of polynomial coefficients), or until some error threshold has been reached. In this manner, the polynomial coefficients are made to more closely fit an increasingly consistent set of data. Although this method requires more computation, it is recommended when noisy data are known to be present.

3.6.4. Least-Squares With Orthogonal Polynomials

The normal equations in Eq. (3.6.10) can be solved if M is greater than K. As K increases, however, solving a large system of equations becomes unstable and inaccurate. Numerical accuracy is further hampered by possible linear dependencies among the equations. As a result, the direct solution of the normal equations is generally not the best way to find the least-squares solution. In this section, we introduce orthogonal polynomials for superior results.

The mapping functions U and V may be rewritten in terms of orthogonal polynomials as

$$u = \sum_{i=1}^{K} a_i P_i(x,y) \qquad (3.6.11)$$

$$v = \sum_{i=1}^{K} b_i P_i(x,y)$$

where a_i and b_i are the unknown coefficients for orthogonal polynomials P_i. As we shall see, introducing orthogonal polynomials allows us to determine the coefficients without solving a linear system of equations. We begin by defining the orthogonality property and then demonstrate how it is used to construct orthogonal polynomials from a set of linearly independent basis functions. The orthogonal polynomials, together with the supplied control points, are combined in a closed-form solution to yield the desired a_i and b_i coefficients.

A set of polynomials $P_1(x,y), P_2(x,y), ..., P_K(x,y)$ is orthogonal over points (x_k,y_k) if

$$\sum_{k=1}^{M} P_i(x_k,y_k) P_j(x_k,y_k) = 0 \qquad i \neq j \qquad (3.6.12)$$

These polynomials may be constructed from a set of linearly independent *basis functions* spanning the space of polynomials. We denote the basis functions here as $h_1(x,y)$,

$h_2(x,y)$, ..., $h_K(x,y)$. The polynomials can be constructed by using the Gram-Schmidt orthogonalization process, a method which generates orthogonal functions by the following incremental procedure.

$$P_1(x,y) = \alpha_{11}h_1(x,y)$$

$$P_2(x,y) = \alpha_{21}P_1(x,y) + \alpha_{22}h_2(x,y)$$

$$P_3(x,y) = \alpha_{31}P_1(x,y) + \alpha_{32}P_2(x,y) + \alpha_{33}h_3(x,y)$$

$$\cdots$$

$$P_K(x,y) = \alpha_{K1}P_1(x,y) + \alpha_{K2}P_2(x,y) + \cdots + \alpha_{KK}h_K(x,y)$$

Basis functions $h_i(x,y)$ are free to be any linearly independent polynomials. For example, the first six basis functions that we shall use are shown below.

$$h_1(x,y) = 1$$

$$h_2(x,y) = x$$

$$h_3(x,y) = y$$

$$h_4(x,y) = x^2$$

$$h_5(x,y) = xy$$

$$h_6(x,y) = y^2$$

The α_{ij} parameters are determined by setting $\alpha_{i1} = 1$ and applying the orthogonality property of Eq. (3.6.12) to the polynomials. That is, the α_{ij}'s are selected such that the product $P_i P_j = 0$ for $i \ne j$. The following results are obtained.

$$\alpha_{ii} = - \frac{\displaystyle\sum_{k=1}^{M} [P_1(x_k,y_k)]^2}{\displaystyle\sum_{k=1}^{M} P_1(x_k,y_k) h_i(x_k,y_k)} \qquad i = 1,...,K \qquad (3.6.13a)$$

$$\alpha_{ij} = -\alpha_{ii} \frac{\displaystyle\sum_{k=1}^{M} P_j(x_k,y_k) h_i(x_k,y_k)}{\displaystyle\sum_{k=1}^{M} [P_j(x_k,y_k)]^2} \qquad \begin{array}{l} i = 1,...,K; \\ j = 1,...,K-1 \end{array} \qquad (3.6.13b)$$

Equations (3.6.13a) and (3.6.13b) are the results of isolating the α_{ij} coefficients once P_i has been multiplied with P_1 and P_j, respectively. This multiplication serves to eliminate all product terms except those associated with the desired coefficient.

Having computed the α_{ij}'s, the orthogonal polynomials are determined. Note that they are simply linear combinations of the basis functions. We must now solve for the a_i

coefficients in Eq. (3.6.11). Using the least-squares approach, the error E is defined as

$$E = \sum_{k=1}^{M} \left[\sum_{i=1}^{K} a_i P_i(x_k, y_k) - u_k \right]^2 \qquad (3.6.14)$$

The coefficients are determined by taking the partial derivatives of E with respect to the coefficients and setting them equal to zero. This results in the following system of linear equations.

$$\sum_{i=1}^{K} a_i \left[\sum_{k=1}^{M} P_i(x_k, y_k) P_j(x_k, y_k) \right] = \sum_{k=1}^{M} u_k P_j(x_k, y_k) \qquad j = 1,...,K \quad (3.6.15)$$

Applying the orthogonal property of Eq. (3.6.12), we obtain the following simplification

$$a_i \sum_{k=1}^{M} [P_i(x_k, y_k)]^2 = \sum_{k=1}^{M} u_k P_i(x_k, y_k) \qquad (3.6.16)$$

The desired coefficients are thereby given as

$$a_i = \frac{\sum_{k=1}^{M} u_k P_i(x_k, y_k)}{\sum_{k=1}^{M} [P_i(x_k, y_k)]^2} \qquad (3.6.17a)$$

A similar procedure is repeated for computing b_i, yielding

$$b_i = \frac{\sum_{k=1}^{M} v_k P_i(x_k, y_k)}{\sum_{k=1}^{M} [P_i(x_k, y_k)]^2} \qquad (3.6.17b)$$

Performing least-squares with orthogonal polynomials offers several advantages worth noting. First, determining coefficients a_i and b_i in Eq. (3.6.11) does not require solving a system of equations. Instead, a closed-form solution is available, as in Eq. (3.6.17). This proves to be a faster and more accurate solution. The computational cost of this method is $O(MK^3)$. Second, additional orthogonal polynomial terms may be added to the mapping function to increase the fitting accuracy of the approximation. For instance, we may define the mean-square error to be

$$E_{ms} = \frac{1}{M} \sum_{k=1}^{M} \left[\sum_{i=1}^{K} a_i P_i(x_k, y_k) - u_k \right]^2 \qquad (3.6.18)$$

If E_{ms} exceeds some threshold, then we may increase the number of orthogonal polynomial terms in the mapping function to reduce the error. The orthogonality property allows these terms to be added *without* recomputation of all the polynomial coefficients. This facilitates a simple means of adaptively determining the degree of the polynomial necessary to satisfy some error bound. Note that this is not true for ordinary polynomials. In that case, as the degree of the polynomial increases the values of all parameters change, requiring the recomputation of all the coefficients.

Numerical accuracy is generally enhanced by normalizing the data before performing the least-square computation. Thus, it is best to translate and scale the data to fit the $[-1,1]$ range. This serves to reduce the ill-conditioning of the problem. In this manner, all the results of the basis function evaluations fall within a narrow range that exploits the numerical properties of floating point computation.

3.6.5. Weighted Least-Squares

One flaw with the least-squares formulation is its global error measure. Nowhere in Eqs. (3.6.7) or (3.6.14) is there any consideration given to the distance between control points (x_k, y_k) and approximation positions (x, y). Intuitively, it is desirable for the error contributions of distant control points to have less influence than those which are closer to the approximating position. This serves to confine local geometric differences, and prevents them from averaging equally over the entire image.

The least-squares method may be localized by introducing a weighting function W_k that represents the contribution of control point (x_k, y_k) on point (x, y).

$$W_k = \frac{1}{\sqrt{\delta + (x - x_k)^2 + (y - y_k)^2}} \qquad (3.6.19)$$

The parameter δ determines the influence of distant control points on approximating points. As δ approaches zero, the distant points have less influence and the approximating mapping function is made to follow the local control points more closely. As δ grows, it dominates the distance measurement and curtails the impact of the weighting function to discriminate between local and distant control points. This serves to smooth the approximated mapping function, making it approach the results of the standard least-squares technique discussed earlier.

The weighted least-squares expression is given as

$$E(x,y) = \sum_{k=1}^{M} [U(x_k, y_k) - u_k]^2 \, W_k(x,y) \qquad (3.6.20)$$

This represents a dramatic increase in computation over the nonweighted method since the error term now becomes a function of position. Notice that each (x, y) point must now recompute the squared-error summation. For ordinary polynomials, Eq. (3.6.10) becomes

$$\sum_{i=0}^{N} \sum_{j=0}^{N-i} a_{ij} \left[\sum_{k=1}^{M} W_k(x,y) x_k^i y_k^j x_k^l y_k^m \right] = \sum_{k=1}^{M} W_k u_k x_k^l y_k^m \qquad (3.6.21)$$

for $l = 0, ..., N$ and $m = 0, ..., N - l$. For orthogonal polynomials, the orthogonality property of Eq. (3.6.12) becomes

$$\sum_{k=1}^{M} W_k(x,y) P_i(x_k, y_k) P_j(x_k, y_k) = 0 \qquad i \neq j \qquad (3.6.22)$$

and the parameters of the orthogonal polynomials are

$$\alpha_{ii} = - \frac{\sum\limits_{k=1}^{M} W_k(x,y)[P_1(x_k,y_k)]^2}{\sum\limits_{k=1}^{M} W_k(x,y)P_1(x_k,y_k)h_i(x_k,y_k)} \qquad i = 1,...,K \qquad (3.6.23a)$$

$$\alpha_{ij} = -\alpha_{ii} \frac{\sum\limits_{k=1}^{M} W_k(x,y)P_j(x_k,y_k)h_i(x_k,y_k)}{\sum\limits_{k=1}^{M} W_k(x,y)[P_j(x_k,y_k)]^2} \qquad \begin{matrix} i = 1,...,K; \\ j = 1,...,K-1 \end{matrix} (3.6.23b)$$

Finally, the desired coefficients for Eq. (3.6.11) as determined by the weighted least-squares method are

$$a_i(x,y) = \frac{\sum\limits_{k=1}^{M} W_k(x,y)u_k P_i(x_k,y_k)}{\sum\limits_{k=1}^{M} W_k(x,y)[P_i(x_k,y_k)]^2} \qquad (3.6.24a)$$

$$b_i(x,y) = \frac{\sum\limits_{k=1}^{M} W_k(x,y)v_k P_i(x_k,y_k)}{\sum\limits_{k=1}^{M} W_k(x,y)[P_i(x_k,y_k)]^2} \qquad (3.6.24b)$$

The computational complexity of weighted least-squares is $O(NMK^3)$. It is N times greater than that of standard least-squares, where N is the number of pixels in the reference image. Although this technique is costlier, it is warranted when the mapping function is to be approximated using information highly in favor of local measurements.

Source code for the weighted least-squares method is provided below. The program, written in C, is based on the excellent exposition of local approximation methods found in [Goshtasby 88]. It expects the user to pass the list of correspondence points in three global arrays: X, Y, and Z. That is, $(X[i],Y[i]) \rightarrow Z[i]$. The full image of interpolated or approximated correspondence values are stored into global array S.

```
#define MXTERMS  10

/* global variables */
int N;
float *X, *Y, *Z, *W, A[MXTERMS][MXTERMS];
float init_alpha(), poly(), coef(), basis();

/*************************************************************************
        Weighted least-squares with orthogonal polynomials
        Input : X, Y, and Z are 3 float arrays for x, y, z=f(x,y) coords
                delta is the smoothing factor (0 is not smooth)
        Output: S <- fitted surface values of points (xsz by ysz)
        (X, Y, Z, and N are passed as global arguments)

        Based on algorithm described by Ardeshir Goshtasby in
        "Image registration by local approximation methods",
        Image and Vision Computing, vol. 6, no. 4, Nov. 1988
*************************************************************************/

wlsq(delta, xsz, ysz, S)
float delta, *S;
int xsz, ysz;
{
        int i, j, k, x, y, t, terms;
        float a, c, f, p, dx2, dy2, *s;

        /* N is already initialized with the number of control points */
        W = (float *) calloc(N, sizeof(float));           /* allocate memory for weights */

        /* determine the number of terms necessary for error < .5 (optional) */
        for(terms=3; terms < MXTERMS; terms++) {
                for(i=0; i<N; i++) {
                        /* init W: the weights of the N control points on x,y */
                        for(j=0; j<N; j++) {
                                dx2  = (X[i]-X[j]) * (X[i]-X[j]);
                                dy2  = (Y[i]-Y[j]) * (Y[i]-Y[j]);
                                W[j] = 1.0 / sqrt(dx2 + dy2 + delta);
                        }
                        /* init A: alpha_jk coeffs of the ortho polynomials */
                        for(j=0; j<terms; j++) A[j][j] = init_alpha(j,j);
                        for(j=0; j<terms; j++) {
                                for(k=0; k<j; k++) A[j][k] = init_alpha(j,k);
                        }
                        /* compute error at each control point over all terms */
                        for(t=f=0; t<terms; t++) {
                                a = coef(t);
                                p = poly(t, X[i], Y[i]);
                                f += (a * p);
                        }
```

```
                                if(ABS(Z[i] - f) > .5) break;
                        }
                        if(i == N) break;        /* found terms such that error < .5 */
                }

                /* perform surface approximation */
                for(y=0; y<ysz; y++) {
                        for(x=0; x<xsz; x++) {
                                /* init W: the weights of the N control points on x,y */
                                for(i=0; i<N; i++) {
                                        dx2  = (x-X[i]) * (x-X[i]);
                                        dy2  = (y-Y[i]) * (y-Y[i]);
                                        W[i] = 1.0 / sqrt(dx2 + dy2 + delta);
                                }
                                /* init A: alpha_jk coeffs of the ortho polynomials */
                                for(j=0; j<terms; j++) A[j][j] = init_alpha(j,j);
                                for(j=0; j<terms; j++) {
                                        for(k=0; k<j; k++) A[j][k] = init_alpha(j,k);
                                }
                                /* evaluate surface at (x,y) over all terms */
                                for(i=f=0; i<terms; i++) {
                                        a = coef(i);
                                        p = poly(i,(float)x,(float)y);
                                        f += (a * p);
                                }
                                *S++ = (float) f;        /* save fitted surface values */
                        }
                }
        }

/***************************************************************************
        Compute parameter alpha (Eq. 3.6.23)
***************************************************************************/

float init_alpha(j,k)
int j, k;
{
        int i;
        float a, h, p, num, denum;

        if(k == 0) a = 1.0;              /* case 0: a_j0 */
        else if(j == k) {               /* case 1: a_jj */
                num = denum = 0;
                for(i=0; i<N; i++) {
                        h = basis(j, X[i], Y[i]);
                        num   += (W[i]);

                        denum += (W[i]*h);
                }
                a = -num / denum;
```

```
        } else {                        /* case 2: a_jk, j!=k */
                num = denum = 0;
                for(i=0; i<N; i++) {
                        h = basis(j, X[i], Y[i]);
                        p = poly(k, X[i], Y[i]);
                        num   += (W[i]*p*h);
                        denum += (W[i]*p*p);
                }
                a = -A[j][j] * num / denum;
        }
        return(a);
}


/***************************************************************
        Find the k th mapping function coefficient (Eq. 3.6.24)
 ***************************************************************/

float coef(k)
int k;
{
        int i;
        float p, num, denum;

        num  = 0;
        denum = 0;
        for(i=0; i<N; i++) {
                p = poly(k, X[i], Y[i]);
                num   += (W[i] * Z[i] * p);
                denum += (W[i] * p*p);
        }
        return(num / denum);
}


/***************************************************************
        Determine the polynomial function at point (x,y)
 ***************************************************************/

float poly(k,x,y)
int k;
float x, y;
{
        int i;
        float p;

        for(i=p=0; i<k; i++)
                p += (A[k][i] * poly(i,x,y));
        p += (A[k][k] * basis(k,x,y));
        return(p);
}
```

```
/*****************************************************************
            Return the (x,y) value of orthogonal basis function f
******************************************************************/


float basis(f, x, y)
int f;
float x, y;
{
        float h;

        switch(f) {
        case 0:        h = 1.0;      break;
        case 1:        h = x;        break;
        case 2:        h = y;        break;
        case 3:        h = x*x;      break;
        case 4:        h = x*y;      break;
        case 5:        h = y*y;      break;
        case 6:        h = x*x*x;    break;
        case 7:        h = x*x*y;    break;
        case 8:        h = x*y*y;    break;
        case 9:        h = y*y*y;    break;
        }
        return(h);

}
```

3.7. PIECEWISE POLYNOMIAL TRANSFORMATIONS

Global polynomial transformations impose a single mapping function upon the whole image. They often do not account for local geometric distortions such as scene elevation, atmospheric turbulence, and sensor nonlinearity. Consequently, piecewise mapping functions have been introduced to handle local deformations [Goshtasby 86, 87].

The study of piecewise interpolation has received much attention in the spline literature. The majority of the work, however, assumes that the data points are available on a rectangular grid. In our application, this is generally not the case. Instead, we must consider the problem of fitting a composite surface to scattered 3-D data [Franke 79].

3.7.1. A Surface Fitting Paradigm for Geometric Correction

The problem of determining functions U and V can be conveniently posed as a surface fitting problem. Consider, for example, knowing M control points labeled (x_k, y_k) in the observed image and (u_k, v_k) in the reference image, where $1 \leq k \leq M$. Deriving mapping functions U and V is equivalent to determining two smooth surfaces: one that passes through points (x_k, y_k, u_k) and the other that passes through (x_k, y_k, v_k) for $1 \leq k \leq M$. Figure 3.13 shows a surface for $U(x, y)$ with control points given at the grid points.

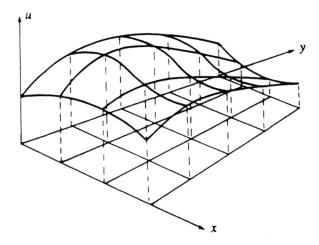

Figure 3.13: Surface $U(x,y)$.

Before an image undergoes geometric distortion, these surfaces are defined to be ramp functions. This follows from the observation that $u_k = x_k$ and $v_k = y_k$ in the absence of any deformation. Introducing geometric distortions will cause these surfaces to deviate from their initial ramp configurations. Note that as long as the surface is monotonically nondecreasing, the resulting image does not fold back upon itself.

Given only sparse control points, it is necessary to interpolate a surface through these points and closely approximate the unknown distortion function. The problem of smooth surface interpolation/approximation from scattered data has been the subject of much attention across many fields. It is of great practical importance to all disciplines concerned with inferring an analytic solution given only a few samples. Traditionally, the solution to this problem is posed in one of two forms: global or local transformations.

A *global transformation* considers all the control points in order to derive the unknown coefficients of the mapping function. Most of the solutions described thus far are global polynomial methods. This chapter devotes a lot of attention to polynomials due to their popularity. Generally, the polynomial coefficients computed from global methods will remain fixed across the entire image. That is, the same polynomial transformation is applied over each pixel.

It is clear that global low-order polynomial mapping functions can only approximate these surfaces. Furthermore, the least-squares technique used to determine the coefficients averages a local geometric difference over the whole image area independent of the position of the difference. As a result, local distortions cannot be handled and they instead contribute to errors at distant locations. We may, instead, interpolate the surface with a global mapping by increasing the degree of the polynomial to match the number of control points. However, the resulting polynomial is likely to exhibit excessive spatial undulations and thereby introduce further artifacts.

Weighted least-squares was introduced as an alternate approach. Although it is a global method that considers all control points, it recomputes the coefficients at each pixel by using a weighting function that is biased in favor of nearby control points. In this manner, it constitutes a hybrid global/local method, computing polynomial coefficients through a global technique, but permitting the coefficients to be spatially-varying. The extent to which the surface is interpolated or approximated is left to a user-specified parameter.

If the control points all lie on a rectangular mesh, as in Fig. 3.13, it is possible to use bicubic spline interpolation. For example, interpolating B-splines or Bezier surface patches can be fitted to the data [Goshtasby 89]. These methods are described globally but remain sensitive to local data. This behavior is contrary to least-squares for fitting polynomials to local data, where a local distortion is averaged out equally over the entire image. With global spline interpolation (see Section 3.8), a local distortion has a global effect on the transformed image, but its effect is vanishingly small on distant points.

A *local transformation* considers only nearby control points in evaluating interpolated values along a surface. In this section, we describe piecewise polynomial transformation, a local technique for computing a surface from scattered points.

3.7.2. Procedure

One general procedure for performing surface interpolation on irregularly-spaced 3-D points consists of the following operations.

1. Partition each image into triangular regions by connecting neighboring control points with noncrossing line segments, forming a planar graph. This process, known as triangulation, serves to delimit local neighborhoods over which surface patches will be defined.

2. Estimate partial derivatives of U (and similarly V) with respect to x and y at each of the control points. This may be done by using a local method, with data values taken from nearby control points, or with a global method using all the control points. Computing the partial derivatives is necessary only if the surface patches are to join smoothly, i.e., for C^1, C^2, or smoother results.[†]

3. For each triangular region, fit a smooth surface through the vertices satisfying the constraints imposed by the partial derivatives. The surface patches are generated by using low-order bivariate polynomials. A linear system of equations must be solved to compute the polynomial coefficients.

4. Those regions lying outside the convex hull of the data points must extrapolate the surface from the patches lying along the boundary.

5. For each point (x,y), determine its enclosing triangle and compute an interpolated value u (similarly for v) by using the polynomial coefficients derived for that triangle. This yields the (u,v) coordinates necessary to resample the input image.

---

† C^1 and C^2 denote continuous first and second derivatives, respectively.

3.7.3. Triangulation

Triangulation is the process of tesselating the convex hull of a set of N distinct points into triangular regions. This is done by connecting neighboring control points with noncrossing line segments, forming a planar graph. Although many configurations are possible, we are interested in achieving a partition such that points inside a triangle are closer to its three vertices than to vertices of any other triangle. This is called the optimal triangulation and it avoids generating triangles with sharp angles and long edges. In this manner, only nearby data points will be used in the surface patch computations that follow. Several algorithms to obtain optimal triangulations are reviewed below.

In [Lawson 77], the author describes how to optimize an arbitrary triangulation initially created from the given data. He gives the following three criteria for optimality.

1. Max-min criterion: For each quadrilateral in the set of triangles, choose the triangulation that maximizes the minimum interior angle of the two obtained triangles. This tends to bias the tesselation against undesirable long thin triangles. Figure 3.14a shows triangle ABC selected in favor of triangle BCD under this criterion. The technique has computational complexity $O(N^{4/3})$.

2. The circle criterion: For each quadrilateral in the set of triangles, pass a circle through three of its vertices. If the fourth vertex lies outside the circle then split the quadrilateral into two triangles by drawing the diagonal that does not pass through the vertex. This is illustrated in Fig. 3.14b.

3. Thessian region criterion: For each quadrilateral in the set of triangles, construct the Thessian regions. In computational geometry, the Thessian regions are also known as Delaunay, Dirichlet, and Voronoi regions. They are the result of intersecting the perpendicular bisectors of the quadrilateral edges, as shown in Fig. 3.14c. This serves to create regions around each control point P such that points in that region are closer to P than to any other control point. Triangulation is obtained by joining adjacent Delaunay regions, a result known as Delaunay triangulation (Fig. 3.15). An $O(N^{3/2})$ triangulation algorithm using this method is described in [Green 78].

An $O(N \log_2 N)$ recursive algorithm that determines the optimal triangulation is given in [Lee 80]. The method recursively splits the data into halves using the x-values of the control points until each subset contains only three or four points. These small subsets are then easily triangulated using any of Lawson's three criteria. Finally, they are merged into larger subsets until all the triangular subsets are consumed, resulting in an optimal triangulation of the control points. Due to its speed and simplicity, this divide-and-conquer technique was used in [Goshtasby 87] to compute piecewise cubic mapping functions. The subject of triangulations and data structures for them is reviewed in [De Floriani 87].

3.7.4. Linear Triangular Patches

Once the triangular regions are determined, the scattered 3-D data (x_i, y_i, u_i) or (x_i, y_i, v_i) are partitioned into groups of three points. Each group is fitted with a low-order bivariate polynomial to generate a surface patch. In this manner, triangulation allows

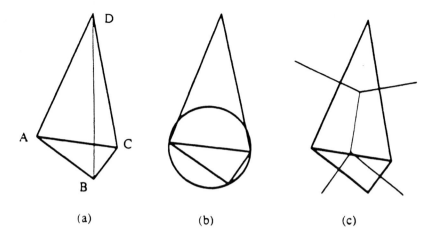

Figure 3.14: Three criteria for optimal triangulation [Goshtasby 86].

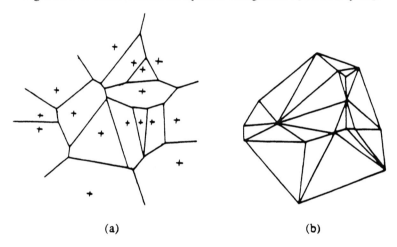

Figure 3.15: (a) Delaunay tesselation; (b) Triangulation [Goshtasby 86].

only nearby control points to influence the surface patch calculations. Together, these patches comprise a composite surface defining the corresponding u or v coordinates at each point in the observed image.

We now consider the case of fitting the triangular patches with a linear interpolant, i.e., a plane. The equation of a plane through three points (x_1, y_1, u_1), (x_2, y_2, u_2), and (x_3, y_3, u_3) is given by

$$Ax + By + Cu + D = 0 \qquad (3.7.1)$$

where

$$A = \begin{vmatrix} y_1 & u_1 & 1 \\ y_2 & u_2 & 1 \\ y_3 & u_3 & 1 \end{vmatrix}; \quad B = -\begin{vmatrix} x_1 & u_1 & 1 \\ x_2 & u_2 & 1 \\ x_3 & u_3 & 1 \end{vmatrix}; \quad C = \begin{vmatrix} x_1 & y_1 & 1 \\ x_2 & y_2 & 1 \\ x_3 & y_3 & 1 \end{vmatrix}; \quad D = -\begin{vmatrix} x_1 & y_1 & u_1 \\ x_2 & y_2 & u_2 \\ x_3 & y_3 & u_3 \end{vmatrix}$$

As seen in Fig. 3.15b, the triangulation covers only the convex hull of the set of control points. In order to extrapolate points outside the convex hull, the planar triangles along the boundary are extended to the image border. Their extents are limited to the intersections of neighboring planes.

3.7.5. Cubic Triangular Patches

Although piecewise linear mapping functions are continuous at the boundaries between neighboring functions, they do not provide a smooth transition across patches. In order to obtain smoother results, the patches must at least use C^1 interpolants. This is achieved by fitting the patches with higher-ordered bivariate polynomials.

This subject has received much attention in the field of computer-aided geometric design. Many algorithms using N-degree polynomials have been proposed. They include $N=2$ [Powell 77], $N=3, 4$ [Percell 76], and $N=5$ [Akima 78]. In this section, we examine the case of fitting triangular regions with cubic patches ($N=3$). A cubic patch f is a third-degree bivariate polynomial of the form

$$f(x,y) = a_1 + a_2 x + a_3 y + a_4 x^2 + a_5 xy + a_6 y^2 + a_7 x^3 + a_8 x^2 y + a_9 xy^2 + a_{10} y^3 \quad (3.7.2)$$

The ten coefficients can be solved by determining ten constraints among them. Three relations are obtained from the coordinates of the three vertices. Six relations are derived from the partial derivatives of the patch with respect to x and y at the three vertices. Smoothly joining a patch with its neighbors requires the partial derivatives of the two patches to be the same in the direction normal to the common edge. This adds three more constraints, yielding a total of twelve relations. Since we have ten unknowns and twelve equations, the system is overdetermined and cannot be solved as given.

The solution lies in the use of the Clough-Tocher triangle, a widely known C^1 triangular interpolant [Clough 65]. Interpolation with the Clough-Tocher triangle requires the triangular region to be divided into three subtriangles. Fitting a surface patch to each subtriangle yields a total of thirty unknown parameters. Since exactly thirty constraints can be derived in this process, a linear system of thirty equations must be solved to compute a surface patch for each region in the triangulation. A full derivation of this method is given in [Goshtasby 87]. A complete review of triangular interpolants can be found in [Barnhill 77].

An interpolation algorithm offering smooth blending across patches requires partial derivative data. Since this is generally not available with the supplied data, it must be estimated. A straightforward approach to estimating the partial derivative at point P consists of fitting a second-degree bivariate polynomial through P and five of its neighbors. This allows us to determine the six parameters of the polynomial and directly compute the partial derivative. More accurate estimates can be obtained by a weighted least-squares technique using more than six points [Lawson 77].

Another approach is given in [Akima 78] where the author uses P and its m nearest points $P_1, P_2, ..., P_m$, to form vector products $V_{ij} = (P - P_i) \times (P - P_j)$ with P_i and P_j being all possible combinations of the points. The vector sum V of all V_{ij}'s is then calculated. Finally, the partial derivatives are estimated from the slopes of a plane that is normal to the vector sum. A similar approach is described in [Klucewicz 78]. Akima later improved this technique by weighting the contribution of each triangle such that small weights were assigned to large or narrow triangles when the vector sum was calculated [Akima 84]. For a comparison of methods, see [Nielson 83] and [Stead 84].

Despite the apparently intuitive formulation of performing surface interpolation by fitting linear or cubic patches to triangulated regions, partitioning a set of irregularly-spaced points into distinct neighborhoods is not straightforward. Three criteria for "optimal" triangulation were described. These heuristics are arbitrary and are not without problems.

In an effort to circumvent the problem of defining neighborhoods, a uniform hierarchical procedure has recently been proposed [Burt 88]. This method fits a polynomial surface to the available data within a local neighborhood of each sample point. The procedure, called hierarchical polynomial fit filtering, yields a multiresolution set of low-pass filtered images, i.e., a pyramid. Finally, the set of blurred images are combined through multiresolution interpolation to form a smooth surface passing through the original data. The recent literature clearly indicates that surface interpolation and approximation from scattered data remains an open problem.

3.8. GLOBAL SPLINES

As is evident, inferring a mapping function given only sparse scattered correspondence points is an important problem. In this section, we examine this problem in terms of a more general framework: surface fitting with sophisticated mapping functions well beyond those defined by polynomials. In particular, we introduce global splines as a general solution. We discuss their definition in terms of basis functions and regularization theory. Research in this area demonstrates that global splines are useful for our purposes, particularly since they provide us a means of imposing constraints on the properties of our inferred mapping functions.

We examine the use of global splines defined in two ways: through basis functions and regularization theory. Although the use of global splines defined through basis functions overlaps with some of the techniques described earlier, we present it here to draw attention to the single underlying mathematical framework. Since they do not depend on any regular structure for the data, they are particularly useful for surface interpolation from scattered data.

3.8.1. Basis Functions

Global splines using basis functions is one of the oldest global interpolation techniques. It consists of the following procedure:

1. Define a set of basis functions $h_i(x,y)$, where $i = 1, ..., K$.
2. Define a set of correspondence points (x_j, y_j, u_j), where $j = 1, ..., M$, and u_j refers to the surface height associated with point (x_j, y_j). In this discussion, we limit ourselves to computing a surface for u. The process must be repeated for v as well.

3. Define the interpolating function to be a linear combination of these basis functions. We refer to the interpolation function as a spline. For example, the expression for mapping function U is a spline that passes through the supplied correspondence points. It is given as

$$U(x,y) = \sum_{i=1}^{K} a_i\, h_i(x,y) \qquad (3.8.1)$$

for some a_i.

4. Determine the unknown a_i coefficients by solving a system of linear equations to ensure that the function interpolates the data. The system of equations is given as $\hat{U} = HA$, or equivalently as

$$\begin{bmatrix} u_1 \\ u_2 \\ \cdots \\ \cdots \\ u_M \end{bmatrix} = \begin{bmatrix} h_1(x_1,y_1) & h_2(x_1,y_1) & \cdots & h_K(x_1,y_1) \\ h_1(x_2,y_2) & h_2(x_2,y_2) & \cdots & h_K(x_2,y_2) \\ & & \cdots & \\ & & \cdots & \\ h_1(x_M,y_M) & h_2(x_M,y_M) & \cdots & h_K(x_M,y_M) \end{bmatrix} \begin{bmatrix} a_1 \\ a_2 \\ \cdots \\ \cdots \\ a_K \end{bmatrix} \qquad (3.8.2)$$

The matrix H is often called the *design matrix* or the *Gram matrix* of the problem.

While the definition of this approach is rather simple, the choice of the basis functions is very nontrivial. Although we present a simple introduction here, a more thorough investigation can be found in [Franke 79], which includes a critical comparison of many global and local methods for scattered interpolation.

A simple choice for the set of basis functions is: $h_i(x,y) = 1, x, y, xy, x^2, y^2$ for $i = 1, ..., 6$. This choice, coincidently, is identical to that used in Eqs. (3.6.1) and (3.6.2) for a second-degree fit. If we were given exactly six data points, it becomes possible to interpolate the data. In that case, the coefficients may be determined by computing $H^{-1}U$, assuming H is nonsingular. Otherwise, if the number of data points exceeds the number of basis functions $(M > K)$, then any approximate solution to the overconstrained linear system can be used. It now no longer becomes possible to interpolate the data unless, of course, the input coincides with a function of order K.

For numerical reasons, it is preferable to compute a decomposition of H rather than compute its inverse. This is sometimes necessary because design matrices are often very ill-conditioned, and care should be taken in solving them. This tends to happen when a clutter of supplied correspondence points may cause several rows in a design matrix to differ only marginally. In such instances, it is suggested that an estimate of the condition number of the particular design matrix be obtained before interpreting the results. Techniques for simultaneously solving a linear system and estimating its condition number can be found in standard linear algebra packages (e.g., LINPACK [Dongarra 79], [NAG 80, IMSL 80]), or directly as the ratio of the largest to smallest nonzero singular value computed through singular value decomposition.

There are obviously many heuristic definitions one could give for the basis functions. While nothing in the definition requires the basis functions to be rotationally symmetric, most heuristic definitions are based on that assumption. In this case, each basis function becomes a function of the radial distance from the respective data point. For example, $h_i(x,y) = g(r)$ for some function g and radial distance $r = \sqrt{(x-x_i)^2 + (y-y_i)^2}$. One of the most popular radially symmetric functions uses the Gaussian basis function

$$g(r) = e^{-r/\sigma} \tag{3.8.3}$$

While it is possible to allow σ to vary with distance, in practice it is held constant. If σ is chosen correctly, this method can provide quite satisfactory results. Otherwise, poor results are obtained. In [Franke 79], it is suggested that $\sigma = 1.008\, d/\sqrt{n}$, where d is the diameter of the point set and n is the number of points.

A second heuristically defined set of radial basis functions suggested by even more researchers uses the popular B-spline. This has the advantage of having basis functions with finite support, i.e., only a small neighborhood of a point needs to be considered when evaluating its interpolated value. However, this method is still global in the sense that there is a chain of interdependencies between all points that must be addressed when evaluating the interpolating B-spline coefficients (see Section 5.2). Nevertheless, the design matrix can be considerably sparser and better conditioned. The basis function can then be taken to be

$$g(r) = 2(1-r/\delta)_+^3 - (1-(2r/\delta))_+^3 \tag{3.8.4}$$

where δ may be taken as $2.4192\, d/\sqrt{n}$, d is the diameter of the point set, and n is the number of points. Note that the $+$ in the subscripts denote that the respective terms are forced to zero when they are negative.

These heuristically defined basis functions have the intuitively attractive property that they fall off with the distance to the data point. This reflects the belief that distant points should exert less influence on local measurements. Again, nothing in the method requires this property. In fact, while it may seem counter-intuitive, fine results have been obtained with basis functions that increase with distance. An example of this behavior is Hardy's multiquadratic splines [Hardy 71, 75]. Here the radial basis functions are taken as

$$g(r) = \sqrt{r^2 + \delta} \tag{3.8.5}$$

for a constant δ. Hardy suggests the value $\delta = 0.815\, m$, where m is the mean squared distance between points. Franke suggests the value $\delta = 2.5\, d/\sqrt{n}$, where d and n take on the same definitions as before. In general, the quality of interpolation is better when the scattered points are approximately evenly distributed. If the points tend to cluster along contours, the results may become unsatisfactory.

Franke reports that of all the global basis functions tested in his research, Hardy's multiquadratic was, overall, the most effective [Franke 79]. One important point to note is that the design matrix for a radial function that increases with distance is generally ill-conditioned. Intuitively, this is true because with the growth of the basis function, the resulting interpolation tends to require large coefficients that delicately cancel out to produce the correct interpolated results.

The methods described thus far suffer from the difficulty of establishing good basis functions. As we shall see, algorithms based on regularization theory define the global splines by their properties rather than by an explicit set of basis functions. In general, this makes it easier to determine the conditions for the uniqueness and existence of a solution. Also, it is often easier to justify the interpolation techniques in terms of their properties rather than their basis functions. In addition, for some classes of functions and dense data, more efficient solutions exist in computing global splines.

3.8.2. Regularization

The term *regularization* commonly refers to the process of finding a unique solution to a problem by minimizing a weighted sum of two terms with the following properties. The first term measures the extent to which the solution conforms to some preconceived conditions, e.g., smoothness. The second term measures the extent to which the solution agrees with the available data. Related techniques have been used by numerical analysts for over twenty years [Atteia 66]. Generalizations are presented in [Duchon 76, 77] and [Meinguet 79a, 79b]. It has only recently been introduced in computer vision [Grimson 81, 83]. We now briefly discuss its use in that field, with a warning that some of this material assumes a level of mathematical sophistication suitable for advanced readers.

Techniques for interpolating (reconstructing) continuous functions from discrete data have been proposed for many applications in computer vision, including visual surface interpolation, inverse visual problems involving discontinuities, edge detection, and motion estimation. All these applications can be shown to be ill-posed because the supplied data is sparse, contains errors, and, in the absence of additional constraints, lies on an infinite number of piecewise smooth surfaces. This precludes any prior guarantee that a solution exists, or that it will be unique, or that it will be stable with respect to measurement errors. Consequently, algorithms based on regularization theory [Tikhonov 77] have been devised to systematically reformulate ill-posed problems into well-posed problems of variational calculus. Unlike the original problems, the variational principle formulations are well-posed in the sense that a solution exists, is unique, and depends continuously on the data.

In practice, these algorithms do not exploit the full power of regularization but rather use one central idea from that theory: an interpolation problem can be made unique by restricting the class of admissible solutions and then requiring the result to minimize a norm or semi-norm. The space of solutions is restricted by imposing global variational principles stated in terms of smoothness constraints. These constraints, known as *stabilizing functionals*, are regularizing terms which stabilize the solution. They are treated together with *penalty functionals*, which bias the solution towards the supplied data points, to form a class of viable solutions.

3.8.2.1. Grimson, 1981

Grimson first applied regularization techniques to the visual surface reconstruction problem [Grimson 81]. Instead of defining a particular surface family (e.g., planes) and then fitting it to the data $z(x,y)$, Grimson proceeded to fit an implicit surface by selecting from among all of the interpolating surfaces $f(x,y)$ the one that minimizes

$$E(f) = S(f) + P(f) \qquad (3.8.6)$$

$$= \left[\iint (f_{xx}^2 + 2f_{xy}^2 + f_{yy}^2)\, dx\, dy \right]^2 + \beta \sum_i \left[f(x_i,y_i) - z(x_i,y_i) \right]^2$$

where E is an energy functional defined in terms of a stabilizing functional S and a penalty functional P. The integral S is a measure of the deviation of the solution f from a desirable smoothness constraint. The form of S given above is based on smoothness properties that are claimed to be consistent with the human visual system, hence the name visual surface reconstruction. The summation P is a measure of the discrepancy between the solution f and the supplied data.

The surfaces that are computable from Eq. (3.8.6) are known in the literature as *thin-plate splines*. Thin-plate interpolating surfaces had been considered in previous work for the interpolation of aircraft wing deflections [Harder 72] and digital terrain maps [Briggs 74]. The stabilizing functional S which these surfaces minimize is known as a second-order Sobolev semi-norm. Grimson referred to it as the *quadratic variation*. This semi-norm has the nice physical analogy of measuring the bending energy in a thin plate. For instance, as S approaches zero, the plate becomes increasingly planar, e.g., no bending.

The penalty measure P is a summation carried over all the data points. It permits the surface to approximate the data $z(x,y)$ in the least-squares sense. The scale parameter β determines the relative importance between a close fit and smoothness. As β approaches zero, the penalty term has greater latitude (since it is suppressed by β) and S becomes more critical to the minimization of E. This results in an approximating surface f that smoothly passes near the data points. Forcing f to become an approximating surface is appropriate when the data is known to contain errors. However, if f is to become an interpolating surface, large values of β should be chosen to fit the data more closely.

This approach is based on minimizing E. If a minimizing solution exists, it will satisfy the necessary condition that the first variation must vanish:

$$\partial E(f) = \partial S(f) + \partial T(f) = 0 \qquad (3.8.7)$$

The resulting partial differential equation is known as the *Euler-Lagrange equation*. It governs the form of the energy-minimizing surface subject to the boundary conditions that correspond to the given sparse data.

The Euler-Lagrange equation does not have an analytic solution in most practical situations. This suggests a numerical solution applied to a discrete version of the problem. Grimson used the conjugate gradient method for approximation and the gradient projection method for interpolation. They are both classical minimization algorithms sharing the following advantages: they are iterative, numerically stable, and have parallel variants which are considered to be biologically feasible, e.g., consistent with the human visual system. In addition, the gradient projection method has the advantage of being a local technique. However, these methods also include the following disadvantages: the rate of convergence is slow, a good criterion for terminating the iteration is lacking, and the use of a grid representation for the discrete approximation precludes a viewpoint-invariant solution.

There are two additional drawbacks to this approach that are due to its formulation. First, the smoothing functional applies to the entire image, regardless of whether there are genuine discontinuities that should not be smoothed. Second, the failure to detect discontinuities gives rise to undesirable overshoots near large gradients. This is a manifestation of a Gibbs or Mach band phenomenon across a discontinuity in the surfaces. These problems are addressed in the work of subsequent researchers.

3.8.2.2. Terzopoulos, 1984

Terzopoulos extended Grimson's results in two important ways: he combined the thin-plate model together with membrane elements to accommodate discontinuities, and he applied multigrid relaxation techniques to accelerate convergence. Terzopoulos finds the unique surface f minimizing E where

$$S(f) = \frac{1}{2} \iint \rho(x,y) \left\{ \tau(x,y)(f_{xx}^2 + 2f_{xy}^2 + f_{yy}^2) + [1 - \tau(x,y)](f_x^2 + f_y^2) \right\} dx \, dy \quad (3.8.8)$$

and

$$P(f) = \frac{1}{2} \sum_i \alpha_i (L_i[f] - L_i[z] - \varepsilon_i)^2 \quad (3.8.9)$$

The stabilizing functional S, now referred to as a *controlled-continuity stabilizer*, is an integral measure that augments a thin-plate spline with a membrane model. The penalty functional P is again defined as a weighted Euclidean norm. It is expressed in terms of the measurement functionals L_i, the associated measurement errors ε_i, and nonnegative real-valued weights α_i. L_i can be used to obtain point values and derivative information.

In Eq. (3.8.8), $\rho(x,y)$ and $\tau(x,y)$ are real-valued weighting functions whose range is [0,1]. They are referred to as *continuity control functions*, determining the local continuity of the surface at any point. An interpretation of τ is surface tension, while that of ρ is surface cohesion. Their correspondence with thin-plate and membrane splines is given below.

$$\lim_{\tau(x,y) \to 0} S(f) \quad \to \quad \text{membrane spline}$$

$$\lim_{\tau(x,y) \to 1} S(f) \quad \to \quad \text{thin-plate spline}$$

$$\lim_{\rho(x,y) \to 0} S(f) \quad \to \quad \text{discontinuous surface}$$

A thin-plate spline is characterized as a C^1 surface which is continuous and has continuous first derivatives. A membrane spline is a C^0 surface that need only be continuous. Membrane splines are introduced to account for discontinuities in orientations, e.g., corners and creases. This reduces the Gibbs phenomena (oscillations) near large gradients by preventing the smoothness condition to apply over discontinuities.

Terzopoulos formulates the discrete problem as a finite element system. Although the finite element method can be solved by using iterative techniques such as relaxation, the process is slow and convergence is not always guaranteed. Terzopoulos used the Gauss-Seidel algorithm, which is a special case of the Successive Over-Relaxation (SOR) algorithm. He greatly enhanced the SOR algorithm by adapting multigrid relaxation techniques developed for solving elliptic partial differential equations. This method computes a coarse approximation to the solution surface, uses it to initiate the iterative solution of a finer approximation, and then uses this finer approximation to refine the coarse estimate. This procedure cycles through to completion at which point we reach a smooth energy-minimizing surface that interpolates (or approximates) the sparse data.

The principle of multigrid operation is consistent with the use of pyramid and multiresolution data structures in other fields. At a single level, the SOR algorithm rapidly smooths away high frequency error, leaving a residual low frequency error to decay slowly. The rate of this decay is dependent on the frequency: high frequencies are removed locally (fast), while low frequencies require long distance propagation taken over many iterations (slow). Dramatic speed improvements are made possible by projecting the low frequencies onto a coarse grid, where they become high frequencies with respect to the new grid. This exploits the fact that neighbor-to-neighbor communication on a coarse grid actually covers much more ground per iteration than on a fine grid. An adaptive scheme switches the relaxation process between levels according to the frequency content of the error signal, as measured by a local Fourier transformation.

We now consider some benefits and drawbacks of Terzopoulos' approach. The advantages include: the methods are far more computationally efficient over those of Grimson, discontinuities (given a priori) can be handled, error can be measured differently at each point, a convenient pyramid structure is used for surface representation, and local computations make this approach biologically feasible.

Some of the disadvantages include: a good criterion for terminating the iteration is lacking, the use of a grid representation for the discrete approximation precludes a viewpoint-invariant solution, there is slower convergence away from the supplied data points and near the grid boundary, and the numerical stability and convergence rates for the multigrid approach are not apparent.

3.8.2.3. Discontinuity Detection

Techniques for increasing the speed and accuracy of this approach have been investigated by Jou and Bovik [Jou 89]. They place emphasis on early localization of surface discontinuities to accelerate the process of minimizing the surface energy with a finite element approximation. Terzopoulos suggests that discontinuities are associated with places of high tension in the thin-plate spline. While this does detect discontinuities, it is prone to error since there is no one-to-one correspondence between the two. For instance, it is possible to have many locations of high tension for a single continuity beacuse of the oscillatory behavior of Gibbs effect. On the other hand, it is possible to miss areas of high tension if the data points around a discontinuity are sparse.

Grimson and Pavlidis describe a method for discontinuity detection based on a hypothesis testing technique. At each point, they compute a planar approximation of the data and use the statistics of the differences between the actual values and the approximations for detection of both steps and creases [Grimson 85]. If the distribution of the residual error appears random, then the hypothesis that there is no discontinuity is accepted. Otherwise, if systematic trends are found, then a discontinuity has been detected.

Blake and Zisserman have developed a technique based on ''weak'' continuity constraints, in which continuity-like constraints are usually enforced but can be broken if a suitable cost is incurred [Blake 87]. Their method is viewpoint-invariant and robustly detects and localizes discontinuities in the presence of noise. Computing the global minimum is difficult because invariant schemes incorporating weak continuity constraints have non-convex cost functions that are not amenable to naive descent algorithms. Furthermore, these schemes do not give rise to linear equations. Consequently, they introduce the Graduated Non-Convexity (GNC) Algorithm as an approximate method for obtaining the global minimum. The GNC approach is a heuristic that minimizes the objective function by a series of convex curves that increasingly refine the approximation of the function near the global minimum. The initial curve localizes the area of the solution and the subsequent curves establish the value precisely.

3.8.2.4. Boult and Kender, 1986

Boult and Kender examine four formalizations of the visual surface reconstruction problem and give alternate realizations for finding a surface that minimizes a functional [Boult 86a]. These methods are:

1. Discretization of the problem using variational principles and then discrete minimization using classical minimization techniques, as in [Grimson 81].

2. Discretization of a partial differential equation formulation of the problem, again using a variational approach, and then use of discrete finite element approximation solved with a multigrid approach, as in [Terzopoulos 84].

3. Direct calculation using semi-reproducing kernel splines, as in [Duchon 76].

4. Direct calculation using quotient reproducing kernel splines, as in [Meinguet 79].

The authors conclude that reproducing kernel splines are a good method for interpolating sparse data. The major computational component of the method is the solution of a dense linear system of equations. They cite the following advantages: the solution of the linear system is well-understood, the algorithm results in functional forms for the surface allowing symbolic calculations (e.g., differentiation or integration), there is no problem with slower convergence away from information points or near the boundary, the algorithm can efficiently allow updating the information (e.g., adding/deleting/modifying data points), no iteration is needed since the computation depends only on the number of information points (not their values), and the method is more efficient than those of Grimson or Terzopoulos for sparse data.

The disadvantages include: the resulting linear system is dense and indefinite which limits the approach with which it can be solved, the reproducing kernels may be difficult to derive, and the method may not be biologically feasible due to the implicit global communication demands. Full details about this method can be found in [Boult 86b]. In what follows, we present a brief derivation of the semi-reproducing kernel spline approach. The name of this method is due to formal mathematical definitions. These details will be suppressed insofar as they lie outside the scope of this book.

The use of semi-reproducing kernel spline allows us to interpolate data by almost the same four steps as used in the computation of global splines defined through heuristic basis functions. This is in contrast to regularization, which employs a discrete minimization method akin to that used by Terzopoulos. Unlike the basis functions suggested earlier, though, the basis functions used to define the semi-reproducing kernel splines compute a surface with minimal energy, as defined in Eq. (3.8.6). To interpolate M data points for mapping function U, the expression is

$$U(x,y) = \sum_{i=1}^{M+3} a_i\, h_i(x,y)$$

where the basis functions h_i are

$$h_i(x,y) = \theta \cdot [(x-x_i)^2 + (y-y_i)^2] \cdot \log[(x-x_i)^2 + (y-y_i)^2], \qquad i=1, ..., M$$

$$h_{M+1}(x,y) = 1$$
$$h_{M+2}(x,y) = x \qquad\qquad\qquad (3.8.10)$$
$$h_{M+3}(x,y) = y$$

for a constant θ (see below).

The above expression for U has more basis functions than data points. These extra terms, h_{M+i}, $i=1, 2, 3$, are called the basis functions of the *null space*, which has dimension $d=3$. They span those functions which can be added to any other function without changing the energy term (Eq. 3.8.6) of that function. They are introduced here because they determine the components of the low-order variations in the data which are not constrained by the smoothness functional (the norm). In our case, since the norm is twice-differentiable in x and y, the low-order variation is a plane and our null space must have dimension $d=3$.

Since we have additional basis functions, the design matrix is insufficient to define the coefficients of the interpolation spline. Instead, the $M+d$ basis function coefficients can be determined from the solution of $(M+d) \times (M+d)$ dense linear system:

$$
\begin{bmatrix}
u_1 \\
u_2 \\
\cdots \\
\cdots \\
u_M \\
0 \\
0 \\
0
\end{bmatrix}
= A
\begin{bmatrix}
a_1 \\
a_2 \\
\cdots \\
\cdots \\
a_M \\
a_{M+1} \\
a_{M+2} \\
a_{M+3}
\end{bmatrix}
\tag{3.8.11}
$$

where

$$
\begin{aligned}
A_{i,j} &= h_i(x_j, y_j) && \text{for } i \le (M+d), \ j \le M, \ i \ne j \\
A_{i,j} &= \beta^{-1} + h_i(x_j, y_j) && \text{for } i = j \le M \\
A_{i,j} &= h_j(x_i, y_i) && \text{for } i \le M, \ M < j \le M+d \\
A_{i,j} &= 0 && \text{for } i > M, \ j > M
\end{aligned}
\tag{3.8.12}
$$

The system can be shown to be nonsingular if the data spans the null space. For this case, the data must contain at least three non-collinear points. Due to the mathematical properties of the basis functions, the above spline is referred to as an interpolating semi-reproducing kernel spline. Note that the β given above corresponds to that used in the expression for the energy functional in Eq. (3.8.6). As it approaches infinity, the system determines the coefficients of the interpolating spline of minimal norm.

One of the most compelling reasons to use this approach over the discrete minimization techniques proposed by Terzopoulos is computational efficiency for very small data sets. The complexity of this approach is $0.33(M+3)^3 + O(MR)$ where M is the number of data points and R is the number of reconstruction points. On the other hand, Terzopoulos' approach has complexity $O(R^2)$ in the worst case, with constant ≥ 30. In the average case, it has cost $O(R^2/M)$. Thus, when M is small compared to R, the semi-reproducing kernel approach can be significantly faster. Since for the problem of warping, the number of known points is small (say $M = 50$), and the resolution of the approximation is high (say 512^2, or $R = 262,144$) the direct approach has significant appeal.

It should be noted that one argument in favor of Terzopoulos' approach over global splines is that the former handles discontinuities while the latter does not. Although this property has particular relevance in computer vision where it is often necessary to model occluding edges and distinct object boundaries, it is less critical in image warping because we usually do not want to introduce discontinuities, or cracks, in the interpolated mapping function. Of course if more dramatic warps are desired, then this property of global splines must be addressed.

3.8.2.5. A Definition of Smoothness

Thus far, our discussion has concentrated on formulations that minimize the energy functional given in Eq. (3.8.6). The term "smoothness" has taken on an implicit meaning which we now seek to express more precisely. This discussion applies to the discrete minimization technique as well as the global splines approach.

If the energy term defined in Eq. (3.8.6) is to be used, the space of functions in which we minimize must be contained in the class referred to as D^2L^2. This is the space of functions such that their second derivatives exist and the integral over all of the real numbers (in the Lebusque sense) of the quadratic variation is bounded. This is the *minimal assumption* necessary for the energy term to be well defined. However, as is generally the case with minimization problems, reducing the space in which one searches for the minimum can have a significant impact on the resulting minimum. This is true even if the same objective function is maintained. Thus, we might ask whether there are alternate classes of functions for which this semi-norm might be minimized. For that matter, we might also ask whether there are other semi-norms to minimize.

An important set of these classes can be parameterized formally as those functions with their m^{th} derivative in H^η, where H^η is the Hilbert space such that if $v \in H^\eta$, then it has a Fourier transform $\tilde{v}$ that satisfies

$$\int\int |\tau|^{2\eta} \cdot |\tilde{v}(\tau)|^2 \, d\tau \quad < \quad \infty \qquad (3.8.13)$$

The class of functions referred to as $D^m H^\eta$ can be equipped with the mth Sobolev semi-norm,

$$\|\cdot\|_{D^m} = \left\{ \int\int \left(\sum_{i+j=m} \begin{bmatrix} m \\ i \end{bmatrix} \left[\frac{\partial^m f}{\partial x^i \partial y^j} \right]^2 \right) \, dx \, dy \right\}^{\frac{1}{2}} \qquad (3.8.14)$$

which results in a semi-Hilbert space if $1 > \eta > 1-m$. Note that if one chooses $m=2$ and $\eta=0$, then using the properties of Fourier transforms, the above definitions yield exactly the space D^2L^2 that was used by Grimson and Terzopoulos.

In order to better understand these classes of functions, the following intuitive definition is offered. First, note that the spaces of functions assume the existence of the mth derivative of the function, *in the distributional sense*. This means that the mth derivative of the functions exist except on sets of measure zero, e.g., at isolated points or lines. Then the classes $D^m H^0$, which are also known as $D^m L^2$, simply assume that the power of these functions is bounded. For the classes $D^m H^\eta$, for $\eta > 0$, we have the squared spectrum of the derivatives going to zero (as the frequency goes to infinity) faster than a specified polynomial of the frequency. This means that the spectrum must taper off quickly. Thus, these functions have less high frequencies and are "smoother" than functions that simply have m derivatives. For the classes $D^m H^\eta$, for $\eta < 0$, we see that the spectrum of the derivatives is bounded away from zero, and that as the frequency goes to infinity, the derivatives go to infinity no faster than a given polynomial. In this case, the spectrum vanishes near zero frequency (DC). Thus, these functions have less low frequencies and are less "smooth" than most functions with m derivatives.

For each member of this family, the surface of minimal norm from the class is as in Eq. (3.8.11) with a different set of basis functions. Those classes which use the mth semi-norm have null spaces spanned by polynomials of total degree $\leq m$. The other basis functions depend on the location of the data points. For the space $D^{-m}H^{\eta}$ the basis function associated with the ith datum is

$$h_i(x,y) = \begin{cases} \theta_m \cdot ((x-x_i)^2 + (y-y_i)^2)^{m/2} \cdot \log((x-x_i)^2 + (y-y_i)^2) & \text{if } m+\eta \text{ is even} \\ \theta_m \cdot ((x-x_i)^2 + (y-y_i)^2)^{(m+\eta)/2} & \text{otherwise} \end{cases} \quad (3.8.15)$$

where

$$\theta_m = \begin{cases} \dfrac{1}{2^{2m-1} \pi (\Gamma(m-1))^2} & \text{if } m \text{ is even} \\ \dfrac{-\Gamma(1-m)}{2^{2m} \pi (m-1)!} & \text{if } m \text{ is odd} \end{cases}$$

where Γ is the gamma function.

It is important to note that while the ith basis spline can be identical for different classes of functions (e.g., for all valid pairs of m and η), the null space depends on the norm and thus reconstructions in the class do differ. One can interpret the interpolation as a combination of least-squares fits to the polynomials which define the null space (a plane, in our case) followed by a minimal energy interpolation of the difference between that surface and the actual data.

3.9. SUMMARY

Spatial transformations are given by mapping functions that relate the coordinates of two images, e.g., the input image and the transformed output. This chapter has focused on various formulations for spatial transformations in common use. Depending on the application, the mapping functions may take on many different forms. In computer graphics, for instance, a general transformation matrix suffices for simple affine and perspective planar mappings. Bilinear transformations are also popular, particularly owing to their computational advantages in terms of separability. However, since they do not preserve straight lines for all orientations, their utility in computer graphics is somewhat undermined with respect to the more predictable results obtained from affine and perspective transformations.

All mappings derived from the general transformation matrix can be expressed in terms of first-order (rational) polynomials. As a result, we introduce a more general class of mappings specified by polynomial transformations of arbitrary degree. Since polynomials of high degree become increasingly oscillatory, we restrict our attention to low-order polynomials. Otherwise, the oscillations would manifest itself as spatial artifacts in the form of undesirable rippling in the warped image.

Polynomial transformations have played a central role in fields requiring geometric correction, e.g., remote sensing. In these applications, we are typically not given the coefficients of the polynomials used to model the transformation. Consequently, numerical techniques are used to solve the overdetermined system of linear equations that relate a set of points in the reference image to their counterparts in the observed (warped) image. We reviewed several methods, including the pseudoinverse solution, least-squares method, and weighted least-squares with orthogonal polynomials.

An alternate approach to global polynomial transformations consists of piecewise polynomial transformations. Rather than defining U and V via a global function, they are expressed as a union of a local functions. In this manner, the interpolated surface is composed of local surface patches, each influenced by nearby control points. This method offers more sensitivity to local deformations than global methods described earlier.

The problem of inferring a mapping function from a set of correspondence points is cast into a broad framework when it is treated as a surface interpolation problem. This framework is clearly consistent with the algebraic methods developed earlier. Consequently, global splines defined through basis functions and regularization methods are introduced for surface interpolation of scattered data. Numerical techniques drawn from numerical analysis, as applied in computer vision for regularization, are described.

The bulk of this chapter has been devoted to the process of inferring a mapping function from a set of correspondence points. Given the various techniques described, it is natural to ask: what algorithm is best-suited for my problem? The answer to this question depends on several factors. If the transformation is known in advance to be adequately modeled by a low-order global polynomial, then it is only necessary to infer the unknown polynomial coefficients. Otherwise, we must consider the number of correspondence points and their distribution.

If the points lie on a quadrilateral mesh, then it is straightforward to fit the data with a tensor product surface, e.g., bicubic patches. When this is not the case, piecewise polynomial transformations offer a reasonable alternative. The user must be aware that this technique is generally not recommended when the points are clustered, leaving large gaps of information that must be extrapolated. In these instances, weighted least-squares might be considered. This method offers several important advantages. It allows the user to adaptively determine the degree of the polynomial necessary to satisfy some error bound. Unlike other global polynomial transformations that can induce undesirable oscillations, the polynomial coefficients in the weighted least-squares approach are allowed to vary at each image position. This expense is often justified if the data is known to contain noise and the mapping function is to be approximated using information biased towards local measurements.

Another class of solutions for inferring mapping functions comes from global splines. Splines defined through heuristic basis functions are one of the oldest global interpolation techniques. They can be shown to be related to some of the earlier techniques. The method, however, is sensitive to a proper choice for the basis functions.

Global splines defined through regularization techniques replace this choice with a formulation that requires the computation of a surface satisfying some property, e.g., smoothness. The surface may be computed by using discrete minimization techniques or basis functions. The latter is best-suited when a small number of correspondence points are supplied. Their computational costs determine when it is appropriate to switch from one method to the other.

In general, the nature of surface interpolation requires a lot of information that is often difficult to quantify. No single solution can be suggested without complete information about the correspondence points and the desired "smoothness" of the interpolated mapping function. Therefore, the reader is encouraged to experiment with the various methods, evaluting the resulting surfaces. Fortunately, this choice can be judged visually rather than on the basis of some mathematical abstraction.

Although the bulk of our discussion on analytic mappings have centered on polynomial transformations, there are other spatial transformations that find wide use in pattern recognition and medical applications. In recent years, there has been renewed interest in the log-spiral (or polar exponential) transform for achieving rotation and scale invariant pattern recognition [Weiman 79]. This transform maps the cartesian coordinate system C to a $(\log r, \theta)$ coordinate system L such that centered scale changes and rotation in C now become horizontal and vertical translations in L, respectively. Among other places, it has found use at the NASA Johnson Space Center where a programmable remapper has been developed in conjunction with Texas Instruments to transform input images so that they may be presented to a shift-invariant optical correlator for object recognition [Fisher 88]. Under the transformation, the location of the peak directly yields the object's rotation and scale change relative to the stored correlation filter. This information is then used to rectify and scale the object for correlation in the cartesian plane.

In related activites, that same hardware has been used to perform quasi-conformal mapping for compensation of human visual field defects [Juday 89]. Many people suffer from retinitis pigmentosa (tunnel vision) and from maculapathy (loss of central field). These are retinal dysfunctions that correspond to damaged parts of the retina in the peripheral and central fields, respectively. By warping the incoming image so that it falls on the viable (working) part of the retina, the effects of these visual defects may be reduced. Conformal mapping is appropriate in these applications because it is consistent with the imaging properties of the human visual system. Analytic and numerical techniques for implementing conformal mappings are given in [Frederick 90].

4

SAMPLING THEORY

4.1. INTRODUCTION

This chapter reviews the principal ideas of digital filtering and sampling theory. Although a complete treatment of this area falls outside the scope of this book, a brief review is appropriate in order to grasp the key issues relevant to the resampling and antialiasing stages that follow. Both stages share the common two-fold problem addressed by sampling theory:

1. Given a continuous input signal $g(x)$ and its sampled counterpart $g_s(x)$, are the samples of $g_s(x)$ sufficient to exactly describe $g(x)$?

2. If so, how can $g(x)$ be reconstructed from $g_s(x)$?

This problem is known as *signal reconstruction*. The solution lies in the frequency domain whereby spectral analysis is used to examine the spectrum of the sampled data.

The conclusions derived from examining the reconstruction problem will prove to be directly useful for resampling and indicative of the filtering necessary for antialiasing. Sampling theory thereby provides an elegant mathematical framework in which to assess the quality of reconstruction, establish theoretical limits, and predict when it is not possible.

In order to better motivate the importance of sampling theory, we demonstrate its role with the following examples. A checkerboard texture is shown projected onto an oblique planar surface in Fig. 4.1. The image exhibits two forms of artifacts: jagged edges and moire patterns. Jagged edges are prominent toward the bottom of the image, where the input checkerboard undergoes magnification. The moire patterns, on the other hand, are noticeable at the top, where minification (compression) forces many input pixels to occupy fewer output pixels.

Figure 4.1: Oblique checkerboard (unfiltered).

Figure 4.1 was generated by projecting the center of each output pixel into the checkerboard and sampling (reading) the value at the nearest input pixel. This point sampling method performs poorly, as is evident by the objectionable results of Fig. 4.1. This conclusion is reached by sampling theory as well. Its role here is to precisely quantify this phenomena and to prescribe a solution. Figure 4.2 shows the same mapping with improved results. This time the necessary steps were taken to preclude artifacts.

4.2. SAMPLING

Consider the imaging system discussed in Section 2.2. For convenience, the images will be taken as one-dimensional signals, i.e., a single scanline image. Recall that the continuous signal, $f(x)$, is presented to the imaging system. Due to the point spread function of the imaging device, the degraded output $g(x)$ is a bandlimited signal with attenuated high frequency components. Since visual detail directly corresponds to spatial frequency, it follows that $g(x)$ will have less detail than its original counterpart $f(x)$. The frequency content of $g(x)$ is given by its spectrum, $G(f)$, as determined by the Fourier transform.

$$G(f) = \int_{-\infty}^{\infty} g(x)\, e^{-i 2\pi f x} dx \qquad (4.2.1)$$

In the discussion that follows, x represents spatial position and f denotes spatial frequency. Note that Chapter 2 used the variable u to refer to frequency in order to avoid

Figure 4.2: Oblique checkerboard (filtered).

confusion with function $f(x)$. Since we will no longer refer to $f(x)$ in this chapter, we return to the more conventional notation of using f for frequency, as adopted in many signal processing textbooks.

The magnitude spectrum of a signal is shown in Fig. 4.3. It shows a concentration of energy in the low-frequency range, tapering off toward the higher frequencies. Since there are no frequency components beyond f_{max}, the signal is said to be *bandlimited* to frequency f_{max}.

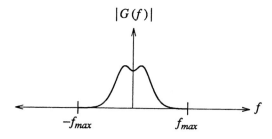

Figure 4.3: Spectrum $G(f)$.

The continuous output $g(x)$ is then digitized by an ideal impulse sampler, the comb function, to get the sampled signal $g_s(x)$. The ideal 1-D sampler is given as

$$s(x) = \sum_{n=-\infty}^{\infty} \delta(x - nT_s) \qquad (4.2.2)$$

where δ is the familiar impulse function and T_s is the sampling period. The running index n is used with δ to define the impulse train of the comb function. We now have

$$g_s(x) = g(x)s(x) \qquad (4.2.3)$$

Taking the Fourier transform of $g_s(x)$ yields

$$G_s(f) = G(f) * S(f) \qquad (4.2.4)$$

$$= G(f) * \left[\sum_{n=-\infty}^{n=\infty} f_s \delta(f - nf_s) \right] \qquad (4.2.5)$$

$$= f_s \sum_{n=-\infty}^{n=\infty} G(f - nf_s) \qquad (4.2.6)$$

where f_s is the sampling frequency and * denotes convolution. The above equations make use of the following well-known properties of Fourier transforms:

1.   Multiplication in the spatial domain corresponds to convolution in the frequency domain. Therefore, Eq. (4.2.3) gives rise to a convolution in Eq. (4.2.4).

2.   The Fourier transform of an impulse train is itself an impulse train, giving us Eq. (4.2.5).

3.   The spectrum of a signal sampled with frequency f_s $(T_s = 1/f_s)$ yields the original spectrum replicated in the frequency domain with period f_s (Eq. 4.2.6).

This last property has important consequences. It yields spectrum $G_s(f)$ which, in response to a sampling period $T_s = 1/f_s$, is *periodic in frequency* with period f_s. This is depicted in Fig. 4.4. Notice then, that a small sampling period is equivalent to a high sampling frequency yielding spectra replicated far apart from each other. In the limiting case when the sampling period approaches zero $(T_s \rightarrow 0, f_s \rightarrow \infty)$, only a single spectrum appears — a result consistent with the continuous case. This leads us, in the next chapter, to answer the central problem posed earlier regarding reconstruction of the original signal from its samples.

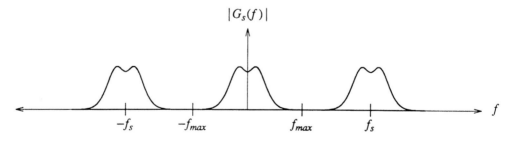

Figure 4.4: Spectrum $G_s(f)$.

4.3. RECONSTRUCTION

The above result reveals that the sampling operation has left the original input spectrum *intact*, merely replicating it periodically in the frequency domain with a spacing of f_s. This allows us to rewrite $G_s(f)$ as a sum of two terms, the low frequency (baseband) and high frequency components. The *baseband* spectrum is exactly $G(f)$, and the high frequency components, $G_{high}(f)$, consist of the remaining replicated versions of $G(f)$ that constitute harmonic versions of the sampled image.

$$G_s(f) = G(f) + G_{high}(f) \qquad (4.3.1)$$

Exact signal reconstruction from sampled data requires us to discard the replicated spectra $G_{high}(f)$, leaving only $G(f)$, the spectrum of the signal we seek to recover. This is a crucial observation in the study of sampled-data systems.

4.3.1. Reconstruction Conditions

The only provision for exact reconstruction is that $G(f)$ be undistorted due to overlap with $G_{high}(f)$. Two conditions must hold for this to be true:

1. The signal must be bandlimited. This avoids spectra with infinite extent that are impossible to replicate without overlap.

2. The sampling frequency f_s must be greater than twice the maximum frequency f_{max}, present in the signal. This minimum sampling frequency, known as the *Nyquist rate*, is the minimum distance between the spectra copies, each with bandwidth f_{max}.

The first condition merely ensures that a sufficiently large sampling frequency exists that can be used to separate replicated spectra from each other. Since all imaging systems impose a bandlimiting filter in the form of a point spread function, this condition is always satisfied for images captured through an optical system.[†] Note that this does not apply to synthetic images, e.g., computer-generated imagery.

The second condition proves to be the most revealing statement about reconstruction. It answers the problem regarding the sufficiency of the data samples to exactly reconstruct the continuous input signal. It states that exact reconstruction is possible only when $f_s > f_{Nyquist}$, where $f_{Nyquist} = 2 f_{max}$. Collectively, these two conclusions about reconstruction form the central message of sampling theory, as pioneered by Claude Shannon in his landmark papers on the subject [Shannon 48, 49]. Interestingly enough, these conditions were first discussed during the early development of television in the landmark 1934 paper by Mertz and Gray [Mertz 34]. In their work, they informally outlined these conditions as a rule-of-thumb for preventing visual artifacts in the reconstructed image.

---

† This does not include the shot noise that may be introduced by digital scanners.

4.3.2. Ideal Low-Pass Filter

We now turn to the second central problem: Given that it is theoretically possible to perform reconstruction, how may it be done? The answer lies with our earlier observation that sampling merely replicates the spectrum of the input signal, generating $G_{high}(f)$ in addition to $G(f)$. Therefore, the act of reconstruction requires us to completely suppress $G_{high}(f)$. This is done by multiplying $G_s(f)$ with $H(f)$, given as

$$H(f) = \begin{cases} 1 & |f| < f_{max} \\ 0 & |f| \geq f_{max} \end{cases} \qquad (4.3.2)$$

$H(f)$ is known as an ideal low-pass filter and is depicted in Fig. 4.5, where it is shown suppressing all frequency components above f_{max}. This serves to discard the replicated spectra $G_{high}(f)$. It is ideal in the sense that the f_{max} cut-off frequency is strictly enforced as the transition point between the transmission and complete suppression of frequency components.

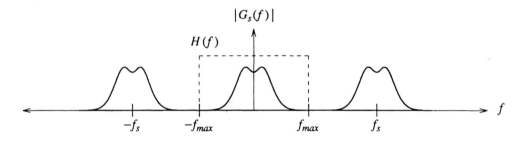

Figure 4.5: Ideal low-pass filter $H(f)$.

In the literature, there appears to be some confusion as to whether it is possible to perform exact reconstruction when sampling at *exactly* the Nyquist rate, yielding an overlap at the highest frequency component f_{max}. In that case, only the frequency can be recovered, but not the amplitude or phase. The only exception occurs if the samples are located at the minimas and maximas of the sinusoid at frequency f_{max}. Since reconstruction is possible in that exceptional instance, some sources in the literature have inappropriately included the Nyquist rate as a sampling rate that permits exact reconstruction. Nevertheless, realistic sampling techniques must sample at rates far above the Nyquist frequency in order to avoid the nonideal elements that enter into the process (e.g., sampling with a narrow pulse rather than an impulse). Therefore, this mistaken point is rather academic for natural images. This has more serious consequences for synthetic images that can indeed be sampled with a perfect comb function.

4.3.3. Sinc Function

In the spatial domain, the ideal low-pass filter is derived by computing the inverse Fourier transform of $H(f)$. This yields the *sinc* function shown in Fig. 4.6. It is defined as

$$sinc(x) = \frac{sin(\pi x)}{\pi x} \tag{4.3.3}$$

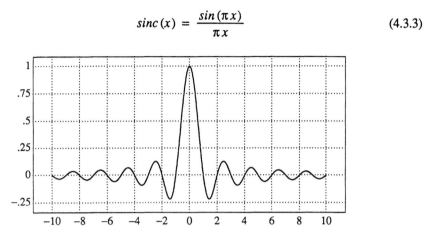

Figure 4.6: The sinc function.

The reader should note the reciprocal relationship between the height and width of the ideal low-pass filter in the spatial and frequency domains. Let A denote the amplitude of the sinc function, and let its zero crossings be positioned at integer multiples of $1/2W$. The spectrum of this sinc function is a rectangular pulse of height $A/2W$ and width $2W$, with frequencies ranging from $-W$ to W. In our example above, $A = 1$ and $W = f_{max} = .5$ cycles/pixel. This value for W is derived from the fact that digital images must not have more than one half cycle per pixel in order to conform to the Nyquist rate.

The sinc function is one instance of a large class of functions known as *cardinal splines*, which are interpolating functions defined to pass through zero at all but one data sample, where they have a value of one. This allows them to compute a continuous function that passes through the uniformly-spaced data samples.

Since multiplication in the frequency domain is identical to convolution in the spatial domain, $sinc(x)$ represents the convolution kernel used to evaluate any point x on the continuous input curve g given only the sampled data g_s.

$$g(x) = sinc(x) * g_s(x) \tag{4.3.4}$$

$$= \int_{-\infty}^{\infty} sinc(\lambda) g_s(x - \lambda) d\lambda$$

Equation (4.3.4) highlights an important impediment to the practical use of the ideal low-pass filter. The filter requires an infinite number of neighboring samples (i.e., an infinite filter support) in order to precisely compute the output points. This is, of course, impossible owing to the finite number of data samples available. However, truncating the sinc function allows for approximate solutions to be computed at the expense of undesirable "ringing", i.e., ripple effects. These artifacts, known as the *Gibbs phenomenon*, are the overshoots and undershoots caused by reconstructing a signal with truncated frequency terms. The two rows in Fig. 4.7 show that truncation in one domain leads to ringing in the other domain. This indicates that a truncated sinc function is actually a poor reconstruction filter because its spectrum has infinite extent and thereby fails to bandlimit the input.

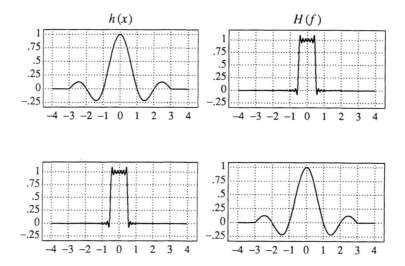

Figure 4.7: Truncation in one domain causes ringing the other domain.

In response to these difficulties, a number of approximating algorithms have been derived, offering a tradeoff between precision and computational expense. These methods permit local solutions that require the convolution kernel to extend only over a small neighborhood. The drawback, however, is that the frequency response of the filter has some undesirable properties. In particular, frequencies below f_{max} are tampered, and high frequencies beyond f_{max} are not fully suppressed. Thus, nonideal reconstruction does not permit us to exactly recover the continuous underlying signal without artifacts. As we shall see, though, there are ways of ameliorating these effects. The problem of nonideal reconstruction receives a great deal of attention in the literature due to its practical significance. We briefly present this problem below, and describe it in more detail in Chapter 5.

4.4. NONIDEAL RECONSTRUCTION

The process of nonideal reconstruction is depicted in Fig. 4.8, which indicates that the input signal satisfies the two conditions necessary for exact reconstruction. First, the signal is bandlimited since the replicated copies in the spectrum are each finite in extent. Second, the sampling frequency exceeds the Nyquist rate since the copies do not overlap. However, this is where our ideal scenario ends. Instead of using an ideal low-pass filter to retain only the baseband spectrum components, a nonideal reconstruction filter is shown in the figure.

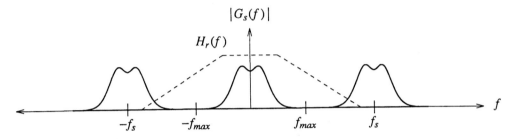

Figure 4.8: Nonideal reconstruction.

The filter response $H_r(f)$ deviates from the ideal response $H(f)$ shown in Fig. 4.5. In particular, $H_r(f)$ does not discard all frequencies beyond f_{max}. Furthermore, that same filter is shown to attenuate some frequencies that should have remained intact. This brings us to the problem of assessing the quality of a filter.

The accuracy of a reconstruction filter can be evaluated by analyzing its frequency domain characteristics. Of particular importance is the filter response in the passband and stopband. In this problem, the *passband* consists of all frequencies below f_{max}. The *stopband* contains all higher frequencies arising from the sampling process.[†]

An ideal reconstruction filter, as described earlier, will completely suppress the stopband while leaving the passband intact. Recall that the stopband contains the offending high frequencies that, if allowed to remain, would prevent us from performing exact reconstruction. As a result, the sinc filter was devised to meet these goals and serve as the ideal reconstruction filter. Its kernel in the frequency domain applies unity gain to transmit the passband and zero gain to suppress the stopband.

The breakdown of the frequency domain into passband and stopband isolates two problems that can arise due to nonideal reconstruction filters. The first problem deals with the effects of imperfect filtering on the passband. Failure to impose unity gain on *all* frequencies in the passband will result in some combination of image smoothing and image sharpening. Smoothing, or blurring, will result when the frequency gains near the cut-off frequency start falling off. Image sharpening results when the high frequency

---

† Note that frequency ranges designated as passbands and stopbands vary among problems.

gains are allowed to exceed unity. This follows from the direct correspondence of visual detail to spatial frequency. Furthermore, amplifying the high passband frequencies yields a sharper transition between the passband and stopband, a property shared by the sinc function.

The second problem addresses nonideal filtering on the stopband. If the stopband is allowed to persist, high frequencies will exist that will contribute to aliasing (described later). Failure to fully suppress the stopband is a condition known as *frequency leakage*. This allows the offending frequencies to fold over into the passband range. These distortions tend to be more serious since they are visually perceived more readily.

Despite the poor performance of nonideal reconstruction filters in the frequency domain, substantial improvements can be made to the output by simply using a higher sampling density. This serves to place further distance between replicated copies of the spectrum, thereby diminishing the extent of frequency leakage. Below we give some examples of the relationship between sampling rate and the quality of reconstruction necessary to avoid artifacts.

A chirp signal $g(x)$, common in FM radio, is shown in Fig. 4.9 alongside its spectrum $G(f)$. The chirp signal in the figure actually consists of 512 regularly spaced samples. These samples are indexed by x, where $0 \leq x < 512$. The spectrum was computed by using the discrete Fourier transform (DFT). As mentioned in Chapter 2, an N-sample input signal can have at most $N/2$ cycles. Therefore, the horizontal axis of $G(f)$ is spatial frequency, ranging from $-N/2$ to $N/2$ cycles (per scanline), where $N = 512$.

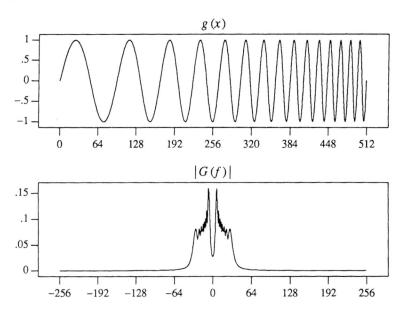

Figure 4.9: (a) Chirp signal and (b) its spectrum.

By inspection, we notice that $G(f)$ tapers to zero at the high frequencies. This means that $g(x)$ is bandlimited, satisfying the first condition necessary for reconstruction. We then uniformly sample $g(x)$ to get $g_s(x)$, as shown in Fig. 4.10. Note that the circles denote the collected samples, spaced four pixels apart. Appropriately, there is a total of four replicated spectra within the range displayed in $G_s(f)$. Each copy is scaled to one-fourth the amplitude of its original counterpart. Again, by inspection, we observe that the sampling frequency exceeds the Nyquist rate since the replicated copies do not overlap.

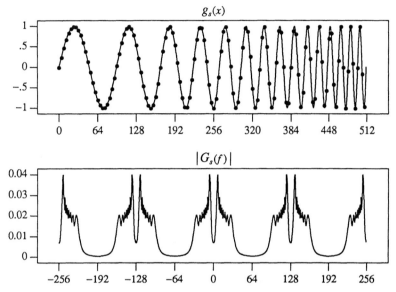

Figure 4.10: Sampled chirp signal.

By applying the ideal low-pass filter to $G_s(f)$, it is possible to recover $g(x)$. In Fig. 4.11, however, a nonideal low-pass filter $G_r(f)$ was applied, generating the output $g_r(x)$. The filter, corresponding to linear interpolation in the spatial domain, permitted some high frequencies to remain. Clearly, $G_r(f)$ is not identical to the original $G(f)$. These high frequencies account for the artifacts in the reconstructed signal. In particular, notice that the left end of $g_s(x)$ is fairly well reconstructed because it is slowly varying. However, as we move towards the right end of the figure, the highly varying sinusoids can no longer be adequately sampled at that same rate.

It is important to note the following subtle point about restoring signals that have not been reconstructed exactly. If the output were to remain a continuous signal, then the original signal may still be recovered by filtering out the undesirable high frequency components by applying an ideal low-pass filter to the degraded output. However, since the poorly reconstructed signal has actually been sampled in this discrete example, the retained samples are corrupted and further low-pass refinements will only serve to further integrate erroneous information.

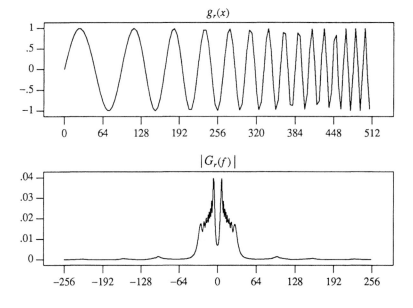

Figure 4.11: Nonideal low-pass filter applied to Fig. (4.10).

4.5. ALIASING

If the two reconstruction conditions outlined in Section 4.3.1 are not met, sampling theory predicts that exact reconstruction is *not* possible. This phenomenon, known as *aliasing*, occurs when signals are not bandlimited or when they are undersampled, i.e., $f_s \leq f_{Nyquist}$. In either case there will be unavoidable overlapping of spectral components, as in Fig. 4.12. Notice that the irreproducible high frequencies fold over into the low frequency range. As a result, frequencies originally beyond f_{max} will, upon reconstruction, appear in the form of much *lower* frequencies. Unlike the spurious high frequencies retained by nonideal reconstruction filters, the spectral components passed due to undersampling are more serious since they actually corrupt the components in the original signal.

Aliasing refers to the higher frequencies becoming aliased, and indistinguishable from, the lower frequency components in the signal if the sampling rate falls below the Nyquist frequency. In other words, undersampling causes high frequency components to appear as spurious low frequencies. This is depicted in Fig. 4.13, where a high frequency signal appears as a low frequency signal after sampling it too sparsely. In digital images, the Nyquist rate is determined by the highest frequency that can be displayed: one cycle every two pixels. Therefore, any attempt to display higher frequencies will produce similar artifacts.

To get a better idea of the effects of aliasing, consider digitizing a page of text into a binary (bilevel) image. If the samples are taken too sparsely, then the digitized image will appear to be a collection of randomly scattered dots, rather than the actual letters.

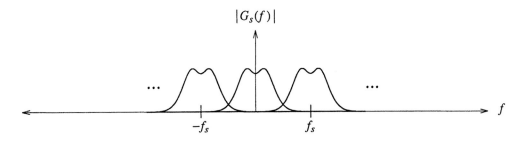

Figure 4.12: Overlapping spectral components give rise to aliasing.

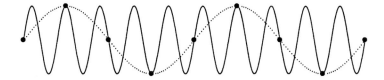

Figure 4.13: Aliasing artifacts due to undersampling.

This form of degradation prevents the output from even closely resembling the input. If the sampling density is allowed to increase, the letters will begin to take shape. At first, the exact spacing of black and white regions is compromised by the poor localization afforded by sparse samples.

In the computer graphics literature there is a misconception that jagged (staircased) edges are always a symptom of aliasing. This is only partially true. Technically, jagged edges arise from high frequencies introduced by inadequate reconstruction. Since these high frequencies are not corrupting the low frequency components, no aliasing is actually taking place. The confusion lies in that the suggested remedy of increasing the sampling rate is also used to eliminate aliasing. Of course, the benefit of increasing the sampling rate is that the replicated spectra are now spaced farther apart from each other. This relaxes the accuracy constraints for reconstruction filters to perform ideally in the stopband where they must suppress all components beyond some specified cut-off frequency. In this manner, the same nonideal filters will produce less objectionable output.

It is important to note that a signal may be densely sampled (far above the Nyquist rate), and continue to appear jagged if a zero-order reconstruction filter is used. Sample-and-hold filters used for pixel replication in real-time hardware zooms are a common example of poor reconstruction filters. In this case, the signal is clearly not aliased but rather poorly reconstructed. The distinction between reconstruction and aliasing artifacts becomes clear when we notice that the appearance of jagged edges is improved by blurring. For example, it is not uncommon to step back from an image exhibiting excessive blockiness in order to see it more clearly. This is a defocusing operation that attenuates the high frequencies admitted through nonideal reconstruction. On the other hand, once a signal is truly undersampled, there is no postprocessing possible to improve its condition. After all, applying an ideal low-pass (reconstruction) filter to a spectrum whose components are already overlapping will only blur the result, not rectify it. This subtlety is made explicit in [Pavlidis 82].

Unfortunately, the terminology in the literature often serves to propagate the confusion regarding the relationship between aliasing, reconstruction, and jagged edges. Some sources refer to undersampling as *prealiasing* and errors due to reconstruction as *postaliasing* [Netravali, Mitchell 88]. These names are used to parallel *prefilter* and *postfilter*, two terms used to mean bandlimiting before sampling, and reconstruction, respectively. In this context, the distinction between aliasing, reconstruction, and jagged edges becomes fuzzy.

Although at first glance it may seem misleading to refer to poor reconstruction as some form of aliasing, the correctness of this claim is actually dependent on whether we are speaking of the continuous or discrete domain. If the reconstructed signal is left in the continuous domain, then clearly poor reconstruction is *not* a form of aliasing since it can be corrected by bandlimiting the signal further. If, instead, we are operating in the discrete domain, then after the signal has been reconstructed it is resampled. It is this discretization that causes the high frequencies that remain from nonideal reconstruction to be folded into the low frequency range after resampling. This *is* aliasing because the continuous signal is no longer properly bandlimited before undergoing sampling.

In practice, most images of interest are not bandlimited, having sharp edges and high visual detail. Computer-generated imagery, in particular, often have step edges that contribute infinitely high frequencies to the spectrum. Furthermore, reconstruction filters are never, in practice, ideal low-pass filters. They tend to extend beyond the cut-off frequency and overlap neighboring spectra copies. Therefore, virtually all output inevitably has some form of degradation due to both aliasing and poor reconstruction. However, careful filter design can keep the errors well within the quantization of the framebuffers that store these images and the monitors that display them.

4.6. ANTIALIASING

The filtering necessary to combat aliasing is known as *antialiasing*. In order to determine corrective action, we must directly address the two conditions necessary for exact signal reconstruction. The first solution calls for low-pass filtering *before* sampling. This method, known as *prefiltering*, bandlimits the signal to levels below f_{max}, thereby eliminating the offending high frequencies. Notice that the frequency at which the signal is to be sampled imposes limits on the allowable bandwidth. This is often necessary when the output sampling grid must be fixed to the resolution of an output device, e.g., screen resolution. Therefore, aliasing is often a problem that is confronted when a signal is forced to conform to an inadequate resolution due to physical constraints. As a result, it is necessary to bandlimit, or narrow, the input spectrum to conform to the allotted bandwidth as determined by the sampling frequency.

The second solution is to point sample at a higher frequency. In doing so, the replicated spectra are spaced farther apart, thereby separating the overlapping spectra tails. This approach theoretically implies sampling at a resolution determined by the highest frequencies present in the signal. Since a surface viewed obliquely can give rise to arbitrarily high frequencies, this method may require extremely high resolution. Whereas the first solution adjusts the bandwidth to accommodate the fixed sampling rate, f_s, the second solution adjusts f_s to accommodate the original bandwidth. Antialiasing by sampling at the highest frequency is clearly superior in terms of image quality. This is, of course, operating under different assumptions regarding the possibility of varying f_s. In practice, antialiasing is performed through a combination of these two approaches. That is, the sampling frequency is increased so as to reduce the amount of bandlimiting to a minimum.

The effects of bandlimiting are shown below. The scanline in Fig. 4.14a is a horizontal cross-section taken from a monochrome version of the Mandrill image. Its frequency spectrum is illustrated in Fig. 4.14b. Since low frequency components often dominate the plots, a log scale is commonly used to display their magnitudes more clearly. In our case, we have simply clipped the zero frequency component to 30, from an original value of 130. This number represents the average input value. It is often referred to as the *DC* (direct current) component, a name derived from the electrical engineering literature.

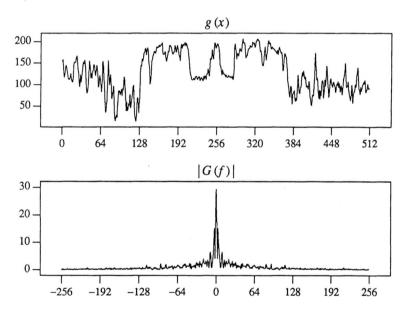

Figure 4.14: (a) A scanline and (b) its spectrum.

If we were to sample that scanline, we would face aliasing artifacts due to the fact that the spectras would overlap. As a result, the samples would not adequately characterize the underlying continuous signal. Consequently, the scanline undergoes blurring so that it may become bandlimited and avoid aliasing artifacts. This reasoning is intuitive since it is logical that a sparse set of samples can only adequately characterize a slowly-varying signal, i.e., one that is blurred. Figures 4.15 through 4.17 show the result of increasingly bandlimiting filters applied to the scanline in Fig. 4.14. They correspond to signals that are immune to aliasing after subsampling one out of every four, eight, and sixteen pixels, respectively.

Antialiasing is an important component to any application that requires high-quality digital filtering. The largest body of antialiasing research stems from computer graphics where high-quality rendering of complicated imagery is the central goal. The developed algorithms have primarily addressed the tradeoff issues of accuracy versus efficiency. Consequently, methods such as supersampling, adaptive sampling, stochastic sampling, pyramids, and preintegrated tables have been introduced. These techniques are described in Chapter 6.

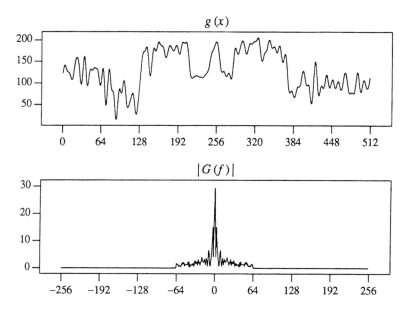

Figure 4.15: Bandlimited scanline appropriate for four-fold subsampling.

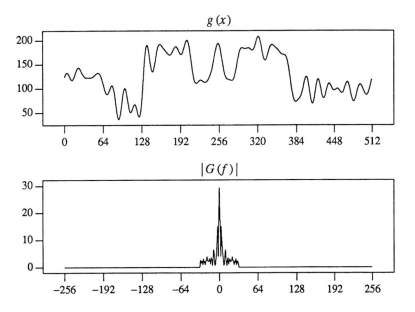

Figure 4.16: Bandlimited scanline appropriate for eight-fold subsampling.

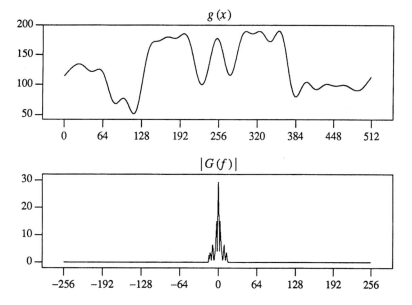

Figure 4.17: Bandlimited scanline appropriate for sixteen-fold subsampling.

4.7. SUMMARY

This chapter has reviewed the basic principles of sampling theory. We have shown that a continuous signal may be reconstructed from its samples if the signal is bandlimited and the sampling frequency exceeds the Nyquist rate. These are the two necessary conditions for image reconstruction to be possible. Since sampling can be shown to replicate a signal's spectrum across the frequency domain, ideal low-pass filtering was introduced as a means of retaining the original spectrum while discarding its copies. Unfortunately, the ideal low-pass filter in the spatial domain is an infinitely wide sinc function. Since this is difficult to work with, nonideal reconstruction filters are introduced to approximate the reconstructed output. These filters are nonideal in the sense that they do not completely attenuate the spectra copies. Furthermore, they contribute to some blurring of the original spectrum. In general, poor reconstruction leads to artifacts such as jagged edges.

Aliasing refers to the phenomenon that occurs when a signal is undersampled. This happens if the reconstruction conditions mentioned above are violated. In order to resolve this problem, one of two actions may be taken. Either the signal can be bandlimited to a range that complies with the sampling frequency, or the sampling frequency can be increased. In practice, some combination of both options are taken, leaving some relatively unobjectionable aliasing in the output.

Examples of the concepts discussed in this chapter are concisely depicted in Figs. 4.18 through 4.20. They attempt to illustrate the effects of sampling and low-pass filtering on the quality of the reconstructed signal and its spectrum. The first row of Fig. 4.18 shows a signal and its spectra, bandlimited to .5 cycle/pixel. For pedagogical purposes, we treat this signal as if it is continuous. In actuality, though, it is really a 256-sample horizontal cross-section taken from the Mandrill image. Since each pixel has 4 samples contributing to it, there is a maximum of two cycles per pixel. The horizontal axes of the spectra account for this fact.

The second row shows the effect of sampling the signal. Since $f_s = 1$ sample/pixel, there are four copies of the baseband spectrum in the range shown. Each copy is scaled by $f_s = 1$, leaving the magnitudes intact. In the third row, the 64 samples are shown convolved with a sinc function in the spatial domain. This corresponds to a rectangular pulse in the frequency domain. Since the sinc function is used here for image reconstruction, it must have an amplitude of unity value in order to interpolate the data. This forces the height of the rectangular pulse in the frequency domain to vary in response to f_s.

A few comments on the reciprocal relationship between the spatial and frequency domains are in order here, particularly as they apply to the ideal low-pass filter. We again refer to the variables A and W as defined in Section 4.3.3. As a sinc function is made broader, the value $1/2W$ is made to change since W is decreasing to accommodate zero crossings at larger intervals. Accordingly, broader sinc functions cause more blurring and their spectra reflect this by reducing the cut-off frequency to some smaller W. Conversely, narrower sinc functions cause less blurring and W takes on some larger value. In either case, the amplitude of the sinc function or its spectrum will change.

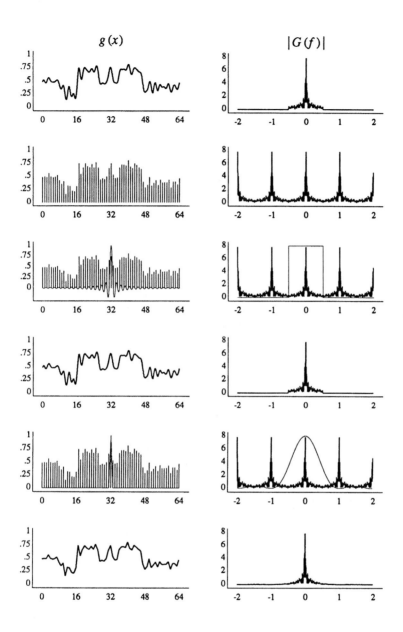

Figure 4.18: Sampling and reconstruction (with an adequate sampling rate). (Created by S. Feiner and G. Wolberg for [Foley 90]. Used with permission.)

That is, we can fix the amplitude of the sinc function so that only the rectangular pulse of the spectrum changes height $A/2W$ as W varies. Alternatively, we can fix $A/2W$ to remain constant as W changes, forcing us to vary A. The choice depends on the application.

When the sinc function is used to interpolate data, it is necessary to fix A to 1. Therefore, as the sampling density changes, the positions of the zero crossings shift, causing W to vary. This makes the amplitude of the spectrum's rectangular pulse change. On the other hand, if the sinc function is applied to bandlimit, not interpolate, the input signal, then it is important to fix $A/2W$ to 1 so that the passband frequencies remain intact. Since W is once again varying, A must change proportionately to keep $A/2W$ constant. Therefore, this application of the ideal low-pass filter requires the amplitude of the sinc function to be responsive to W.

In the examples presented below, our objective is to interpolate (reconstruct) the input and so $A=1$ regardless of the sampling density. Consequently, the height of the spectrum of the reconstruction filter changes. To make the Fourier transforms of the filters easier to see, we have not drawn the frequency response of the reconstruction filters to scale. Therefore, the rectangular pulse function in the third row of Fig. 14.18 actually has height $A/2W=1$. The fourth row of the figure shows the result after applying the ideal low-pass filter. As sampling theory predicts, the output is identical to the original signal. The last two rows of the figure illustrate the consequences of nonideal reconstruction filtering. Instead of using a sinc function, a triangle function corresponding to linear interpolation was applied. In the frequency domain this corresponds to the square of the sinc function. Not surprisingly, the spectrum of the reconstructed signal suffers in both the passband and the stopband.

The identical sequence of filtering operations is performed in Fig. 4.19. In this figure, though, the sampling rate has been lowered to $f_s=.5$, meaning that only one sample is collected for every two output pixels. Consequently, the replicated spectra are multiplied by .5, leaving the magnitudes at 4. Unfortunately, this sampling rate causes the replicated spectra to overlap. This, in turn, gives rise to aliasing, as depicted in the fourth row of the figure. Applying the triangle function to perform linear interpolation also yields poor results.

In order to combat these artifacts, the input signal must be bandlimited to accommodate the low sampling rate. This is shown in the second row of Fig. 14.20 where we see that all frequencies beyond $W=.25$ are truncated. This causes the input signal to be blurred. In this manner we have traded aliasing for blurring, a far less objectionable artifact. Sampling this function no longer causes the replicated copies to overlap. Convolving with an ideal low-pass filter now properly isolates the bandlimited spectrum.

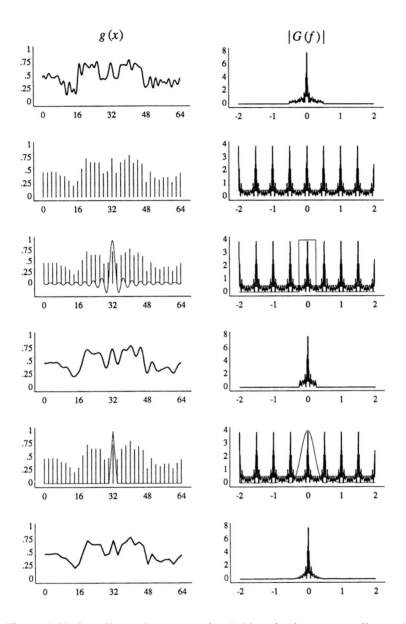

Figure 4.19: Sampling and reconstruction (with an inadequate sampling rate). (Created by S. Feiner and G. Wolberg for [Foley 90]. Used with permission.)

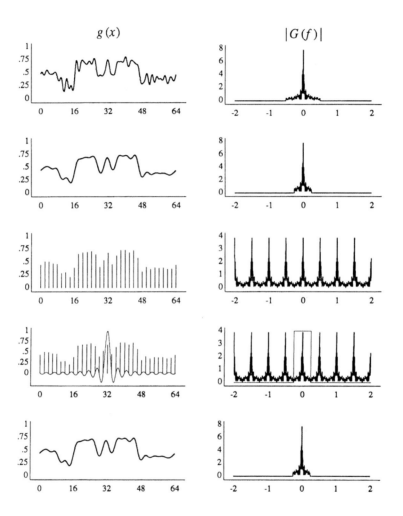

Figure 4.20: Antialiasing filtering, sampling, and reconstruction stages. (Created by S. Feiner and G. Wolberg for [Foley 90]. Used with permission.)

5

IMAGE RESAMPLING

5.1. INTRODUCTION

Image resampling is the process of transforming a sampled image from one coordinate system to another. The two coordinate systems are related to each other by the mapping function of the spatial transformation. This permits the output image to be generated by the following straightforward procedure. First, the inverse mapping function is applied to the output sampling grid, projecting it onto the input. The result is a *resampling grid*, specifying the locations at which the input is to be resampled. Then, the input image is sampled at these points and the values are assigned to their respective output pixels.

The resampling process outlined above is hindered by one problem. The resampling grid does not generally coincide with the input sampling grid, taken to be the integer lattice. This is due to the fact that the range of the continuous mapping function is the set of real numbers, a superset of the integer grid upon which the input is defined. The solution therefore requires a match between the domain of the input and the range of the mapping function. This can be achieved by converting the discrete image samples into a continuous surface, a process known as *image reconstruction*. Once the input is reconstructed, it can be resampled at any position.

Conceptually, image resampling is comprised of two stages: image reconstruction followed by sampling. Although resampling takes its name from the sampling stage, image reconstruction is the implicit component in this procedure. It is achieved through an interpolation procedure, and, in fact, the terms reconstruction and interpolation are often used interchangeably.

The image resampling process is depicted in Fig. 5.1 for the 1-D case. A discrete input (squares) is shown passing through the image reconstruction module, yielding a continuous input signal (solid curve). Reconstruction is performed by convolving the discrete input signal with a continuous interpolating function. The reconstructed input is then modulated (multiplied) with a resampling grid (dashed arrows). Note that the resampling grid is the result of projecting the output grid onto the input through a spatial

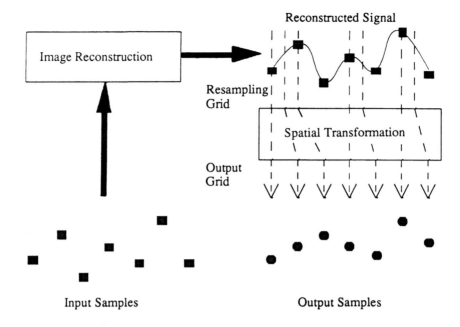

Figure 5.1: Image resampling.

transformation. After the reconstructed signal is sampled by the resampling grid, the samples (circles) are assigned to the uniformly spaced output image.

Image magnification and minification are two typical instances of image resampling. These operations are known by many different names. For instance, stretching, zooming, scaling up, interpolation, and upsampling are all informal terms used to describe magnification. Similarly, minification,[†] compression, shrinking, scale reduction, decimation, and downsampling are all terms that describe the process of reducing the size of an image. These two processes are illustrated in Fig. 5.2. In the top half of the figure, the interval between two adjacent black and white pixels must be reconstructed in order to generate five output points. A ramp is fitted between these points and uniformly sampled at five locations to yield the intensity gradation appearing at the output. In the bottom half of the figure, a scale reduction is shown. This was achieved by discarding points, a method prone to aliasing. Later we shall review antialiasing algorithms that use prefilters to bandlimit the input *before* resampling the continuous warped signal. Prefilters will be shown to be related to the interpolation functions used in reconstruction.

---

† This term originated in the computer graphics literature [Smith 83].

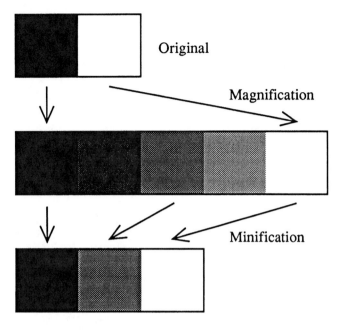

Figure 5.2: Image magnification and minification.

The two topics of reconstruction and antialiasing must be coupled in order to perform accurate image resampling. This chapter focuses on interpolation functions useful in reconstructing a continuous function from sampled image data. Before proceeding to image reconstruction, we briefly present an overview of ideal resampling. Although somewhat theoretical, the presentation should serve to identify the roles of reconstruction and prefiltering in their proper context. Together, they are used to define the ideal resampling filter.

5.2. IDEAL IMAGE RESAMPLING

There are four basic elements to ideal image resampling: reconstruction, warping, prefiltering, and sampling [Smith 83, Heckbert 89]. They are depicted in Fig. 5.3, and outlined in Table 5.1.

The progression begins with $f(u)$, the discrete input defined over integer values of u. It is reconstructed into $f_c(u)$ through convolution with reconstruction filter $r(u)$. From sampling theory, we know that the ideal reconstruction filter is the sinc function. The continuous input $f_c(u)$ is then warped according to mapping function m. The forward map is given as $x = m(u)$ and the inverse map is $u = m^{-1}(x)$. In this case, the warp is defined as an inverse mapping. It is also possible to formulate this as a forward mapping instead. The spatial transformation leaves us with $g_c(x)$, the continuous warped

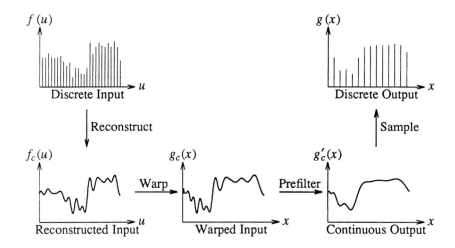

Figure 5.3: Ideal resampling [Heckbert 89].

Stage	Mathematical definition
Discrete Input	$f(u), \quad u \in Z$
Reconstructed Input	$f_c(u) = f(u) * r(u) = \sum_{k \in Z} f(k) r(u-k)$
Warped Signal	$g_c(x) = f_c(m^{-1}(x))$
Continuous Output	$g'_c(x) = g_c(x) * h(x) = \int g_c(t) h(x-t) dt$
Discrete Output	$g(x) = g'_c(x) s(x)$

Table 5.1: Elements of ideal resampling.

output. Depending on the inverse mapping function $m^{-1}(x)$, $g_c(x)$ may have arbitrarily high frequencies. Therefore, it is bandlimited by function $h(x)$ to conform to the Nyquist rate of the output. The bandlimited result is $g'_c(x)$. This function is sampled by $s(x)$, the output sampling grid, to produce the discrete output $g(x)$. Note that $s(x)$, often referred to as the comb function, is not required to sample the output at the same density as that of the input.

There are only two filtering components to the entire resampling process: reconstruction and prefiltering. We may cascade them into a single filter, derived as follows:

$$g(x) = g'_c(x) \qquad \text{for } x \in Z$$

$$= \int f_c(m^{-1}(t)) \, h(x-t) \, dt$$

$$= \int \left[\sum_{k \in Z} f(k) \, r(m^{-1}(t)-k) \right] h(x-t) \, dt \tag{5.2.1}$$

$$= \sum_{k \in Z} f(k) \, \rho(x,k)$$

where

$$\rho(x,k) = \int r(m^{-1}(t)-k) \, h(x-t) \, dt \tag{5.2.2}$$

is the *resampling filter* that specifies the weight of the input sample at location k for an output sample at location x [Heckbert 89].

Assuming that $m^{-1}(x)$ is invertible, we can express $\rho(x,k)$ in terms of an integral in the input space, rather than the output space. Substituting $t = m(u)$, we have

$$\rho(x,k) = \int r(u-k) \, h(x-m(u)) \left| \frac{\partial m}{\partial u} \right| du \tag{5.2.3}$$

where $\left| \partial m / \partial u \right|$ is the determinant of the Jacobian matrix interrelating the input and output coordinate systems. In one dimension,

$$\left| \frac{\partial m}{\partial u} \right| = \frac{dm}{du} \tag{5.2.4}$$

In two dimensions,

$$\left| \frac{\partial m}{\partial u} \right| = \begin{vmatrix} x_u & x_v \\ y_u & y_v \end{vmatrix} \tag{5.2.5}$$

where $x_u = \partial x / \partial u$, and similar notation holds for the other partial derivatives.

Either the input-space or output-space integral can be used to define the resampling filter. In the input-space form, ρ is expressed in terms of a reconstruction filter and a warped prefilter. This can be readily justified by noting that the reconstruction filter is applied *before* the warp and therefore it can be applied directly to the input. The prefilter, however, is applied *after* the warp and so its domain, still defined in terms of u, must undergo the geometric transformation. Since equal increments in u do not generally correspond to identical increments in $m(u)$, the prefilter is warped. This formulation of the resampling filter is depicted in Fig. 5.4. A similar case holds for the output-space form of ρ, which is written in terms of a warped reconstruction filter and a prefilter. Therefore, the actual warping is incorporated into either the reconstruction filter or prefilter, but not both.

The resampling filter takes on a simple form for space-invariant linear warps. In that case, the resampling filter can be shown to be equivalent to the convolution of the reconstruction filter and prefilter [Heckbert 89]. Expressed in input-space form, we have

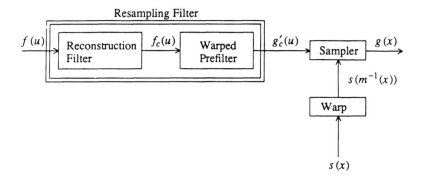

Figure 5.4: Ideal resampling with input-space resampling filter [Heckbert 89].

$$\rho(x,k) = \rho'(m^{-1}(x)-k) \qquad (5.2.6)$$
$$= h'(u) * r(u)$$
$$= \left[|J| h(uJ) \right] * r(u)$$

where J is the Jacobian matrix. This formulation is suitable for linear warps defined in terms of forward mapping functions, i.e., $m(u) = uJ$.

In the special case of magnification, we may ignore the prefilter altogether, treating it instead as an impulse function. This is due to the fact that no high frequencies are introduced into the output upon magnification. Conversely, minification introduces high frequencies and does not require any reconstruction of the input image. Consequently, we can ignore the reconstruction filter and treat it simply as an impulse function. Therefore,

$$\rho_{mag}(x,k) = r(u) \qquad (5.2.7a)$$
$$\rho_{min}(x,k) = |J| h(uJ) \qquad (5.2.7b)$$

Equations (5.2.7a) and (5.2.7b) lead us to an important observation about the shape of reconstruction filters and prefilters for linear warps. According to Eq. (5.2.7a), the shape of the reconstruction filter does not change in response to the mapping function. Indeed, magnification is achieved by selecting a reconstruction filter and directly convolving it across the input. Its shape remains fixed independently of the magnification scale factor. A similar procedure is taken in minification, whereby a reconstruction filter is replaced by a prefilter. The prefilter is selected on the basis of some desired frequency response characteristics. Unlike reconstruction filters, though, the actual shape must be scaled by an amount linearly related to the minification factor. As the input is increasingly decimated, the prefilter must become broader and shorter. It becomes broader in order to average more neighboring pixels together, thereby further bandlimiting the input. Since larger neighborhoods are used to compute each output pixel, the normalized weights applied to the input decrease to reflect the diminishing impact of each input sample. As a result, the prefilter grows shorter.

This observation is a direct consequence of the reciprocal relationship between the spatial and frequency domains. Due to the importance of this property, a proof is presented below. We start by writing the expression for the Fourier transform of $h(u)$.

$$h(u) \longleftrightarrow \int h(u) e^{-i2\pi fu} du \qquad (5.2.8)$$

Note that we use the symbol $\longleftrightarrow$ here to denote a transform pair. After we warp the input $h(u)$ through mapping function $m(u)$, we get

$$h(m(u)) \longleftrightarrow \int h(m(u)) e^{-i2\pi fu} du \qquad (5.2.9)$$

Letting $x = au = m(u)$ and $dx = \left| \dfrac{\partial m}{\partial u} \right| du$, we have

$$h(m(u)) \longleftrightarrow \int h(x) e^{-i2\pi fm^{-1}(x)} \frac{dx}{\left| \dfrac{\partial m}{\partial u} \right|} \qquad (5.2.10)$$

where $m^{-1}(x) = 1/a$ and $|\partial m / \partial u| = |J| = a$. This gives us

$$h(au) \longleftrightarrow \frac{1}{a} \int h(x) e^{-i2\pi f/a} dx \qquad (5.2.11)$$

or simply

$$h(au) \longleftrightarrow \frac{1}{a} H\left[\frac{f}{a} \right] \qquad (5.2.12)$$

This equation expresses the reciprocal relationship between the spatial and frequency domains. Notice that multiplying the spatial axis by a factor of a results in dividing the frequency axis and the spectrum values by that same factor.

This proves to be a fundamental result in linear filtering theory that bears significant consequences. For instance, we would ideally like to use narrow filters in the spatial domain. In this manner, each output pixel can be computed by weighting only a small number of input samples. However, the reciprocal relationship tells us that narrow filters in the spatial domain correspond to wide frequency spectrums. This, however, is undesirable as it hinders our attempts to avoid aliasing due to spectral overlaps. On the other hand, wide spatial filters are costly, but they do permit us to perform more effective bandlimiting. This tradeoff between narrow filters in the spatial domain and good filter response in the frequency domain is at the heart of filter design.

The remainder of this chapter focuses on interpolation for reconstruction, the central component of image resampling. This area has received extensive treatment due to its practical significance in numerous applications. Although theoretical limits on image reconstruction are derived by sampling theory, the algorithms proposed in this chapter address tradeoff issues in accuracy and complexity.

5.3. INTERPOLATION

Interpolation is the process of determining the values of a function at positions lying between its samples. It achieves this process by fitting a continuous function through the discrete input samples. This permits input values to be evaluated at arbitrary positions in the input, not just those defined at the sample points. While sampling generates an infinite bandwidth signal from one that is bandlimited, interpolation plays an opposite role: it reduces the bandwidth of a signal by applying a low-pass filter to the discrete signal. That is, interpolation reconstructs the signal lost in the sampling process by smoothing the data samples with an interpolation function.

For equally spaced data, interpolation can be expressed as

$$f(x) = \sum_{k=0}^{K-1} c_k h(x - x_k) \tag{5.3.1}$$

where h is the interpolation kernel weighted by coefficients c_k and applied to K data samples, x_k. Equation (5.3.1) formulates interpolation as a convolution operation. In practice, h is nearly always a symmetric kernel, i.e., $h(-x) = h(x)$. We shall assume this to be true in the discussion that follows. Furthermore, in all but one case that we will consider, the c_k coefficients are the data samples themselves.

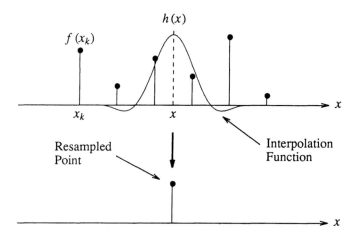

Figure 5.5: Interpolation of a single point.

The computation of one interpolated point is illustrated in Fig. 5.5. The interpolating function is centered at x, the location of the point to be interpolated. The value of that point is equal to the sum of the values of the discrete input scaled by the corresponding values of the interpolation kernel. This follows directly from the definition of convolution.

The interpolation function shown in the figure extends over four points. If x is offset from the nearest point by distance d, where $0 \le d < 1$, we sample the kernel at

$h(-d)$, $h(-1-d)$, $h(1-d)$, and $h(2-d)$. Since h is symmetric, it is defined only over the positive interval. Therefore, $h(d)$ and $h(1+d)$ are used in place of $h(-d)$ and $h(-1-d)$, respectively. Note that if the resampling grid is uniformly spaced, only a fixed number of points on the interpolation kernel must be evaluated. Large performance gains can be achieved by precomputing these weights and storing them in lookup tables for fast access during convolution. This approach will be described in more detail later in this chapter.

Although interpolation has been posed in terms of convolution, it is rarely implemented this way. Instead, it is simpler to directly evaluate the corresponding interpolating polynomial at the resampling positions. Why then is it necessary to introduce the interpolation kernel and the convolution process into the discussion? The answer lies in the ability to compare interpolation algorithms. Whereas evaluation of the interpolation polynomial is used to implement the interpolation, analysis of the kernel is used to determine the numerical accuracy of the interpolated function. This provides us with a quantitative measure which facilitates a comparison of various interpolation methods [Schafer 73].

Interpolation kernels are typically evaluated by analyzing their performance in the passband and stopband. Recall that an ideal reconstruction filter will have unity gain in the passband and zero gain in the stopband in order to transmit and suppress the signal's spectrum in these respective frequency ranges. Ideal filters, as well as superior nonideal filters, generally have wide extent in the spatial domain. For instance, the sinc function has infinite extent. As a result, they are categorized as *infinite impulse reponse filters* (IIR). It should be noted, however, that sinc functions are not physically realizable IIR filters. That is, they can only be realized approximately. The physically realizable IIR filters must necessarily use a finite number of computational elements. Such filters are also known as *recursive filters* due to their structure: they always have feedback, where the output is fed back to the input after passing through some delay element.

An alternative is to use filters with finite support that do not incorporate feedback, called *finite impulse response filters* (FIR). In FIR filters, each output value is computed as the weighted sum of a finite number of neighboring input elements. Note that they are not functions of past output, as is the case with IIR filters. Although IIR filters can achieve superior results over FIR filters for a given number of coefficients, they are difficult to design and implement. Consequently, FIR filters find widespread use in signal and image processing applications. Commonly used FIR filters include the box, triangle, cubic convolution kernel, cubic B-spline, and windowed sinc functions. They serve as the interpolating functions, or kernels, described below.

5.4. INTERPOLATION KERNELS

The numerical accuracy and computational cost of interpolation algorithms are directly tied to the interpolation kernel. As a result, interpolation kernels are the target of design and analysis in the creation and evaluation of interpolation algorithms. They are subject to conditions influencing the tradeoff between accuracy and efficiency.

In this section, the analysis is applied to the 1-D case. Interpolation in 2-D will be shown to be a simple extension of the 1-D results. In addition, the data samples are assumed to be equally spaced along each dimension. This restriction imposes no serious problems since images tend to be defined on regular grids. We now review the interpolation schemes in the order of their complexity.

5.4.1. Nearest Neighbor

The simplest interpolation algorithm from a computational standpoint is the *nearest neighbor* algorithm, where each interpolated output pixel is assigned the value of the nearest sample point in the input image. This technique, also known as the *point shift* algorithm, is given by the following interpolating polynomial.

$$f(x) = f(x_k) \qquad \frac{x_{k-1} + x_k}{2} < x \le \frac{x_k + x_{k+1}}{2} \qquad (5.4.1)$$

It can be achieved by convolving the image with a one-pixel width rectangle in the spatial domain. The interpolation kernel for the nearest neighbor algorithm is defined as

$$h(x) = \begin{cases} 1 & 0 \le |x| < .5 \\ 0 & .5 \le |x| \end{cases} \qquad (5.4.2)$$

Various names are used to denote this simple kernel. They include the *box filter*, *sample-and-hold function*, and *Fourier window*. The kernel and its Fourier transform are shown in Fig. 5.6. The reader should note that the figure refers to frequency f in $H(f)$, not function f.

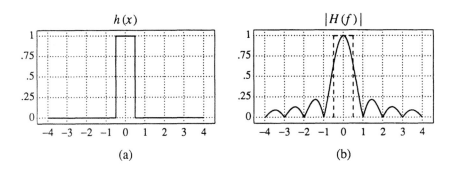

Figure 5.6: Nearest neighbor: (a) kernel, (b) Fourier transform.

Convolution in the spatial domain with the rectangle function h is equivalent in the frequency domain to multiplication with a sinc function. Due to the prominent side lobes and infinite extent, a sinc function makes a poor low-pass filter. Consequently, the nearest neighbor algorithm has a poor frequency domain response relative to that of the ideal low-pass filter.

The technique achieves magnification by pixel replication, and minification by sparse point sampling. For large-scale changes, nearest neighbor interpolation produces images with a blocky appearance. In addition, shift errors of up to one-half pixel are possible. These problems make this technique inappropriate when sub-pixel accuracy is required.

One notable property of this algorithm is that, except for the shift error, the resampled data exactly reproduce the original data if the resampling grid has the same spacing as that of the input. This means that the frequency spectra of the original and resampled images differ only by a pure linear phase shift. In general, the nearest neighbor algorithm permits zero-degree reconstruction and yields exact results only when the sampled function is piecewise constant.

Nearest neighbor interpolation was first used in remote sensing at a time when the processing time limitations of general purpose computers prohibited more sophisticated algorithms. It was found to simplify the entire mapping problem because each output point is a function of only one input sample. Furthermore, since the majority of problems involved only slight distortions with a scale factor near one, the results were considered adequate.

Currently, this method has been superceded by more elaborate interpolation algorithms. Dramatic improvements in digital computers account for this transition. Nevertheless, the nearest neighbor algorithm continues to find widespread use in one area: frame buffer hardware zoom functions. By simply diminishing the rate at which to sample the image and by increasing the cycle period in which the sample is displayed, pixels are easily replicated on the display monitor. This scheme is known as a sample-and-hold function. Although it generates images with large blocky patches, the nearest neighbor algorithm derives its primary use as a means for real-time magnification. For more sophisticated algorithms, this has only recently become realizable with the use of special-purpose hardware.

5.4.2. Linear Interpolation

Linear interpolation is a first-degree method that passes a straight line through every two consecutive points of the input signal. Given an interval (x_0, x_1) and function values f_0 and f_1 for the endpoints, the interpolating polynomial is

$$f(x) = a_1 x + a_0 \qquad (5.4.3)$$

where a_0 and a_1 are determined by solving

$$[f_0 \, f_1] = [a_1 \, a_0] \begin{bmatrix} x_0 & x_1 \\ 1 & 1 \end{bmatrix}$$

This gives rise to the following interpolating polynomial.

$$f(x) = f_0 + \left\lfloor \frac{x - x_0}{x_1 - x_0} \right\rfloor (f_1 - f_0) \qquad (5.4.4)$$

Not surprisingly, we have just derived the equation of a line joining points (x_0, f_0) and (x_1, f_1). In order to evaluate this method of interpolation, we must examine the frequency response of its interpolation kernel.

In the spatial domain, linear interpolation is equivalent to convolving the sampled input with the following interpolation kernel.

$$h(x) = \begin{cases} 1 - |x| & 0 \le |x| < 1 \\ 0 & 1 \le |x| \end{cases} \qquad (5.4.5)$$

Kernel h is referred to as a *triangle filter, tent filter, roof function, Chateau function*, or *Bartlett window*.

This interpolation kernel corresponds to a reasonably good low-pass filter in the frequency domain. As shown in Fig. 5.7, its response is superior to that of the nearest neighbor interpolation function. In particular, the side lobes are far less prominent, indicating improved performance in the stopband. Nevertheless, a significant amount of spurious high-frequency components continue to leak into the passband, contributing to some aliasing. In addition, the passband is moderately attenuated, resulting in image smoothing.

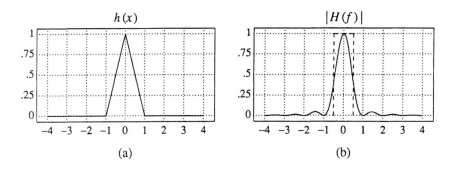

Figure 5.7: Linear interpolation: (a) kernel, (b) Fourier transform.

Linear interpolation offers improved image quality above nearest neighbor techniques by accommodating first-degree fits. It is the most widely used interpolation algorithm for reconstruction since it produces reasonably good results at moderate cost. Often, though, higher fidelity is required and thus more sophisticated algorithms have been formulated.

Although second-degree interpolating polynomials appear to be the next step in the progression, it was shown that their filters are space-variant with phase distortion [Schafer 73]. These problems are shared by all polynomial interpolators of even-degree.

This is attributed to the fact that the number of sampling points on each side of the inter-polated point always differ by one. As a result, interpolating polynomials of even-degree are not considered.

5.4.3. Cubic Convolution

Cubic convolution is a third-degree interpolation algorithm originally suggested by Rifman and McKinnon [Rifman 74] as an efficient approximation to the theoretically optimum sinc interpolation function. Its interpolation kernel is derived from constraints imposed on the general cubic spline interpolation formula. The kernel is composed of piecewise cubic polynomials defined on the unit subintervals $(-2,-1)$, $(-1,0)$, $(0,1)$, and $(1,2)$. Outside the interval $(-2,2)$, the interpolation kernel is zero.[†] As a result, each interpolated point is a weighted sum of four consecutive input points. This has the desir-able symmetry property of retaining two input points on each side of the interpolating region. It gives rise to a symmetric, space-invariant, interpolation kernel of the form

$$h(x) = \begin{cases} a_{30}|x|^3 + a_{20}|x|^2 + a_{10}|x| + a_{00} & 0 \le |x| < 1 \\ a_{31}|x|^3 + a_{21}|x|^2 + a_{11}|x| + a_{01} & 1 \le |x| < 2 \\ 0 & 2 \le |x| \end{cases} \qquad (5.4.6)$$

The values of the coefficients can be determined by applying the following set of con-straints to the interpolation kernel.

1. $h(0) = 1$ and $h(x) = 0$ for $|x| = 1$ and 2.
2. h must be continuous at $|x| = 0$, 1, and 2.
3. h must have a continuous first derivative at $|x| = 0$, 1, and 2.

The first constraint states that when h is centered on an input sample, the interpola-tion function is independent of neighboring samples. This permits f to actually pass through the input points. In addition, it establishes that the c_k coefficients in Eq. (5.3.1) are the data samples themselves. This follows from the observation that at data point x_j,

$$f(x_j) = \sum_{k=0}^{K-1} c_k h(x_j - x_k) \qquad (5.4.7)$$

$$= \sum_{k=j-2}^{j+2} c_k h(x_j - x_k)$$

According to the first constraint listed above, $h(x_j - x_k) = 0$ unless $j = k$. Therefore, the right-hand side of Eq. (5.4.7) reduces to c_j. Since this equals $f(x_j)$, we see that all c_k coefficients must equal the data samples in the four-point interval.

The first two constraints provide four equations for these coefficients:

---

† We again assume that our data points are located on the integer grid.

$$1 = h(0) = a_{00} \tag{5.4.8a}$$

$$0 = h(1^-) = a_{30} + a_{20} + a_{10} + a_{00} \tag{5.4.8b}$$

$$0 = h(1^+) = a_{31} + a_{21} + a_{11} + a_{01} \tag{5.4.8c}$$

$$0 = h(2^-) = 8a_{31} + 4a_{21} + 2a_{11} + a_{01} \tag{5.4.8d}$$

Three more equations are obtained from constraint (3):

$$-a_{10} = h'(0^-) = h'(0^+) = a_{10} \tag{5.4.8e}$$

$$3a_{30} + 2a_{20} + a_{10} = h'(1^-) = h'(1^+) = 3a_{31} + 2a_{21} + a_{11} \tag{5.4.8f}$$

$$12a_{31} + 4a_{21} + a_{11} = h'(2^-) = h'(2^+) = 0 \tag{5.4.8g}$$

The constraints given above have resulted in seven equations. However, there are eight unknown coefficients. This requires another constraint in order to obtain a unique solution. By allowing $a = a_{31}$ to be a free parameter that may be controlled by the user, the family of solutions given below may be obtained.

$$h(x) = \begin{cases} (a+2)|x|^3 - (a+3)|x|^2 + 1 & 0 \le |x| < 1 \\ a|x|^3 - 5a|x|^2 + 8a|x| - 4a & 1 \le |x| < 2 \\ 0 & 2 \le |x| \end{cases} \tag{5.4.9}$$

Additional knowledge about the shape of the desired result may be imposed upon Eq. (5.4.9) to yield bounds on the value of a. The heuristics applied to derive the kernel are motivated from properties of the ideal reconstruction filter, the sinc function. By requiring h to be concave upward at $|x| = 1$, and concave downward at $x = 0$, we have

$$h''(0) = -2(a+3) < 0 \quad \rightarrow \quad a > -3 \tag{5.4.10a}$$

$$h''(1) = -4a > 0 \quad \rightarrow \quad a < 0 \tag{5.4.10b}$$

Bounding a to values between -3 and 0 makes h resemble the sinc function. In [Rifman 74], the authors use the constraint that $a = -1$ in order to match the slope of the sinc function at $x = 1$. This choice results in some amplification of the frequencies at the high-end of the passband. As stated earlier, such behavior is characteristic of image sharpening.

Other choices for a include $-.5$ and $-.75$. Keys selected $a = -.5$ by making the Taylor series approximation of the interpolated function agree in as many terms as possible with the original signal [Keys 81]. He found that the resulting interpolating polynomial will exactly reconstruct a second-degree polynomial. Finally, $a = -.75$ is used to set the second derivatives of the two cubic polynomials in h to 1 [Simon 75]. This allows the second derivative to be continuous at $x = 1$.

Of the three choices for a, the value -1 is preferable if visually enhanced results are desired. That is, the image is sharpened, making visual detail perceived more readily.

However, the results are not mathematically precise, where precision is measured by the order of the Taylor series. To maximize this order, the value $a = -.5$ is preferable. The kernel and spectrum of a cubic convolution kernel with $a = -.5$ is shown in Fig. 5.8.

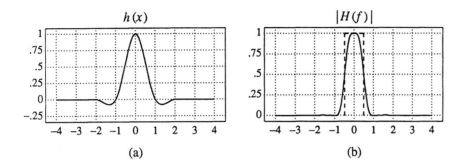

Figure 5.8: Cubic convolution: (a) kernel ($a = -.5$), (b) Fourier transform.

In a recent paper [Maeland 88], Maeland showed that at the Nyquist frequency the spectrum attains a value that is independent of the free parameter a. The value is equal to $(48/\pi^4)f_s$, while the value at the zero frequency is $H(0) = f_s$. This result implies that adjusting a can alter the cut-off rate between the passband and stopband, but not the attenuation at the Nyquist frequency. In comparing the effect of varying a, Maeland points out that cubic convolution with $a = 0$ is superior to the simple linear interpolation method when a strictly positive kernel is necessary. The role of a has also been studied in [Park 83], where a discussion is given on its optimal selection based on the frequency content of the image.

It is important to note that in the general case cubic convolution can give rise to values outside the range of the input data. Consequently, when using this method in image processing it is necessary to properly clip or rescale the results into the appropriate range for display.

5.4.4. Two-Parameter Cubic Filters

In [Mitchell 88], Mitchell and Netravali describe a variation of cubic convolution in which two parameters are used to describe a family of cubic reconstruction filters. Through a different set of constraints, the number of free parameters in Eq. (5.4.6) are reduced from eight to two. The constraints they use are:

1. $h(x) = 0$ for $|x| = 2$.

2. $h'(x) = 0$ for $|x| = 0$ and 2.

3. h must be continuous at $|x| = 1$. That is, $h(1^-) = h(1^+)$.

4. h must have a continuous first derivative at $|x| = 1$. That is, $h'(1^-) = h'(1^+)$.

5. $\displaystyle\sum_{n=-\infty}^{\infty} h(x-n) = 1$.

The first four constraints ensure that the interpolation kernel is flat at $|x| = 0$ and 2, and has continuous first derivatives at $|x| = 1$. They result in five equations for the unknown coefficients. The last constraint enforces a *flat-field response*, meaning that if the digital image has constant pixel values, then the reconstructed image will also have constant value. This yields the sixth of eight equations needed to solve for the unknown coefficients in Eq. (5.4.6). That leaves us with the following two-parameter family of solutions.

$$h(x) = \frac{1}{6} \begin{cases} (-9b-6c+12)|x|^3 + (12b+6c-18)|x|^2 + (-2b+6) & 0 \le |x| < 1 \\ (-b-6c)|x|^3 + (6b+30c)|x|^2 + (-12b-48c)|x| + K & 1 \le |x| < 2 \\ 0 & 2 \le |x| \end{cases} \quad (5.4.11)$$

where $K = 8b + 24c$. Several well-known cubic filters are derivable from Eq. (5.4.11) through an appropriate choice of values for (b,c). For instance, $(0,-c)$ corresponds to the cubic convolution kernel in Eq. (5.4.9) and $(1,0)$ is the cubic B-spline given later in Eq. (5.4.18).

The evaluation of these parameters is performed in the spatial domain, using the visual artifacts described in [Schreiber 85] as the criteria for judging image quality. In order to better understand the behavior of (b,c), the authors partitioned the parameter space into regions characterizing different artifacts, including blur, anisotropy, and ringing. As a result, the parameter pair $(.33,.33)$ is found to offer superior image quality. Another suggestion is $(1.5,-.25)$, corresponding to a band-reject, or notch, filter. This suppresses the signal energy near the Nyquist frequency that is most responsible for conspicuous moire patterns.

Despite the added flexibility made possible by a second free parameter, the benefits of the method for reconstruction fidelity are subject to scrutiny. In a recent paper [Reichenbach 89], the frequency domain analysis developed in [Park 82] was used to show that the additional parameter beyond that of the one-parameter cubic convolution does not improve the reconstruction fidelity. That is, the optimal two-parameter convolution kernel is identical to the optimal kernel for the traditional one-parameter algorithm, where optimality is taken to mean the minimization of the squared error at low spatial frequencies. It is then reasonable to ask whether this optimality criterion is useful. If so, why might images reconstructed with other interpolation kernels be preferred in a subjective test? Ultimately, any quantity that represents reconstruction error must necessarily conform to the subjective properties of the human visual system. This suggests that merging image restoration with reconstruction can yield significant improvements in the quality of reconstruction filters.

Further improvements in reconstruction are possible when derivative values can be given along with the signal amplitude. This is possible for synthetic images where this information may be available. In that case, Eq. (5.3.1) can be rewritten as

$$f(x) = \sum_{k=0}^{K-1} \left[f_k g(x-x_k) + f'_k h(x-x_k) \right] \quad (5.4.12)$$

where

$$g(x) = \frac{\sin^2 \pi x}{\pi^2 x^2} \qquad (5.4.13a)$$

$$h(x) = \frac{\sin^2 \pi x}{\pi^2 x} \qquad (5.4.13b)$$

An approximation to the resulting reconstruction formula can be given by Hermite cubic interpolation.

$$g(x) = \begin{cases} 2|x|^3 - 3|x|^2 + 1 & 0 \le |x| < 1 \\ 0 & 1 \le |x| \end{cases} \qquad (5.4.14a)$$

$$h(x) = \begin{cases} |x|^3 - 2x|x| + x & 0 \le |x| < 1 \\ 0 & 1 \le |x| \end{cases} \qquad (5.4.14b)$$

5.4.5. Cubic Splines

The next reconstruction technique we describe is the method of cubic spline interpolation. A *cubic spline* is a piecewise continuous third-degree polynomial. Given n points labeled (x_k, y_k) for $0 \le k < n$, the interpolating cubic spline consists of $n-1$ cubic polynomials. They pass through the supplied points, which are also known as *control points*.

We now derive the piecewise interpolating polynomials. The kth polynomial piece, f_k, is defined to pass through two consecutive input points in the fixed interval (x_k, x_{k+1}). Furthermore, f_k are joined at x_k (for $k = 1, \dots, n-2$) such that f_k, f_k', and f_k'' are continuous (Fig. 5.9). The interpolating polynomial f_k is given as

$$f_k(x) = a_3(x - x_k)^3 + a_2(x - x_k)^2 + a_1(x - x_k) + a_0 \qquad (5.4.15)$$

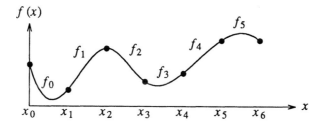

Figure 5.9: A spline consisting of 6 piecewise cubic polynomials.

The four coefficients of f_k can be defined in terms of the data points and their first (or second) derivatives. Assuming that the data samples are on the integer lattice, each spaced one unit apart, then the coefficients, defined in terms of the data samples and their first derivatives, are given below.

$$a_0 = y_k \tag{5.4.16a}$$

$$a_1 = y'_k \tag{5.4.16b}$$

$$a_2 = 3\Delta y_k - 2y'_k - y'_{k+1} \tag{5.4.16c}$$

$$a_3 = -2\Delta y_k + y'_k + y'_{k+1} \tag{5.4.16d}$$

where $\Delta y_k = y_{k+1} - y_k$.

Although the derivatives are not supplied with the data, they are derived by solving the following system of linear equations.

$$
\begin{bmatrix}
2 & 4 & & & & & \\
1 & 4 & 1 & & & & \\
& 1 & 4 & 1 & & & \\
& & & \cdot & & & \\
& & & & \cdot & & \\
& & & & 1 & 4 & 1 \\
& & & & & 4 & 2
\end{bmatrix}
\begin{bmatrix}
y'_0 \\
y'_1 \\
y'_2 \\
\cdot \\
\cdot \\
y'_{n-2} \\
y'_{n-1}
\end{bmatrix}
=
\begin{bmatrix}
-5y_0 + 4y_1 + y_2 \\
3(y_2 - y_0) \\
3(y_3 - y_1) \\
\cdot \\
\cdot \\
3(y_{n-1} - y_{n-3}) \\
-y_{n-3} - 4y_{n-2} + 5y_{n-1}
\end{bmatrix}
\tag{5.4.17}
$$

The not-a-knot boundary condition [de Boor 78] was used above, as reflected in the first and last rows of the matrices. It is superior to the artificial boundary conditions commonly reported in the literature, such as the natural or cyclic end conditions, which have no relevance in our application. Note that the need to solve a linear system of equations arises from global dependencies introduced by the constraints for continuous first and second derivatives at the knots. A complete derivation is given in Appendix 2.

In order to compare interpolating cubic splines with other methods, we must analyze the interpolation kernel. Thus far, however, the piecewise interpolating polynomials have been derived without any reference to an interpolation kernel. We seek to express the interpolating cubic spline as a convolution in a manner similar to the previous algorithms. This can be done with the use of cubic B-splines as interpolation kernels [Hou 78].

5.4.5.1. B-Splines

A *B-spline* of degree n is derived through n convolutions of the box filter, B_0. Thus, $B_1 = B_0 * B_0$ denotes a B-spline of degree 1, yielding the familiar triangle filter shown in Fig. 5.7a. Interpolation by B_1 consists of a sequence of straight lines joined at the knots continuously. This is equivalent to linear interpolation.

The second-degree B-spline B_2 is produced by convolving $B_0 * B_1$. Using B_2 to interpolate data yields a sequence of parabolas that join at the knots continuously together with their slopes. The span of B_2 is limited to three points.

The *cubic B-spline* B_3 is generated from convolving $B_0 * B_2$. That is, $B_3 = B_0 * B_0 * B_0 * B_0$. The interpolation with B_3 is composed of a series of cubic polynomials that join at the knots continuously together with their slopes and curvatures, i.e., their first and second derivatives. Figure 5.10 summarizes the shapes of these low-order B-splines.

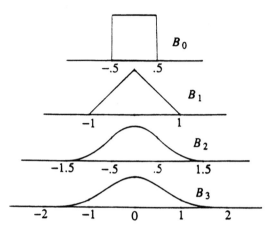

Figure 5.10: Low-order B-splines are derived from repeated box filters.

Denoting the cubic B-spline interpolation kernel as h, we have the following piecewise cubic polynomials defining the kernel.

$$h(x) = \frac{1}{6} \begin{cases} 3|x|^3 - 6|x|^2 + 4 & 0 \le |x| < 1 \\ -|x|^3 + 6|x|^2 - 12|x| + 8 & 1 \le |x| < 2 \\ 0 & 2 \le |x| \end{cases} \qquad (5.4.18)$$

This kernel is sometimes called the *Parzen window*.

There are several properties of cubic B-splines worth noting. As in the cubic convolution method, the extent of the cubic B-spline is over four points. This allows two points on each side of the central interpolated region to be used in the convolution. Consequently, the cubic B-spline is shift-invariant as well.

Unlike cubic convolution, however, the cubic B-spline kernel is not interpolatory since it does not satisfy the necessary constraint that $h(0) = 1$ and $h(1) = h(2) = 0$. Instead, it is an approximating function that passes near the points but not necessarily through them. This is due to the fact that the kernel is strictly positive.

The positivity of the cubic B-spline kernel is actually attractive for our image processing application. When using kernels with negative lobes, (e.g., the cubic convolution and windowed sinc functions), it is possible to generate negative values while interpolating positive data. Since negative intensity values are meaningless for display, it is desirable to use strictly positive interpolation kernels to guarantee the positivity of the interpolated image.

There are problems, however, in directly interpolating the data with kernel h, as given in Eq. (5.4.18). Due to the low-pass (blur) characteristics of h, the image undergoes considerable smoothing. This is evident by examining its frequency response where the stopband is effectively suppressed at the expense of additional attenuation in the passband. This leads us to the development of an interpolation method built upon the local support of the cubic B-spline.

5.4.5.2. Interpolating B-Splines

Interpolating with cubic B-splines requires that at data point x_j, we again satisfy Eq. (5.4.7). Namely,

$$f(x_j) = \sum_{k=j-2}^{j+2} c_k h(x_j - x_k) \tag{5.4.19}$$

From Eq. (5.4.18), we have $h(0) = 4/6$, $h(-1) = h(1) = 1/6$, and $h(-2) = h(2) = 0$. This yields

$$f(x_j) = \frac{1}{6}(c_{j-1} + 4c_j + c_{j+1}) \tag{5.4.20}$$

Since this must be true for all data points, we have a chain of global dependencies for the c_k coefficients. The resulting linear system of equations is similar to that obtained for the derivatives of the cubic interpolating spline algorithm. We thus have,

$$
\begin{bmatrix} f_0 \\ f_1 \\ f_2 \\ \cdot \\ \cdot \\ f_{n-2} \\ f_{n-1} \end{bmatrix}
=
\begin{bmatrix}
4 & 1 & & & & \\
1 & 4 & 1 & & & \\
& 1 & 4 & 1 & & \\
& & & \cdot & & \\
& & & & 1 & 4 & 1 \\
& & & & & 1 & 4
\end{bmatrix}
\begin{bmatrix} c_0 \\ c_1 \\ c_2 \\ \cdot \\ \cdot \\ c_{n-2} \\ c_{n-1} \end{bmatrix}
\tag{5.4.21}
$$

Labeling the three matrices above as F, K, and C, respectively, we have

$$F = KC \tag{5.4.22}$$

The coefficients in C may be evaluated by multiplying the known data points F with the inverse of the tridiagonal matrix K.

$$C = FK^{-1} \tag{5.4.23}$$

The inversion of tridiagonal matrix K has an efficient algorithm that is solvable in linear time [Press 88]. In [Lee 83], the matrix inversion step is modified to introduce high-frequency emphasis. This serves to compensate for the undesirable low-pass filter imposed by the point-spread function of the imaging system.

In all the previous methods, the coefficients c_k were taken to be the data samples themselves. In the cubic spline interpolation algorithm, however, the coefficients must be determined by solving a tridiagonal matrix problem. After the interpolation coefficients have been computed, cubic spline interpolation has the same computational cost as cubic convolution.

5.4.6. Windowed Sinc Function

Sampling theory establishes that the sinc function is the ideal interpolation kernel. Although this interpolation filter is exact, it is not practical since it is an IIR filter defined by a slowly converging infinite sum. Nevertheless, it is perfectly reasonable to consider the effects of using a truncated, and therefore finite, sinc function as the interpolation kernel.

The results of this operation are predicted by sampling theory, which demonstrates that truncation in one domain leads to ringing in the other domain. This is due to the fact that truncating a signal is equivalent to multiplying it with a rectangle function $Rect(x)$, defined as

$$Rect(x) = \begin{cases} 1 & 0 \le |x| < .5 \\ 0 & .5 \le |x| \end{cases} \qquad (5.4.24)$$

Since multiplication in one domain is convolution in the other, truncation amounts to convolving the signal's spectrum with a sinc function, the transform pair of $Rect(x)$. We have already seen an example of this in Fig. 4.7. Since the stopband is no longer eliminated, but rather attenuated by a ringing filter (i.e., a sinc), the input is not bandlimited and aliasing artifacts are introduced. The most typical problems occur at step edges, where the Gibbs phenomena becomes noticeable in the form of undershoots, overshoots, and ringing in the vicinity of edges. In [Ratzel 80], the author found this method to perform poorly.

The $Rect$ function above served as a *window*, or kernel, that weighs the input signal. In Fig. 5.11a, we see the $Rect$ window extended over three pixels on each side of its center, i.e., $Rect(6x)$ is plotted. The corresponding windowed sinc function $h(x)$ is shown in Fig. 5.11b. This is simply the product of the sinc function with the window function, i.e., $sinc(x)Rect(6x)$. Its spectrum, shown in Fig. 5.11c, is nearly an ideal low-pass filter. Although it has a fairly sharp transition from the passband to the stopband, it is plagued by ringing. In order to more clearly see the values in the spectrum, we use a logarithmic scale for the vertical axis of the spectrum in Fig. 5.11d. The next few figures will be illustrated by using this same four-part format.

Ringing can be mitigated by using a different windowing function exhibiting smoother fall-off than the rectangle. The resulting windowed sinc function can yield

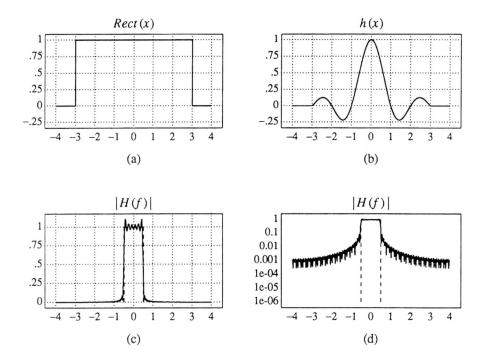

Figure 5.11: (a) Rectangular window; (b) Windowed sinc; (c) Spectrum; (d) Log plot.

better results. However, since slow fall-off requires larger windows, the computation remains costly.

Aside from the rectangular window mentioned above, the most frequently used window functions are: Hann,[†] Hamming, Blackman, and Kaiser [Antoniou 79]. These filters identify a quantity known as the ripple ratio, defined as the ratio of the maximum side-lobe amplitude to the main-lobe amplitude. Good filters will have small ripple ratios to achieve effective attenuation in the stopband. A tradeoff exists, however, between ripple ratio and main-lobe width. Therefore, as the ripple ratio is decreased, the main-lobe width is increased. This is consistent with the reciprocal relationship between the spatial and frequency domains, i.e., narrow bandwidths correspond to wide spatial functions.

In general, though, each of these smooth window functions is defined over a small finite extent. This is tantamount to multiplying the smooth window with a rectangle function. While this is better than the *Rect* function alone, there will inevitably be some form of aliasing. Nevertheless, the window functions described below offer a good compromise between ringing and blurring.

---

† Due to Julius von Hann. It is often mistakenly referred to as the Hanning window.

5.4.6.1. Hann and Hamming Windows

The Hann and Hamming windows are defined as

$$
Hann/Hamming(x) = \begin{cases} \alpha + (1-\alpha)\cos \dfrac{2\pi x}{N-1} & |x| < \dfrac{N-1}{2} \\ 0 & \text{otherwise} \end{cases} \tag{5.4.25}
$$

where N is the number of samples in the windowing function. The two windowing functions differ in the choice of α. In the Hann window $\alpha = 0.5$, and in the Hamming window $\alpha = 0.54$. Since they both amount to a scaled and shifted cosine function, they are also known as the *raised cosine window*.

The spectra for the Hann and Hamming windows can be shown to be the sum of a sinc, the spectrum of *Rect* (x), with two shifted counterparts: a sinc shifted to the right by $2\pi/(N-1)$, as well as one shifted to the left by the same amount. This serves to cancel the right and left side lobes in the spectrum of *Rect* (x). As a result, the Hann and Hamming windows have reduced side lobes in their spectra as compared to those of the rectangular window. The Hann window is illustrated in Fig. 5.12.

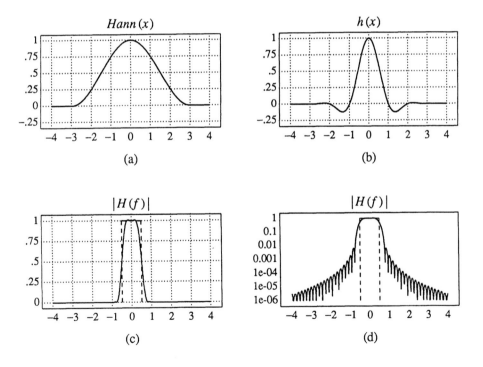

Figure 5.12: (a) Hann window; (b) Windowed sinc; (c) Spectrum; (d) Log plot.

Notice that the passband is only slightly attenuated, but the stopband continues to retain high frequency components in the stopband, albeit less than that of *Rect*(*x*). It performs somewhat better in the stopband than the Hamming window, as shown in Fig. 5.13. This is partially due to the fact that the Hamming window is discontinuous at its ends, giving rise to "kinks" in the spectrum.

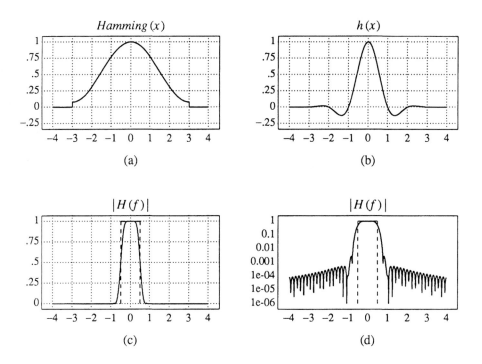

Figure 5.13: (a) Hamming window; (b) Windowed sinc; (c) Spectrum; (d) Log plot.

5.4.6.2. Blackman Window

The Blackman window is similar to the Hann and Hamming windows. It is defined as

$$Blackman(x) = \begin{cases} 0.42 + 0.5\cos\dfrac{2\pi x}{N-1} + 0.08\cos\dfrac{4\pi x}{N-1} & |x| < \dfrac{N-1}{2} \\ 0 & \text{otherwise} \end{cases} \qquad (5.4.26)$$

The purpose of the additional cosine term is to further reduce the ripple ratio. This window function is shown in Fig. 5.14.

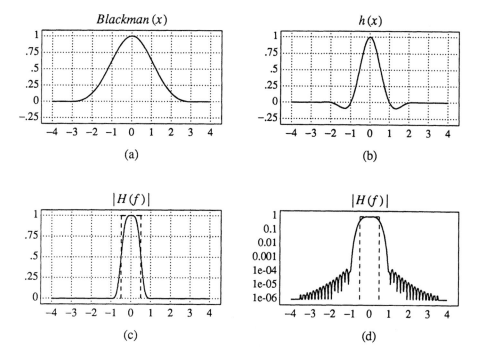

Figure 5.14: (a) Blackman window; (b) Windowed sinc; (c) Spectrum; (d) Log plot.

5.4.6.3. Kaiser Window

The Kaiser window is defined as

$$Kaiser(x) = \begin{cases} \dfrac{I_0(\beta)}{I_0(\alpha)} & |x| < \dfrac{N-1}{2} \\ \\ 0 & \text{otherwise} \end{cases} \tag{5.4.27}$$

where α is a free parameter and

$$\beta = \alpha \left[1 - \left(\frac{2x}{N-1} \right)^2 \right]^{1/2} \tag{5.4.28}$$

I_0 is the zeroth-order Bessel function of the first kind. This can be evaluated to any desired degree of accuracy by using the rapidly converging series

$$I_0(n) = 1 + \sum_{k=1}^{\infty} \left[\frac{1}{k!} \left(\frac{n}{2} \right)^k \right]^2 \tag{5.4.29}$$

The Kaiser window leaves the filter designer much flexibility in controlling the ripple ratio by adjusting the parameter α. As α is incremented, the level of sophistication of the window function grows as well. Therefore, the rectangular window corresponds to a Kaiser window with $\alpha = 0$, while more sophisticated windows such as the Hamming window correspond to $\alpha = 5$. This formulation facilitates a tradeoff between ringing and edge softening.

5.4.6.4. Lanczos Window

Windowed sinc functions are notorious for producing ringing artifacts near edges. Although they are an improvement over truncated sinc functions, they retain a fairly sharp transition from passband to stopband. Superior filters can be designed by imposing further constraints on the filter response in the frequency domain.

Reasonable constraints to impose on the kernel include: unity gain in the low-pass region with cut-off at frequency f_1, zero gain at high frequencies beyond f_2, and linear fall-off in the transition range between f_1 and f_2. This frequency response can be expressed as the convolution of two boxes. In the spatial domain, this corresponds to the multiplication of two sinc functions, yielding a function known as the *Lanczos window*. The widths of the two sinc functions determine the extent of the transition range.

The two-lobed Lanczos window function is defined as

$$Lanczos\,2(x) = \begin{cases} \dfrac{sinc(\pi x/2)}{\pi x/2} & 0 \le |x| < 2 \\ \\ 0 & 2 \le |x| \end{cases} \tag{5.4.30}$$

The *Lanczos* 2 window function is the central lobe of a sinc function. It is wide enough to extend over two lobes of the ideal low-pass filter, i.e., a second sinc function. The windowed sinc function is therefore given by the product $sinc\,(x)\,Lanczos\,2(x)$. This can be rewritten as $sinc\,(x)\,sinc\,(x/2)\,Rect(x/4)$, where the first term is the ideal low-pass filter, the second term is $Lanczos\,2(x)$, and $Rect(x/4)$ is the rectangular function that truncates $Lanczos\,2$ past $x = 2$. Note that its abscissa is $x/4$ because $Rect$ is defined over $-.5 \le x \le .5$. The spectrum of this product is $Rect(f)*Rect(2f)*sinc\,(4f)$, where $*$ is convolution. The $Lanczos\,2(x)$ window function is shown in Fig. 5.15.

This formulation can be generalized to an N-lobed window function by replacing the value 2 in Eq. (5.4.30) to the value N. For instance, the 3-lobed Lanczos window is defined as

$$Lanczos\,3(x) = \begin{cases} \dfrac{sinc(\pi x/3)}{\pi x/3} & 0 \le |x| < 3 \\ \\ 0 & 3 \le |x| \end{cases} \tag{5.4.31}$$

The $Lanczos\,3(x)$ window function is shown in Fig. 5.16. As we let more more lobes pass under the Lanczos window, then the spectrum of the windowed sinc function becomes $Rect(f)*Rect(N f)*sinc\,(2N f)$. This proves to be a superior frequency

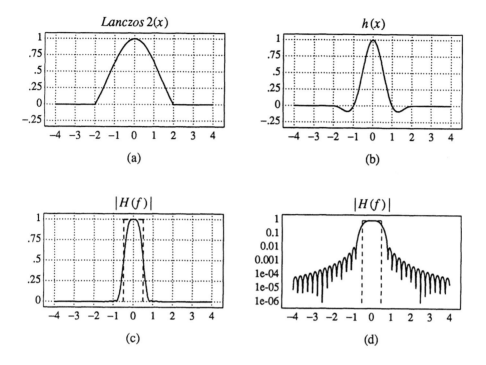

Figure 5.15: (a) Lanczos2 window; (b) Windowed sinc; (c) Spectrum; (d) Log plot.

response than that of the 2-lobed Lanczos window because the $Rect(Nf)$ term causes faster fall-off in the transition region and $sinc(2Nf)$ is a narrower sinc function that produces less deleterious ringing artifacts.

5.4.6.5. Gaussian Window

The Gaussian function is defined as

$$Gauss(x) = \frac{1}{\sqrt{2\pi}\,\sigma}\,e^{-x^2/2\sigma^2} \qquad (5.4.32)$$

where σ is the standard deviation. Gaussians have the nice property that their spectrum is also a Gaussian. They can be used to directly smooth the input for prefiltering purposes or to smooth the sinc function for windowed sinc reconstruction filters. Furthermore, since the tails of a Gaussian diminish rapidly, they may be truncated and still produce results that are not plagued by excessive ringing. The rate of fall-off is determined by σ, with low values of σ resulting in faster decay.

The general form of the Gaussian may be expressed more conveniently as

$$Gauss_\sigma(x) = 2^{-(x/\sigma)^2} \qquad (5.4.33)$$

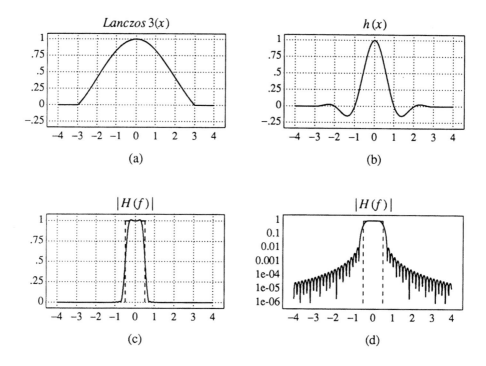

Figure 5.16: (a) Lanczos3 window; (b) Windowed sinc; (c) Spectrum; (d) Log plot.

Two plots of this function are shown in Fig. 5.17 with $\sigma = 1/2$ and $1/\sqrt{2}$. The latter is a wider Gaussian than the first, and its magnitude doesn't become negligible until two samples away from the center. The benefit of these choices of σ is that many of their coefficients for commonly used resampling ratios are scaled powers of two, which makes way for fast computation. In [Turkowski 88a], these two functions have been examined for use in image resampling.

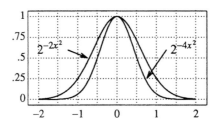

Figure 5.17: Two Gaussian functions.

5.4.7. Exponential Filters

A superior class of reconstruction filters can be derived using exponential functions. Consider, for instance, the hyberbolic tangent function *tanh* defined in Eq. (5.4.34).

$$tanh(x) = \frac{e^x - e^{-x}}{e^x + e^{-x}} \tag{5.4.34}$$

This function has several desirable properties. First, it converges quickly to ± 1. Second, its transition from -1 to 1 is sharp. We can sharpen the transition even further by scaling the domain, i.e., use $tanh(kx)$ for $k \geq 1$. In addition, this function is infinitely differentiable everywhere, i.e., it satisfies an important smoothness constraint. These properties are readily apparent in Fig. 5.18, which illustrates $tanh(kx)$ for $k = 1$, 4, and 10. Notice that the function quickly approximates *Rect* for larger values of k.

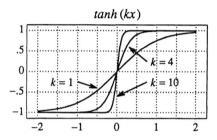

Figure 5.18: Scaled hyperbolic tangent function.

Given $tanh(kx)$ as our starting point, we can define a new function that resembles the ideal low-pass filter $Rect(f)$, i.e., a box in the frequency domain. This is done by treating *tanh* as one half of *Rect*, and then merely compositing that with a mirror image of itself. Since *tanh* lies between -1 and 1, some care must be taken to normalize the expression so that it yields a box of unity height. The resulting function is given as

$$H_k(f) = \left[\frac{tanh(k(x + f_c)) + 1}{2} \right] \left[\frac{tanh(k(-x + f_c)) + 1}{2} \right] \tag{5.4.35}$$

where f_c is the cut-off frequency. In our examples, we shall use $f_c = .5$ to conform to the Nyquist rate. The purpose of the addition and division operations is to normalize $H_k(f)$ so that $0 \leq H_k(f) \leq 1$.

Function H_k is treated as the desired spectrum of our reconstruction filter. By varying k, we can control the shape of the spectrum. For low values of k, H_k is smooth and resembles a Gaussian function. As k is made larger, H_k will have increasingly sharper corners, eventually approximating a *Rect* function. Figure 5.19 shows $H_k(f)$ for $k = 1$, 4, and 10.

Having established H_k to be the desired spectrum of our interpolation kernel, the actual kernel is derived by computing the inverse Fourier transform of Eq. (5.4.35). This

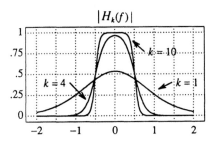

Figure 5.19: Spectrum $H_k(f)$ is a function of $tanh\,(kx)$.

gives us $h_k(x)$, as shown in Fig. 5.20. Not surprisingly, it has infinite extent. However, unlike the sinc function that decays at a rate of $1/x$, h_k decays exponentially fast. This is readily verified by inspecting the log plots in Fig. 5.20.[†] This means that we may truncate it with negligible penalty in reconstruction quality. The truncation is, in effect, implicit in the decay of the filter. In practice, a 7-point kernel (with 3 points on each side of the center) yields excellent results [Massalin 90].

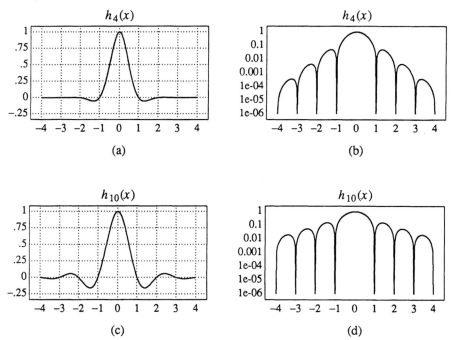

Figure 5.20: Interpolation kernels derived from $H_4(f)$ and $H_{10}(f)$.

---

† Note that a linear fall-off in log scale corresponds to an exponential function.

5.5. COMPARISON OF INTERPOLATION METHODS

The quality of the popular interpolation kernels are ranked in ascending order as follows: nearest neighbor, linear, cubic convolution, cubic spline, and sinc function. These interpolation methods are compared in many sources, including [Andrews 76, Parker 83, Maeland 88, Ward 89]. Below we give some examples of these techniques for the magnification of the Star and Madonna images. The Star image helps show the response of the filters to a high contrast image with edges oriented in many directions. The Madonna image is typical of many natural images with smoothly varying regions (skin), high frequency regions (hair), and sharp transitions on curved boundaries (cheek). Also, a human face (especially one as famous as this) comes with a significant amount of a priori knowledge, which may affect the subjective evaluation of quality. Only monochrome images are used here to avoid obscuring the results over three color channels.

In Fig. 5.21, a small 50×50 section was taken from the center of the Star image, and magnified to 500×500 by using the following interpolation methods: nearest neighbor, linear interpolation, cubic convolution (with $A = -1$), and cubic convolution (with $A = -.5$). Figure 5.22 shows the same image magnified by the following interpolation methods: cubic spline, Lanczos2 windowed sinc function, Hamming windowed sinc, and the exponential filter derived from the tanh function.

The algorithms are rated according to the passband and stopband performances of their interpolation kernels. If an additional process is required to compute coefficients used together with the kernel, its effect must be evaluated as well. In [Parker 83], the authors failed to consider this when they erroneously concluded that cubic convolution is superior to cubic spline interpolation. Their conclusion was based on an inappropriate comparison of the cubic B-spline kernel with that of the cubic convolution. The fault lies in neglecting the effect of computing the coefficients in Eq. (5.3.1). Had the data samples been directly convolved with the cubic B-spline kernel, then the analysis would have been correct. However, in performing a matrix inversion to determine the coefficients, a certain periodic filter must be multiplied together with the spectrum of the cubic B-spline in order to produce the interpolation kernel. The resulting kernel can be easily demonstrated to be of infinite support and oscillatory, sharing the same properties as the Cardinal spline (sinc) kernel [Maeland 88]. This is reasonable, considering the recursive nature of the interpolation kernel. By a direct comparison, cubic spline interpolation performs better than cubic convolution, albeit at slightly greater computational cost.

It is important to note that high quality interpolation algorithms are not always warranted for adequate reconstruction. This is due to the natural relationship that exists between the rate at which the input is sampled and the interpolation quality necessary for accurate reconstruction. If a bandlimited input is densely sampled, then its replicating spectra are spaced far apart. This diminishes the role of frequency leakage in the degradation of the reconstructed signal. Consequently, we can relax the accuracy of the interpolation kernel in the stopband. Therefore, the stopband performance necessary for adequate reconstruction can be made a function of the input sampling rate. Low sampling rates require the complexity of the sinc function, while high rates allow simpler algorithms. Although this result is intuitively obvious, it is reassuring to arrive at the same

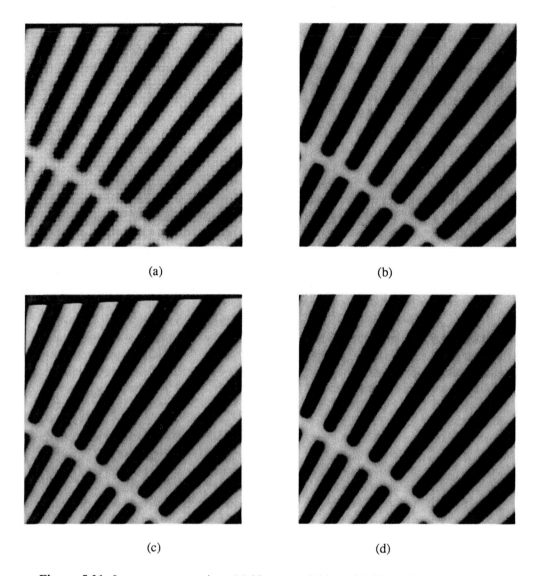

(a)   (b)

(c)   (d)

Figure 5.21: Image reconstruction. (a) Nearest neighbor; (b) Linear interpolation; (c) Cubic convolution ($A = -1$); (d) Cubic convolution ($A = -.5$).

conclusion from an interpretation in the frequency domain.

The above discussion has focused on reconstructing gray-scale (color) images. Complications emerge when the attention is restricted to bi-level (binary) images. In [Abdou 82], the authors analyze several interpolation schemes for bi-level image applications. This is of practical importance for the geometric transformation of images of black-and-white documents. Subtleties are introduced due to the nonlinear elements that

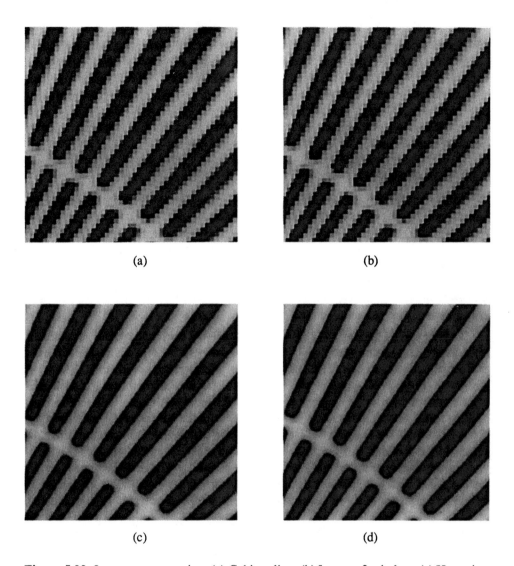

(a)   (b)

(c)   (d)

Figure 5.22: Image reconstruction. (a) Cubic spline; (b) Lanczos2 window; (c) Hamming window; (d) Exponential filter.

enter into the imaging process: quantization and thresholding. Since binary signals are not bandlimited and the nonlinear effects are difficult to analyze in the frequency domain, the analysis is performed in the spatial domain. Their results confirm the conclusions already derived regarding interpolation kernels. In addition, they arrive at useful results relating the errors introduced in the tradeoff between sampling rate and quantization.

5.6. IMPLEMENTATION

In this section, we present two methods to speed up the image resampling stage. The first approach addresses the computational bottleneck of evaluating the interpolation function at any desired position. These computed values are intended for use as weights applied to the input. The second method we describe is a fast 1-D resampling algorithm that combines image reconstruction with antialiasing to perform image resampling in scanline order. This algorithm, as originally proposed, implements reconstruction using linear interpolation, and implements antialiasing using a box filter. It is ideally suited for hardware implementation for use in nonlinear image warping.

5.6.1. Interpolation with Coefficient Bins

When implementing image resampling, it is necessary to weigh the input samples with appropriate values taken from the interpolation kernel. Depending on the inverse mapping, the interpolation kernel may be centered anywhere in the input. The weights applied to the neighboring input pixels must be evaluated by sampling the centered kernel at positions coinciding with the input samples. By making some assumptions about the allowable set of positions at which we can sample the interpolation kernel, we can accelerate the resampling operation by precomputing the input weights and storing them in lookup tables for fast access during convolution [Ward 89].

In [Ward 89], image resampling is done by mapping each output point back into the input image, i.e., inverse mapping. The distance between input pixels is divided into a number of intervals, or bins, each having a set of precomputed coefficients. The set of coefficients for each bin corresponds to samples of the interpolation function positioned at the center of the bin. Computing each new pixel then requires quantization of its input position to the nearest bin and a table lookup to obtain the corresponding set of weights that are applied to the respective input samples. The advantage of this method is that the calculation of coefficients, which requires evalution of the interpolation function at positions corresponding to the original samples, is replaced by a table lookup operation. A mean-squared error analysis with this method shows that the quantization effects due to the use of coefficient bins can be made the same as integer roundoff if 17 bins are used. More coefficient bins yields a higher density of points for which the interpolation function is accurately computed, yielding more precise output values. Ward shows that a lookup table of 65 coefficient bins adds virtually no error to that due to roundoff.

This approach is demonstrated below for the special case of 1-D magnification. The function *magnify_*1D takes *IN*, an input array of *INlen* pixels, and magnifies it to fill *OUTlen* entries in *OUT*. For convenience, we will assume that the symmetric convolution kernel extends over seven pixels (three on each side), i.e., a 7-point kernel. The kernel is oversampled at a rate of *Oversample* samples per pixel. Therefore, if we wish to center the kernel at any of, say, 512 subpixel positions, then we would choose *Oversample* = 512 and initialize *kern* with 7×512 kernel samples. Note that the 7-point kernel in the code is included to make the program more efficient and readable. A more general version of the code would permit kernels of arbitrary width.

```
#define KernShift    12                    /* 12-bit kernel integers */
#define KernHalf     (1 << (KernShift-1))  /* 1/2 in kernel's notation */
#define Oversample512                      /* subdivisions per pixel */

magnify_1D(IN, OUT, INlen, OUTlen, kern)
unsigned char *IN, *OUT;
int INlen, OUTlen, *kern;
{
        int x, i, ii, dii, ff, dff, len;
        long val;

        len = OUTlen;
        OUTlen--;
        INlen-- ;
        ii = 0;                            /* ii indexes into bin */
        ff = OUTlen / 2;                   /* ff is fractional remainder */
        x = INlen * Oversample;
        dii = x / OUTlen;                  /* dii is ii increment */
        dff = x % OUTlen;                  /* dff is ff increment */

        /* compute all output pixels */
        for(x=0; x<len; x++) {
                /* compute convolution centered at current position */
                val    = (long) IN[-2] * kern[2*Oversample + ii]
                       + (long) IN[-1] * kern[1*Oversample + ii]
                       + (long) IN[ 0] * kern[ii]
                       + (long) IN[ 1] * kern[1*Oversample - ii]
                       + (long) IN[ 2] * kern[2*Oversample - ii]
                       + (long) IN[ 3] * kern[3*Oversample - ii];
                if(ii == 0)
                        val += (long) IN[-3] * kern[3*Oversample + ii];

                /* roundoff and restore into 8-bit number */
                val = (val + KernHalf) >> KernShift;
                if(val < 0) val = 0;               /* clip from below */
                if(val > 0xFF) val = 0xFF;         /* clip from above */
                OUT[x] = val;                      /* save result */

                /* Bresenham-like algorithm to recenter kernel */
                if((ff += dff) >= OUTlen) {        /* check if fractional part overflows */
                        ff -= OUTlen;              /* normalize */
                        ii++;                      /* increment integer part */
                }
                if((ii += dii) >= Oversample) {    /* check if integer part overflows */
                        ii -= Oversample;          /* normalize */
                        IN++;                      /* increment input pointer */
                }
        }
}
```

The function *magnify_1D* above operates exclusively using integer arithmetic. This proves to be efficient for those applications in which a floating point accelerator is not available. We choose to represent kernel samples as integers scaled by 4096 for 12-bit accuracy. Note that the sum of 7 products, each of which is 20 bits long (8-bit intensity and 12-bit kernel), can be stored in 23 bits. This leaves plenty of space after packing the results into 32-bit integers, a common size used in most machines. Although more bits can be devoted to the precision of the kernel samples, higher quality results must necessarily use larger values of *Oversample* as well.

A second motivation for using integer arithmetic comes from the desire to circumvent division while recentering the kernel. The most direct way of computing where to center the kernel in *IN* is by evaluating $(x)(INlen / OUTlen)$, where x is the index into *OUT*. An alternate, and cheaper, approach is to compute this positional information incrementally using rational numbers in mixed radix notation [Massalin 90]. We identify three variables of interest: *IN*, *ii*, and *ff*. *IN* is the current input pixel in which the kernel center resides. It is subdivided into *Oversample* bins. The variable *ii* indexes into the proper bin. Since the true floating point precision has been quantized in this process, *ff* is used to maintain the position *within* the bin. Therefore, as we scan the input, the new positions can be determined by adding *di* to *ii* and *dff* to *ff*. When doing so, *ff* may overflow beyond *OUTlen* and *ii* may overflow beyond *Oversample*. In these cases, the appropriate roundoff and normalizations must be made. In particular, *ii* is incremented if *ff* is found to exceed *OUTlen* and *IN* is incremented if *ii* is found to exceed *Oversample*. It should be evident that *ii* and *ff*, taken together, form a fractional component that is added *IN*. Although only *ii* is needed to determine which coefficient bin to select, *ff* is needed to prevent the accrual of error during the incremental computation. Together, these three variables form the following pointer *P* into the input:

$$P = IN + \frac{ii + ff \, / \, OUTlen}{Oversample} \qquad (5.6.1)$$

where $0 \le ii < Oversample$ and $0 \le ff < OUTlen$.

A few additional remarks are in order here. Since the convolution kernel can extend beyond the image boundary, we assume that *IN* has been padded with a 3-pixel border on both sides. For minimal border artifacts, their values should taken to be that of the image boundary, i.e., $IN[0]$ and $IN[INlen-1]$, respectively. After each output pixel is computed, the convolution kernel is shifted by an amount $(INlen-1)/(OUTlen-1)$. The value -1 enters into the calculation because this is necessary to guarantee that the boundary values will remain fixed. That is, for resampling operations such as magnification, we generally want $OUT[0]=IN[0]$ and $OUT[OUTlen-1]=IN[INlen-1]$.

Finally, it should be mentioned that the approach taken above is a variant of the Bresenham line-drawing algorithm. Division is made unnecessary by use of rational arithmetic where the integer numerator and denominator are maintained exactly. In Eq. (5.6.1), for instance, *ii* can be used directly to index in the kernel without any additional arithmetic. A mixed radix notation is used, such that *ii* and *ff* cannot exceed *Oversample* and *OUTlen*, respectively. These kinds of incremental algorithms can be viewed in terms of mixed radix arithmetic. An intuitive example of this notation is our system for telling time: the units of days, hours, minutes, and seconds are not mutually related by the same factor. That is, 1 day = 24 hours, 1 hour = 60 minutes, etc. Performing arithmetic above is similar to performing calculations involving time. A significant difference, however, is that whereas the scales of time are fixed, the scales used in this magnification example are derived from *INlen* and *OUTlen*, two data-dependent parameters. A similar approach to the algorithm described above can be taken to perform minification. This problem is left as an exercise for the reader.

5.6.2. Fant's Resampling Algorithm

The central benefit of separable algorithms is the reduction in complexity of 1-D resampling algorithms. When the input is restricted to be one-dimensional, efficient solutions are made possible for the image reconstruction and antialiasing components of resampling. Fant presents a detailed description of such an algorithm that is well-suited for hardware implementation [Fant 86]. Related patents on this method include [Graf 87, Fant 89].

The process treats the input and output as streams of pixels that are consumed and generated at rates determined by the spatial mapping. The input is assumed to be mapped onto the output along a single direction, i.e., with no folds. As each input pixel arrives, it is weighted by its partial contribution to the current output pixel and integrated into an accumulator. In terms of the input and output streams, one of three conditions is possible:

1. The current input pixel is entirely consumed without completing an output pixel.

2. The input is entirely consumed while completing the output pixel.

3. The output pixel will be completed without entirely consuming the current input pixel. In this case, a new input value is interpolated from the neighboring input pixels at the position where the input was no longer consumed. It is used as the next element in the input stream.

If conditions (2) or (3) apply, the output computation is complete and the accumulator value is stored into the output array. The accumulator is then reset to zero in order to receive new input contributions for the next output pixel. Since the input is unidirectional, a one-element accumulator is sufficient. The process continues to cycle until the entire input stream is consumed.

The algorithm described in [Fant 86] is a principal 1-D resampling method used in separable transformations defined in terms of forward mapping functions. Like the example given in the preceding section, this method can be shown to use a variant of the Bresenham algorithm to step through the input and output streams. It is demonstrated in the example below. Consider the input arrays shown in Fig. 5.23. The first array specifies the values of $F_v(u)$ for $u = 0, 1, ..., 4$. These represent new x-coordinates for their respective input pixels. For instance, the leftmost pixel will start at $x = 0.6$ and terminate at $x = 2.3$. The next input pixel begins to influence the output at $x = 2.3$ and proceeds until $x = 3.2$. This continues until the last input pixel is consumed, filling the output between $x = 3.3$ and $x = 3.9$.

The second array specifies the distribution range that each input pixel assumes in the output. It is simply the difference between adjacent coordinates. Note that this requires the first array to have an additional element to define the length of the last input pixel. Large values correspond to stretching, and small values reflect compression. They determine the rate at which input is consumed to generate the output stream.

The input intensity values are given in the third array. Their contributions to the output stream is marked by connecting segments. The output values are labeled A_0 through A_3 and are defined below. For clarity, the following notation is used: interpolated input values are written within square brackets ([]), weights denoting contributions to output pixels are written within an extra level of parentheses, and input intensity values are printed in boldface.

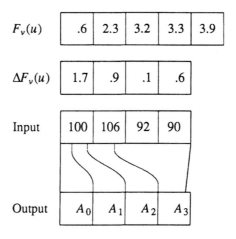

| $F_v(u)$ | .6 | 2.3 | 3.2 | 3.3 | 3.9 |

| $\Delta F_v(u)$ | 1.7 | .9 | .1 | .6 |

| Input | 100 | 106 | 92 | 90 |

| Output | A_0 | A_1 | A_2 | A_3 |

Figure 5.23: Resampling example.

$$A_0 = (\mathbf{100})((.4)) = 40$$

$$A_1 = \left[(\mathbf{100})\left[1 - \frac{.4}{1.7}\right] + (\mathbf{106})\left[\frac{.4}{1.7}\right] \right] ((1)) = 101$$

$$A_2 = \left[(\mathbf{100})\left[1 - \frac{1.4}{1.7}\right] + (\mathbf{106})\left[\frac{1.4}{1.7}\right] \right] ((.3)) + (\mathbf{106})((.7)) = 106$$

$$A_3 = \left[(\mathbf{106})\left[1 - \frac{.7}{.9}\right] + (\mathbf{92})\left[\frac{.7}{.9}\right] \right] ((.2)) + (\mathbf{92})((.1)) + (\mathbf{90})((.6)) = 82$$

The algorithm demonstrates both image reconstruction and antialiasing. When we are not positioned at pixel boundaries in the input stream, linear interpolation is used to reconstruct the discrete input. When more than one input element contributes to an output pixel, the weighted results are integrated in an accumulator to achieve antialiasing. These two cases are both represented in the above equations, as denoted by the expressions between square brackets and double parentheses, respectively.

Fant presents several examples of this algorithm on images that undergo magnification, minification, and a combination of these two operations. The resampling algorithm is illustrated in Fig. 5.24. It makes references to the following variables. *SIZFAC* is the multiplicative scale factor from the input to the output. For example, *SIZFAC* = 2 denotes two-fold magnification. *INSFAC* is 1/*SIZFAC*, or the inverse size factor. It indicates how many input pixels contribute to each output pixel. Thus, in the case of two-fold magnification, only one half of an input pixel is needed to fill an entire output pixel. *INSEG* indicates how much of the current input pixel remains to contribute to the next output pixel. This ranges from 0 (totally consumed) to 1 (entirely available). Analogously, *OUTSEG* indicates how much of the current output pixel remains to be filled. Finally, *Pixel* is the intensity value of the current input pixel.

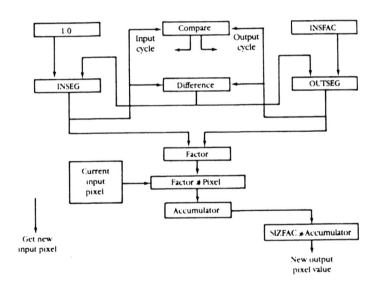

Figure 5.24: Fant's resampling algorithm [Fant 86].

The following C code implements the algorithm as described above. Input intensities are found in *IN*, an array containing *INlen* entries. The output is stored in *OUT*, an array of *OUTlen* elements. In this example, we assume that a constant scale factor, *OUTlen/INlen*, applies to each pixel.

```
resample(IN, OUT, INlen, OUTlen)
unsigned char *IN, *OUT;
int INlen, OUTlen;
{
        int u, x;
        double acc, intensity, INSFAC, SIZFAC, INSEG, OUTSEG;

        SIZFAC = (double) OUTlen / INlen;        /* scale factor */
        INSFAC = 1.0 / SIZFAC;                   /* inverse scale factor */
        OUTSEG = INSFAC;                         /* # input pixels that map onto 1 output pixel */
        INSEG  = 1.0;                            /* entire input pixel is available */
        acc = 0.;                                /* clear accumulator */

        /* compute all output pixels */
        for(x = u = 0; x < OUTlen; ) {
                /* use linear interpolation for reconstruction */
                intensity = (INSEG * IN[u])  + ((1-INSEG) * IN[u+1]);

                /* INSEG < OUTSEG: input pixel is entirely consumed before output pixel */
                if(INSEG < OUTSEG) {
                        acc += (intensity * INSEG);      /* accumulate weighted contribution */
                        OUTSEG -= INSEG;                 /* INSEG portion has been filled */
                        INSEG = 1.0;                     /* new input pixel will be available */
                        u++;                             /* index into next input pixel */
                }

                /* INSEG >= OUTSEG: input pixel is not entirely consumed before output pixel */
                else {
                        acc += (intensity * OUTSEG);     /* accumulate weighted contribution */
                        OUT[x] = acc * SIZFAC;           /* init output with normalized accumulator */
                        acc = 0.;                        /* reset accumulator for next output pixel */

                        INSEG -= OUTSEG;                 /* OUTSEG portion of input has been used */
                        OUTSEG = INSFAC;                 /* restore OUTSEG */
                        x++;                             /* index into next output pixel */
                }
        }
}
```

The code given above is restricted to transformations characterized by a constant scale factor, i.e., affine transformations. It can be modified to handle nonlinear image warping, where the scale factor is made to vary from pixel to pixel. The more sophisticated mappings involved in this general case can be conveniently stored in a forward mapping address table F. The elements of this table are point samples of the forward mapping function — it contains the output coordinates for each input pixel. Consequently, F is made to have the same dimensions as IN, the array containing the input image values.

```
resample_gen(F, IN, OUT, INlen, OUTlen)
double *F;
unsigned char *IN, *OUT;
int INlen, OUTlen;
{
      int u, x;
      double acc, intensity, inpos[2048], INSFAC, INSEG, OUTSEG;

      /* precompute input index for each output pixel */
      for(u = x = 0; x < OUTlen; x++) {
            while(F[u+1] < x) u++;
            inpos[x] = u + (double) (x-F[u]) / (F[u+1]-F[u]);
      }


      INSEG  = 1.0;                    /* entire input pixel is available */
      OUTSEG = inpos[1];               /* # input pixels that map onto 1 output pixel */
      INSFAC = OUTSEG;                 /* inverse scale factor */
      acc = 0.;                        /* clear accumulator */

      /* compute all output pixels */
      for(x = u = 0; x < OUTlen; ) {
            /* use linear interpolation for reconstruction */
            intensity = (INSEG * IN[u]) + ((1-INSEG) * IN[u+1]);

            /* INSEG < OUTSEG: input pixel is entirely consumed before output pixel */
            if(INSEG < OUTSEG) {
                  acc += (intensity * INSEG);      /* accumulate weighted contribution */
                  OUTSEG -= INSEG;                 /* INSEG portion has been filled */
                  INSEG = 1.0;                     /* new input pixel will be available */
                  u++;                             /* index into next input pixel */
            }

            /* INSEG >= OUTSEG: input pixel is not entirely consumed before output pixel */
            else {
                  acc += (intensity * OUTSEG);     /* accumulate weighted contribution */
                  OUT[x] = acc / INSFAC;           /* init output with normalized accumulator */
                  acc = 0.;                        /* reset accumulator for next output pixel */
                  INSEG -= OUTSEG;                 /* OUTSEG portion of input has been used */
                  x++;                             /* index into next output pixel */
                  INSFAC = inpos[x+1] - inpos[x];  /* init spatially-varying INSFAC */
                  OUTSEG = INSFAC;                 /* init spatially-varying SIZFAC */
            }
      }
}
```

The version of Fant's resampling algorithm given above will always produce antialiased images as long as the scale change does not exceed the precision of *INSFAC*. That is, an eight bit *INSFAC* is capable of scale factors no greater than 255. The algorithm, however, is more sensitive in the spatial domain, i.e., spatial position inaccuracies.

This may become manifest in the form of spatial jitter between consecutive rows on the right edge of the output line (assuming that the row was processed from left to right). These errors are due to the continued mutual subtraction of *INSEG* and *OUTSEG*. Examples are given in [Fant 86].

The sensitivity to spatial jitter is due to incremental errors. This can be mitigated using higher precision for the intermediate computation. Alternatively, the problem can be resolved by separately treating each interval spanned by the input pixels. Although such a method may appear to be less elegant than that presented above, it serves to decouple errors made among intervals. We demonstrate its operation in the hope of further illustrating the manner in which forward mappings are conducted. Since it is less tightly coupled, it is perhaps easier to follow.

The approach is based on the fact that the input pixel can either lie fully embedded in an output pixel, or it may straddle several output pixels. In the first case, the input is weighted by its partial contribution to the output pixel, and then that value is deposited to an accumulator. The accumulator will ultimately be stored in the output array only when the input interval passes across its rightmost boundary (assuming that the algorithm proceeds from left to right). In the second case, the input pixel actually crosses, or straddles, at least one output pixel boundary. A single input pixel may give rise to a "left straddle" if it occupies only a partial output pixel before it crosses its first output boundary from the left side. As long as the input pixel continues to fully cover output pixels, it is said to be in the "central interval." Finally, the last partial contribution to an output pixel on the right side is called a "right straddle."

Note that not all three types of coverage must result upon resampling. For instance, if an input pixel is simply translated by .6, then it has a left straddle of .4, no central straddle, and a right straddle of .6. The following code serves to demonstrate this approach. It assumes that processing proceeds from left to right, and no foldovers are allowed. That is, the forward mapping function is strictly nondecreasing. Again, *IN* contains *INlen* input pixels that must be resampled to *OUTlen* entries stored in *OUT*. As before, only a one-element accumulator is necessary. For simplicity, we let *OUT* accumulate partial contributions instead of using a separate *acc* accumulator. In order to do this accurately, *OUT* is made to have double precision. As in *resample_gen*, *F* is the sampled forward mapping function that facilitates spatially-varying scale factors.

```
resample_intervals(F, IN, OUT, INlen, OUTlen)
double *F, *IN, *OUT;
int INlen, OUTlen;
{
        int u, x, x0, x1, ix0, ix1;
        double intensity, dl;

        /* clear output array (also used to accumulate intermediate results) */
        for(x = 0; x <= OUTlen; x++) OUT[x] = 0;

        /* visit all input pixels (IN) and compute resampled output (OUT) */
        for(u = 0; u < INlen; u++) {
                /* input pixel u stretches in the output from x0 to x1 */
                x0 = F[u];
                x1 = F[u+1];
                ix0 = (int) x0;        /* for later use as integer index */
                ix1 = (int) x1;        /* for later use as integer index */

                /* check if interval is embedded in output pixel */
                if(ix0 == ix1) {
                        intensity = IN[u] * (x1-x0);        /* weighted intensity */
                        OUT[ix1] += intensity;              /* accumulate contributions */
                        continue;                           /* go on to next pixel */
                }
                /* if we got this far, input straddles more than one output pixel */

                /* left straddle */
                intensity = IN[u] * (ix0+1 - x0);        /* weighted intensity */
                OUT[ix0] += intensity;                   /* accumulate contribution */

                /* central interval */
                dl = (IN[u+1] - IN[u]) / (x1 - x0);        /* for linear interpolation */
                for(x=ix0+1; x<ix1; x++)                   /* visit all pixels in central interval */
                        OUT[x] = IN[u] + dl*(x-x0);        /* init output pixel */

                /* right straddle */
                if(x1 != ix1) {
                        /* partial output pixel remains: accumulate its contribution in OUT */
                        intensity = (IN[u] + dl*(ix1-x0)) * (x1 - ix1);
                        OUT[ix1] += intensity;
                }
        }
}
```

The 1-D interpolation algorithms described above generalize quite simply to 2-D. This is accomplished by performing 1-D interpolation in each dimension. For example, the horizontal scanlines are first processed, yielding an intermediate image which then undergoes a second pass of interpolation in the vertical direction. The result is independent of the order: processing the vertical lines before the horizontal lines gives the same results. Each of the two passes are elements of a separable transformation that allow a reconstruction filter $h(x,y)$ to be replaced by the product $h(x)h(y)$.

In 2-D, the nearest neighbor and bilinear interpolation algorithms use a 2×2 neighborhood about the desired location. The separable transform result is identical to computing these methods directly in 2-D. The proof for bilinear interpolation was given in Chapter 3. In cubic convolution, a 4×4 neighborhood is used to achieve an approximation to the radially symmetric 2-D sinc function. Note that this is not equivalent to the result obtained through direct computation. This can be easily verified by observing that the zeros are all aligned along the rectangular grid instead of being distributed along concentric circles. Nevertheless, separable transforms provide a substantial reduction in computational complexity from $O(N^2 M^2)$ to $O(NM^2)$ for an $M \times M$ image and an $N \times N$ filter kernel.

5.7. DISCUSSION

Image reconstruction plays a critical role in image resampling because geometric transformations often require image values at points that do not coincide with the input lattice. Therefore, some form of interpolation is necessary to reconstruct the continuous image from its samples. This chapter has described various image reconstruction algorithms for resampling. It is certainly easy to be overwhelmed with the many different goals and assumptions that lie embedded in these techniques. In this section, we attempt to clarify some of the underlying similarities and differences between these methods. This should also serve to indicate when certain algorithms are more appropriate than others.

To better evaluate the different reconstruction algorithms, we review the goals of image reconstruction and then we evaluate the described techniques in terms of these objectives. Ideally, we want a reconstruction kernel with a small neighborhood in the spatial domain and a narrow transition region in the frequency domain. The use of small neighborhoods allow us to produce the output with less computation. Narrow transition regions reflect the sharp cut-off between passband and stopband that is necessary to minimize blurring and aliasing. These two goals, however, are mutually incompatible as a consequence of the reciprocal relationship between the spatial and frequency domains. Instead, we attempt to accommodate a tradeoff. Unfortunately, these tradeoffs contribute to ringing artifacts, as well as some combination of blurring and aliasing.

The simplest functions we described are the box filter and the triangle filter. They were used for nearest neighbor and linear interpolation, respectively. Their formulation was based solely on characteristics in the spatial domain. Assuming that the input data is accurately modeled as piecewise constant or piecewise linear functions, these two respective approaches can exactly reconstruct the data. Similarly, cubic splines can exactly reconstruct the samples assuming the data is accurately modeled as a cubic function.

The method of cubic convolution, however, had different origins. Instead of defining its kernel by assuming that we can model the input, the cubic convolution kernel is defined by approximating the truncated sinc function with a piecewise cubic polynomials. The motivation for this approach is to approximate the infinite sinc function with a finite representation. In this manner, an approximation to the ideal reconstruction filter can be applied to the data without any need to place restrictions on the input model. A free parameter is available for the user to fine-tune the response of the filter. Properties of the sinc function are often used as heuristics to select the free parameter.

In a related approach, windowed sinc functions have been introduced to directly apply a finite approximation of the sinc function to the input. Instead of approximating the sinc with piecewise cubic polynomials, the sinc function is multiplied with a smooth window so that truncation does not produce excessive ringing. Various window functions have been proposed: Hann, Hamming, Blackman, Kaiser, Lanczos, and Gaussian windows. They are all motivated by different goals. The Hann, Hamming, and Blackman windows use the cosine function to generate a smooth fall-off. The spectrum of these windows can be shown to be related to the summation of shifted sinc functions. Proper choice of parameters allows the side lobes to delicately cancel out.

The Lanczos window uses the central lobe of a sinc function to taper the tails of the ideal low-pass filter. The rationale here is best understood by considering the frequency domain. Since the spectrum of the ideal filter is a box, then windowing will cause it to be convolved with another spectrum. If that spectrum is chosen to be another box, then the passband and stopband can continue to have ideal performance. Only a transition band needs to be introduced. The problem, however, is that the suggested window is itself a sinc function. Since that too must be truncated, there will be some additional ringing.

Superior results were derived with a new class of filters introduced in this chapter. We began by abandoning the premise that the starting point must be an ideal filter. Instead, we formulated an analytic function with a free parameter that could be tuned to produce a desired transition width between the passband and stopband. The analytic function we used in our example was defined in terms of the hyperbolic tangent. This function was chosen because its corresponding kernel, although still of infinite extent, exhibits exponential fall-off. The success of this method hinges on this important property. As a result, we could simply truncate the kernel as soon as its response fell below the desired accuracy, i.e., quantization error. Response accuracy beyond the quantization error is wasteful because the augmented fidelity cannot be noticed. This observation can be exploited to design cheaper filters.

6

ANTIALIASING

The geometric transformation of digital images is inherently a sampling process. As with all sampled data, digital images are susceptible to aliasing artifacts. This chapter reviews the antialiasing techniques developed to counter these deleterious effects. The largest contribution to this area stems from work in computer graphics and image processing, where visually complex images containing high spatial frequencies must be rendered onto a discrete array. In particular, antialiasing has played a critical role in the quality of texture-mapped and ray-traced images. Remote sensing and medical imaging, on the other hand, typically do not deal with large scale changes that warrant sophisticated filtering. They have therefore neglected this stage of the processing.

6.1. INTRODUCTION

Aliasing occurs when the input signal is undersampled. There are two solutions to this problem: raise the sampling rate or bandlimit the input. The first solution is ideal but may require a display resolution which is too costly or unavailable. The second solution forces the signal to conform to the low sampling rate by attenuating the high frequency components that give rise to the aliasing artifacts. In practice, some compromise is reached between these two solutions [Crow 77, 81].

6.1.1. Point Sampling

The naive approach for generating an output image is to perform *point sampling*, where each output pixel is a single sample of the input image taken independently of its neighbors (Fig. 6.1). It is clear that information is lost between the samples and that aliasing artifacts may surface if the sampling density is not sufficiently high to characterize the input. This problem is rooted in the fact that intermediate intervals between samples, which should have some influence on the output, are skipped entirely.

The Star image is a convenient example that overwhelms most resampling filters due to the infinitely high frequencies found toward the center. Nevertheless, the extent of the artifacts are related to the quality of the filter and the actual spatial transformation.

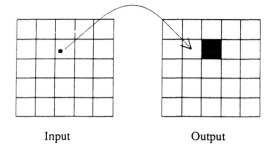

Input   Output

Figure 6.1: Point sampling.

Figure 6.2 shows two examples of the moire effects that can appear when a signal is undersampled using point sampling. In Fig. 6.2a, one out of every two pixels in the Star image was discarded to reduce its dimension. In Fig. 6.2b, the artifacts of undersampling are more pronounced as only one out of every four pixels are retained. In order to see the small images more clearly, they are magnified using cubic spline reconstruction. Clearly, these examples show that point sampling behaves poorly in high frequency regions.

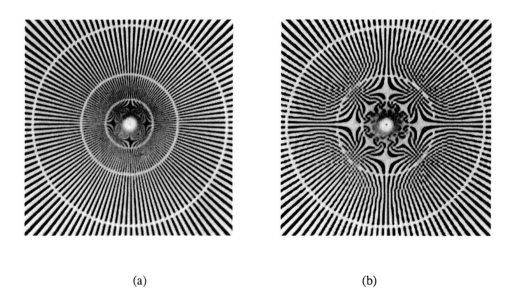

(a)   (b)

Figure 6.2: Aliasing due to point sampling. (a) 1/2 and (b) 1/4 scale.

There are some applications where point sampling may be considered acceptable. If the image is smoothly-varying or if the spatial transformation is mild, then point sampling can achieve fast and reasonable results. For instance, consider the following example in Fig. 6.3. Figures 6.3a and 6.3b show images of a hand and a flag, respectively. In Fig. 6.3c, the hand is made to appear as if it were made of glass. This effect is achieved by warping the underlying image of the flag in accordance with the principles of refraction. Notice, for instance, that the flag is more warped near the edge of the hand where high curvature in the fictitious glass would cause increasing refraction.

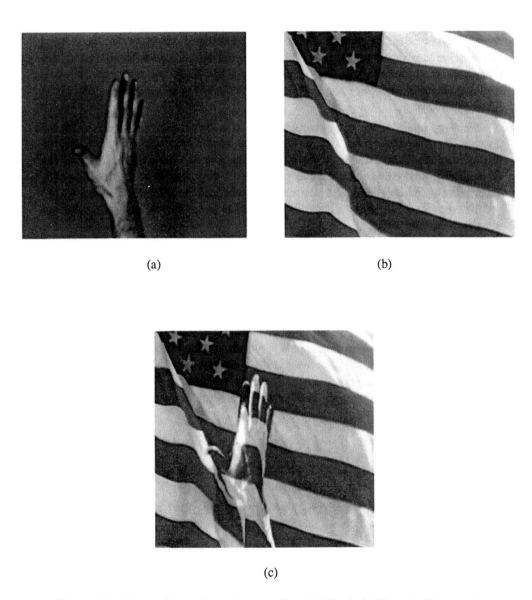

(a)   (b)

(c)

Figure 6.3: Visual effect using point sampling. (a) Hand; (b) Flag; (c) Glass hand.

The procedure begins by isolating the hand pixels from the blue background in Fig. 6.3a. A spatial transformation is derived by evaluating the distance of these pixels from the edge of the hand. Once the distance values are normalized, they serve as a displacement function that is used to perturb the current positions. This yields a new set of coordinates to sample the flag image. Those pixels which are far from the edge sample nearby flag pixels. Pixels that lie near the edge sample more distant flag pixels. Of course, the normalization process must smooth the distance values so that the warping function does not appear too ragged. Although close inspection reveals some point sampling artifacts, the result rivals that which can be achieved by ray-tracing without even requiring an actual model of a hand. This is a particularly effective use of image warping for visual effects.

Aliasing can be reduced by point sampling at a higher resolution. This raises the Nyquist limit, accounting for signals with higher bandwidths. Generally, though, the display resolution places a limit on the highest frequency that can be displayed, and thus limits the Nyquist rate to one cycle every two pixels. Any attempt to display higher frequencies will produce aliasing artifacts such as moire patterns and jagged edges. Consequently, antialiasing algorithms have been derived to bandlimit the input *before* resampling onto the output grid.

6.1.2. Area Sampling

The basic flaw in point sampling is that a discrete pixel actually represents an area, not a point. In this manner, each output pixel should be considered a window looking onto the input image. Rather than sampling a point, we must instead apply a low-pass filter (LPF) upon the projected area in order to properly reflect the information content being mapped onto the output pixel. This approach, depicted in Fig. 6.4, is called *area sampling* and the projected area is known as the *preimage*. The low-pass filter comprises the *prefiltering* stage. It serves to defeat aliasing by bandlimiting the input image prior to resampling it onto the output grid. In the general case, prefiltering is defined by the convolution integral

$$g(x,y) = \int\int f(u,v) h(x-u,y-v) \, du \, dv \qquad (6.1.1)$$

where f is the input image, g is the output image, h is the filter kernel, and the integration is applied to all $[u,v]$ points in the preimage.

Images produced by area sampling are demonstrably superior to those produced by point sampling. Figure 6.5 shows the Star image subjected to the same downsampling transformation as that in Fig. 6.2. Area sampling was implemented by applying a box filter (i.e., averaging) the Star image before point sampling. Notice that antialiasing through area sampling has traded moire patterns for some blurring. Although there is no substitute to high resolution imagery, filtering can make lower resolution less objectionable by attenuating aliasing artifacts.

Area sampling is akin to direct convolution except for one notable exception: independently projecting each output pixel onto the input image limits the extent of the

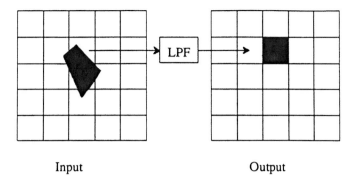

Input   Output

Figure 6.4: Area sampling.

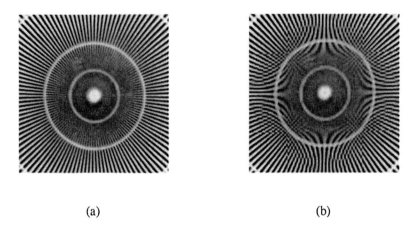

(a)   (b)

Figure 6.5: Aliasing due to area sampling. (a) 1/2 and (b) 1/4 scale.

filter kernel to the projected area. As we shall see, this constraint can be lifted by considering the bounding area which is the smallest region that completely bounds the pixel's convolution kernel. Depending on the size and shape of convolution kernels, these areas may overlap. Since this carries extra computational cost, most area sampling algorithms limit themselves to the restrictive definition which, nevertheless, is far superior to point sampling. The question that remains open is the manner in which the incoming data is to be filtered. There are various theoretical and practical considerations to be addressed.

6.1.3. Space-Invariant Filtering

Ideally, the sinc function should be used to filter the preimage. However, as discussed in Chapters 4 and 5, an FIR filter or a physically realizable IIR filter must be used instead to form a weighted average of samples. If the filter kernel remains constant as it scans across the image, it is said to be *space-invariant*.

Fourier convolution can be used to implement space-invariant filtering by transforming the image and filter kernel into the frequency domain using an FFT, multiplying them together, and then computing the inverse FFT. For wide space-invariant kernels, this becomes the method of choice since it requires $O(N \log_2 N)$ operations instead of $O(MN)$ operations for direct convolution, where M and N are the lengths of the filter kernel and image, respectively. Since the cost of Fourier convolution is independent of the kernel width, it becomes practical when $M > \log_2 N$. This means, for example, that scaling an image can best be done in the frequency domain when excessive magnification or minification is desired. An excellent tutorial on the theory supporting digital filtering in the frequency domain can be found in [Smith 83]. The reader should note that the term "excessive" is taken to mean any scale factor beyond the processing power of fast hardware convolvers. For instance, current advances in pipelined hardware make direct convolution reasonable for filter neighborhoods as large as 17×17.

6.1.4. Space-Variant Filtering

In most image warping applications, however, *space-variant* filters are required, where the kernel varies with position. This is necessary for many common operations such as perspective mappings, nonlinear warps, and texture mapping. In such cases, space-variant FIR filters are used to convolve the preimage. Proper filtering requires a large number of preimage samples in order to compute each output pixel. There are various sampling strategies used to collect these samples. They can be broadly categorized into two classes: regular sampling and irregular sampling.

6.2. REGULAR SAMPLING

The process of using a regular sampling grid to collect image samples is called *regular sampling*. It is also known as *uniform sampling*, which is slightly misleading since an irregular sampling grid can also generate a uniform distribution of samples. Regular sampling includes point sampling, as well as the supersampling and adaptive sampling techniques described below.

6.2.1. Supersampling

The process of using more than one regularly-spaced sample per pixel is known as *supersampling*. Each output pixel value is evaluated by computing a weighted average of the samples taken from their respective preimages. For example, if the supersampling grid is three times denser than the output grid (i.e., there are nine grid points per pixel area), each output pixel will be an average of the nine samples taken from its projection in the input image. If, say, three samples hit a green object and the remaining six samples hit a blue object, the composite color in the output pixel will be one-third green and two-thirds blue, assuming a box filter is used.

Supersampling reduces aliasing by bandlimiting the input signal. The purpose of the high-resolution supersampling grid is to refine the estimate of the preimages seen by the output pixels. The samples then enter the prefiltering stage, consisting of a low-pass filter. This permits the input to be resampled onto the (relatively) low-resolution output grid without any offending high frequencies introducing aliasing artifacts. In Fig. 6.6 we see an output pixel subdivided into nine subpixel samples which each undergo inverse mapping, sampling the input at nine positions. Those nine values then pass through a low-pass filter to be averaged into a single output value.

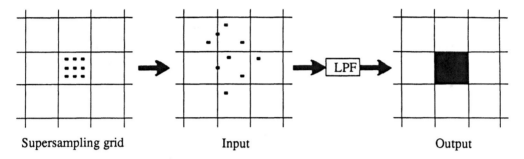

Supersampling grid   Input   Output

Figure 6.6: Supersampling.

The impact of supersampling is easily demonstrated in the following example of a checkerboard projected onto an oblique plane. Figure 6.7 shows four different sampling rates used to perform an inverse mapping. In Fig. 6.7a, only one checkerboard sample per output pixel is used. This contributes to the jagged edges at the bottom of the image and to the moire patterns at the top. They directly correspond to poor reconstruction and antialiasing, respectively. The results are progressively refined as more samples are used to compute each output pixel.

There are two problems associated with straightforward supersampling. The first problem is that the newly designated high frequency of the prefiltered image continues to be fixed. Therefore, there will always be sufficiently higher frequencies that will alias. The second problem is cost. In our example, supersampling will take nine times longer than point sampling. Although there is a clear need for the additional computation, the dense placement of samples can be optimized. Adaptive supersampling is introduced to address these drawbacks.

6.2.2. Adaptive Supersampling

In *adaptive supersampling*, the samples are distributed more densely in areas of high intensity variance. In this manner, supersamples are collected only in regions that warrant their use. Early work in adaptive supersampling for computer graphics is described in [Whitted 80]. The strategy is to subdivide areas between previous samples

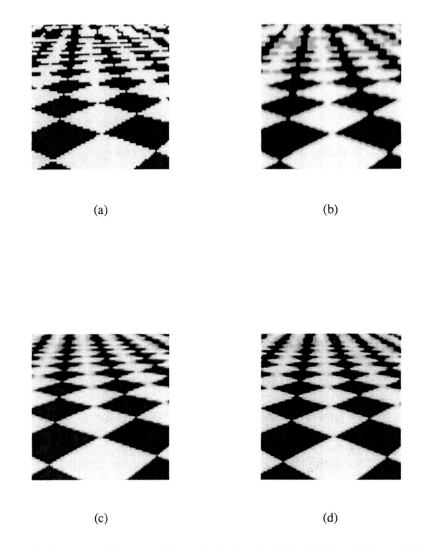

(a)   (b)

(c)   (d)

Figure 6.7: Supersampling an oblique checkerboard. (a) 1; (b) 4; (c) 16; and (d) 256 samples per output pixel.

when an edge, or some other high frequency pattern, is present. Two approaches to adaptive supersampling have been described in the literature. The first approach allows sampling density to vary as a function of local image variance [Lee 85, Kajiya 86]. A second approach introduces two levels of sampling densities: a regular pattern for most areas and a higher-density pattern for regions demonstrating high frequencies. The regular pattern simply consists of one sample per output pixel. The high density pattern involves local supersampling at a rate of 4 to 16 samples per pixel. Typically, these rates are adequate for suppressing aliasing artifacts.

A strategy is required to determine where supersampling is necessary. In [Mitchell 87], the author describes a method in which the image is divided into small square super-sampling cells, each containing eight or nine of the low-density samples. The entire cell is supersampled if its samples exhibit excessive variation. In [Lee 85], the variance of the samples are used to indicate high frequency. It is well-known, however, that variance is a poor measure of visual perception of local variation. Another alternative is to use contrast, which more closely models the nonlinear response of the human eye to rapid fluctuations in light intensities [Caelli 81]. Contrast is given as

$$ C \ = \ \frac{I_{max} - I_{min}}{I_{max} + I_{min}} \tag{6.2.1} $$

Adaptive sampling reduces the number of samples required for a given image quality. The problem with this technique, however, is that the variance measurement is itself based on point samples, and so this method can fail as well. This is particularly true for sub-pixel objects that do not cross pixel boundaries. Nevertheless, adaptive sampling presents a far more reliable and cost-effective alternative to supersampling.

An example of the effectiveness of adaptive supersampling is shown in Fig. 6.8. The image, depicting a bowl on a wooden table, is a computer-generated picture that made use of bilinear interpolation for reconstruction and box filtering for antialiasing. Higher sampling rates were chosen in regions of high variance. For each output pixel, the following operations were taken. First, the four pixel corners were projected into the input. The average of these point samples was computed. If any of the corner values differed from that average by more than some user-specified threshold, then the output pixel was subdivided into four subpixels. The process repeats until the four corners satisfy the uniformity condition. Each output pixel is the average of all the computed input values that map onto it.

6.2.3. Reconstruction from Regular Samples

Each output pixel is evaluted as a linear combination of the preimage samples. The low-pass filters shown in Figs. 6.4 and 6.6 are actually reconstruction filters used to interpolate the output point. They share the identical function of the reconstruction filters discussed in Chapter 5: they bandlimit the sampled signal (suppress the replicated spectra) so that the resampling process does not itself introduce aliasing. The careful reader will notice that reconstruction serves two roles:

Figure 6.8: A ray-traced image using adaptive supersampling.

1) Reconstruction filters interpolate the input samples to compute values at nonintegral positions. These values are the preimage samples that are assigned to the supersampling grid.

2) The very same filters are used to interpolate a new value from the dense set of samples collected in step (1). The result is applied to the output pixel.

When reconstruction filters are applied to interpolate new values from regularly-spaced samples, errors may appear as observable derivative discontinuities across pixel boundaries. In antialiasing, reconstruction errors are more subtle. Consider an object of constant intensity which is entirely embedded in pixel p, i.e., a sub-pixel sized object. We will assume that the popular triangle filter is used as the reconstruction kernel. As the object moves away from the center of p, the computed intensity for p decreases as it moves towards the edge. Upon crossing the pixel boundary, the object begins to contribute to the adjacent pixel, no longer having an influence on p. If this motion were animated, the object would appear to flicker as it crossed the image. This artifact is due to the limited range of the filter. This suggests that a wider filter is required, in order to reflect the object's contribution to neighboring pixels.

One ad hoc solution is to use a square pyramid with a base width of 2×2 pixels. This approach was used in [Blinn 76], an early paper on texture mapping. In general, by varying the width of the filter a compromise is reached between passband transmission and stopband attenuation. This underscores the need for high-quality reconstruction filters to prevent aliasing in image resampling.

Despite the apparent benefits of supersampling and adaptive sampling, all regular sampling methods share a common problem: information is discarded in a coherent way. This produces coherent aliasing artifacts that are easily perceived. Since spatially correlated errors are a consequence of the regularity of the sampling grid, the use of irregular sampling grids has been proposed to address this problem.

6.3. IRREGULAR SAMPLING

Irregular sampling is the process of using an irregular sampling grid in which to sample the input image. This process is also referred to as *nonuniform sampling* and *stochastic sampling*. As before, the term nonuniform sampling is a slight misnomer since irregular sampling can be used to produce a uniform distribution of samples. The name stochastic sampling is more appropriate since it denotes the fact that the irregularly-spaced locations are determined probabilistically via a Monte Carlo technique.

The motivation for irregular sampling is that coherent aliasing artifacts can be rendered incoherent, and thus less conspicuous. By collecting irregularly-spaced samples, the energies of the offending high frequencies are made to appear as featureless noise of the correct average intensity, an artifact that is much less objectionable than aliasing. This claim is supported by evidence from work in color television encoding [Limb 77], image noise measurement [Sakrison 77], dithering [Limb 69, Ulichney 87], and the distribution of retinal cells in the human eye [Yellott 83].

6.3.1. Stochastic Sampling

Although the mathematical properties of stochastic sampling have received a great deal of attention, this technique has only recently been advocated as a new approach to antialiasing for images. In particular, it has played an increasing role in ray tracing where the rays (point samples) are now stochastically distributed to perform a Monte Carlo evaluation of integrals in the rendering equation. This is called *distributed ray tracing* and has been used with great success in computer graphics to simulate motion blur, depth of field, penumbrae, gloss, and translucency [Cook 84, 86].

There are three common forms of stochastic sampling discussed in the literature: Poisson sampling, jittered sampling, and point-diffusion sampling.

6.3.2. Poisson Sampling

Poisson sampling uses an irregular sampling grid that is stochastically generated to yield a uniform distribution of sample points. This approximation to uniform sampling can be improved with the addition of a minimum-distance constraint between sample points. The result, known as the *Poisson-disk distribution*, has been suggested as the optimal sampling pattern to mask aliasing artifacts. This is motivated by evidence that the Poisson-disk distribution is found among the sparse retinal cells outside the foveal region of the eye. It has been suggested that this spatial organization serves to scatter aliasing into high-frequency random noise [Yellott 83].

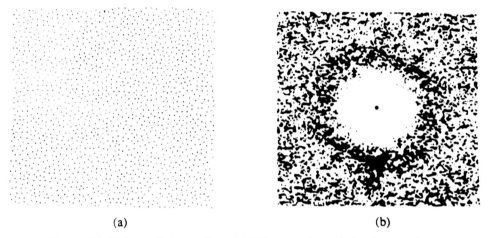

(a)   (b)

Figure 6.9: Poisson-disk sampling: (a) Point samples; (b) Fourier transform.

A Poisson-disk sampling pattern and its Fourier transform are shown in Fig. 6.9. Theoretical arguments can be given in favor of this sampling pattern, in terms of its spectral characteristics. An ideal sampling pattern, it is argued, should have a broad noisy spectrum with minimal low-frequency energy. A perfectly random pattern such as white noise is an example of such a signal where all frequency components have equal magnitude. This is equivalent to the "snow", or random dot patterns, that appear on a television with poor reception. Such a pattern exhibits no coherence which can give rise to structured aliasing artifacts.

Dot-patterns with low-frequency noise often give rise to clusters of dots that coalesce to form clumps. Such granular appearances are undesirable for a uniformly distributed sampling pattern. Consequently, low-frequency attenuation is imposed to concentrate the noise energy into the higher frequencies which are not readily perceived. These properties have direct analog to the Poisson and Poisson-disk distributions, respectively. That is, white noise approximates the Poisson distribution while the low-frequency attenuation approximates the minimal-distance constraint necessary for the Poisson-disk distribution.

Distributions which satisfy these conditions are known as *blue noise*. Similar constraints have been applied towards improving the quality of dithered images. These distinct applications share the same problem of masking undesirable artifacts under the guise of less objectionable noise. The solution offered by Poisson-disk sampling is appealing in that it accounts for the response of the human visual system in establishing the optimal sampling pattern.

Poisson-disk sampling patterns are difficult to generate. One possible implementation requires a large lookup table to store random sample locations. As each new random sample location is generated, it is tested against all locations already chosen to be on the sampling pattern. The point is added onto the pattern unless it is found to be closer than a certain distance to any previously chosen point. This cycle iterates until the sampling region is full. The pattern can then be replicated to fill the image provided that care is taken to prevent regularities from appearing at the boundaries of the copies.

In practice, this costly algorithm is approximated by cheaper alternatives. Two such methods are jittered sampling and point-diffusion sampling.

6.3.3. Jittered Sampling

Jittered sampling is a class of stochastic sampling introduced to approximate a Poisson-disk distribution. A jittered sampling pattern is created by randomly perturbing each sample point on a regular sampling pattern. The result, shown in Fig. 6.10, is inferior to that of the optimal Poisson-disk distribution. This is evident in the granularity of the distribution and the increased low-frequency energy found in the spectrum. However, since the magnitude of the noise is directly proportional to the sampling rate, improved image quality is achieved with increased sample density.

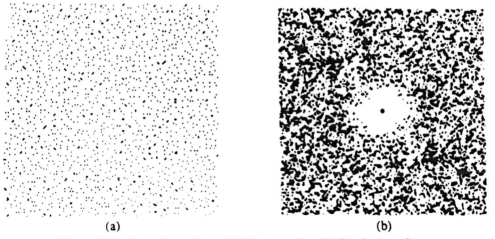

(a)   (b)

Figure 6.10: Jittered sampling: (a) Point samples; (b) Fourier transform.

6.3.4. Point-Diffusion Sampling

The *point-diffusion* sampling algorithm has been proposed by Mitchell as a computationally efficient technique to generate Poisson-disk samples [Mitchell 87]. It is based on the Floyd-Steinberg error-diffusion algorithm used for dithering, i.e., to convert gray-scale images into bitmaps. The idea is as follows. An intensity value, g, may be converted to black or white by thresholding. However, an error e is made in the process:

$$e = \text{MIN}(\text{white}-g,\ g-\text{black}) \qquad (6.3.1)$$

This error can be used to refine successive thresholding decisions. By spreading, or diffusing, e within a small neighborhood we can compensate for previous errors and provide sufficient fluctuation to prevent a regular pattern from appearing at the output.

These fluctuations are known as dithering signals. They are effectively high-frequency noise that are added to an image in order to mask the false contours that inevitably arise in the subsequent quantization (thresholding) stage. Regularities in the form of textured patterns, for example, are typical in ordered dithering where the dither signal is predefined and replicated across the image. In contrast, the Floyd-Steinberg algorithm is an adaptive thresholding scheme in which the dither signal is generated on-the-fly based on errors collected from previous thresholding decisions.

The success of this method is due to the pleasant distribution of points it generates to simulate gray scale. It finds use in stochastic sampling because the distribution satisfies the blue-noise criteria. In this application, the point samples are selected from a supersampling grid that is four times denser than the display grid (i.e., there are 16 grid points per pixel area). The diffusion coefficients are biased to ensure that an average of about one out of 16 grid points will be selected as sample points. The pattern and its Fourier transform are shown in Fig. 6.11.

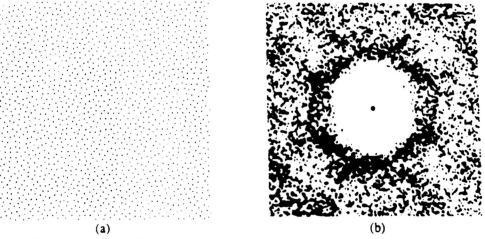

(a)   (b)

Figure 6.11: Point-diffusion sampling: (a) Point samples; (b) Fourier transform.

The Floyd-Steinberg algorithm was first introduced in [Floyd 75] and has been described in various sources [Jarvis 76, Stoffel 81, Foley 90], including a recent dissertation analyzing the role of blue-noise in dithering [Ulichney 87].

6.3.5. Adaptive Stochastic Sampling

Supersampling and adaptive sampling, introduced earlier as regular sampling methods, can be applied to irregular sampling as well. In general, irregular sampling requires rather high sampling densities and thus adaptive sampling plays a natural role in this process. It serves to dramatically reduce the noise level while avoiding needless computation.

As before, an initial set of samples is collected in the neighborhood about each pixel. If the sample values are very similar, then a smooth region is implied and a lower sampling rate is adequate. However, if these samples are very dissimilar, then a rapidly varying region is indicated and a higher sampling rate is warranted. Suggestions for an error estimator, error bound, and initial sampling rate can be found in [Dippe 85a, 85b].

6.3.6. Reconstruction from Irregular Samples

Once the irregularly-spaced samples are collected, they must pass through a reconstruction filter to be resampled at the display resolution. Reconstruction is made difficult by the irregular distribution of the samples. One common approach is to use *weighted-average* filters:

$$f(x) = \frac{\sum\limits_{k=-\infty}^{\infty} h(x-x_k)f(x_k)}{\sum\limits_{k=-\infty}^{\infty} h(x-x_k)} \tag{6.3.2}$$

The value of each pixel $f(x)$ is the sum of the values of the nearby sample points $f(x_k)$ multiplied by their respective filter values $h(x-x_k)$. This total is then normalized by dividing by the sum of the filter values. This technique, however, can be shown to fail upon extreme variation in sampling density.

Mitchell proposes a *multi-stage filter* [Mitchell 87]. Bandlimiting is achieved through repeated application of weighted-average filters with ever-narrowing low-pass cutoff. The strategy is to compute normalized averages over dense clusters of supersamples before combining them with nearby values. Since averaging is done over a dense grid (16 supersamples per pixel area), a crude box filter is used for efficiency. Ideally, the sophistication of the applied filters should increase with every iteration, thereby refining the shape of the bandlimited spectrum.

Various other filtering suggestions are given in [Dippe 85a, 85b], including Wiener filtering and the use of the raised cosine function. The raised cosine function, often used in image restoration, is recommended as a reconstruction kernel to reduce Gibb's phenomenon and guarantee strictly positive results. The filter is given by

$$h(x) = \frac{1}{W} \left[\cos \frac{2\pi}{W} |x| \right] + 1 \qquad |x| < W \qquad (6.3.3)$$

where W is the radius of kernel h, and x is the distance from its center.

6.4. DIRECT CONVOLUTION

Whether regular or irregular sampling is used, direct convolution requires fast space-variant filtering. Most of the work in antialiasing research has focused on this problem. They have generally addressed approximations to the convolution integral of Eq. (6.1.1).

In the general case, a preimage can be of arbitrary shape and the kernel can be an arbitrary filter. Solutions to this problem have typically achieved performance gains by adding constraints. For example, most methods approximate a curvilinear preimage by a quadrilateral. In this manner, techniques discussed in Chapter 3 can be used to locate points in the preimage. Furthermore, simple kernels are often used for computational efficiency. Below we summarize several direct convolution techniques. For consistency with the texture mapping literature from which they are derived, we shall refer to the input and output coordinate systems as *texture space* and *screen space*, respectively.

6.4.1. Catmull, 1974

The earliest work in texture mapping is rooted in Catmull's dissertation on subdivision algorithms for curved surfaces [Catmull 74]. For every screen pixel, his subdivision patch renderer computed an unweighted average (i.e., box filter convolution) over the corresponding quadrilateral preimage. An accumulator array was used to properly integrate weighted contributions from patch fragments at each pixel.

6.4.2. Blinn and Newell, 1976

Blinn and Newell extended Catmull's results by using a triangle filter. In order to avoid the artifacts mentioned in Sec. 6.2.3, the filter formed overlapping square pyramids two pixels wide in screen space. In this manner, the 2×2 region surrounding the given output pixel is inverse mapped to the corresponding quadrilateral in the input. The input samples within the quadrilateral are weighted by a pyramid distorted to fit the quadrilateral. Note that the pyramid is a 2-D separable realization of the triangle filter. The sum of the weighted values is then computed and assigned to the output pixel [Blinn 76].

6.4.3. Feibush, Levoy, and Cook, 1980

High-quality filtering was advanced in computer graphics by Feibush, Levoy, and Cook in [Feibush 80]. Their method is summarized as follows. At each output pixel, the bounding rectangle of the kernel is transformed into texture space where it is mapped into an arbitrary quadrilateral. All input samples contained within the bounding rectangle of this quadrilateral are then mapped into the output. The extra points selected in this procedure are eliminated by clipping the transformed input points against the bounding rectangle of the kernel mask in screen space. A weighted average of the selected (remaining) samples is then computed and assigned to the respective output pixel.

The method is distinct in that the filter weights are stored in a lookup table and indexed by each sample's location within the pixel. Since the kernel is in a lookup table, any high-quality filter of arbitrary shape can be stored at no extra cost. Typically, circularly symmetric (isotropic) kernels are used. In [Feibush 80], the authors used a Gaussian filter. Since circles in the output can map into ellipses in the input, more refined estimates of the preimage are possible. This method achieves a good discrete approximation of the convolution integral.

6.4.4. Gangnet, Perny, and Coueignoux, 1982

The technique described in [Gangnet 82] is similar to that introduced in [Feibush 80], with pixels assumed to be circular and overlapping. The primary difference is that in [Gangnet 82], supersampling is used to refine the preimage estimates. The supersampling density is determined by the length of the major axis of the ellipse in texture space. This is approximated by the length of the longest diagonal of the parallelogram approximating the texture ellipse. Supersampling the input requires image reconstruction to evaluate the samples that do not coincide with the input sampling grid. Note that in [Feibush 80] no image reconstruction is necessary because the input samples are used directly. The collected supersamples are then weighted by the kernel stored in the lookup table. The authors used bilinear interpolation for image reconstruction and a truncated sinc function (2 pixels wide) as the convolution kernel.

This method is superior to [Feibush 80] because the input is sampled at a rate determined by the span of the inverse projection. Unfortunately, the supersampling rate is excessive along the minor axis of the ellipse in texture space. Nevertheless, the additional supersampling mechanism in [Gangnet 82] makes the technique superior, and more costly, to that in [Feibush 80].

6.4.5. Greene and Heckbert, 1986

A variation to the last two filtering methods, called the elliptical weighted average (EWA) filter, was proposed by Greene and Heckbert in [Greene 86]. As before, the filter assumes overlapping circular pixels in screen space which map onto arbitrary ellipses in texture space, and kernels continue to be stored in lookup tables. However, in [Feibush 80] and [Gangnet 82], the input samples were all mapped back onto screen space for weighting by the circular kernel. This mapping is a costly operation which is avoided in EWA. Instead, the EWA distorts the circular kernel into an ellipse in texture space where the weighting can be computed directly.

An elliptic paraboloid Q in texture space is defined for every circle in screen space

$$Q(u,v) = Au^2 + Buv + Cv^2 \tag{6.4.1}$$

where $u = v = 0$ is the center of the ellipse. The parameters of the ellipse can be determined from the directional derivatives

$$A = V_x^2 + V_y^2$$
$$B = -2(U_x V_x + U_y V_y)$$
$$C = U_x^2 + U_y^2$$

where

$$(U_x, V_x) = \left[\frac{\partial u}{\partial x}, \frac{\partial v}{\partial x}\right]$$

$$(U_y, V_y) = \left[\frac{\partial u}{\partial y}, \frac{\partial v}{\partial y}\right]$$

Once the ellipse parameters are determined, samples in the texture space may be tested for point-inclusion in the ellipse by incrementally computing Q for new values of u and v. In texture space the contours of Q are concentric ellipses. Points inside the ellipse satisfy $Q(u,v) < F$ for some threshold F.

$$F = U_x V_y - U_y V_x \qquad (6.4.2)$$

This means that point-inclusion testing for ellipses can be done with one function evaluation rather than the four needed for quadrilaterals (four line equations).

If a point is found to satisfy $Q < F$, then the sample value is weighted with the appropriate lookup table entry. In screen space, the lookup table is indexed by r, the radius of the circle upon which the point lies. In texture space, though, Q is related to r^2. Rather than indexing with r, which would require us to compute $r = \sqrt{Q}$ at each pixel, the kernel values are stored into the lookup table so that they may be indexed by Q directly. Initializing the lookup table in this manner results in large computational efficiency. Thus, instead of determining which concentric circle the texture point maps onto in screen space, we determine which concentric ellipse the point lies upon in texture space and use it to index the appropriate weight in the lookup table.

Explicitly treating preimages as ellipses permits the function Q to take on a dual role: point-inclusion testing and lookup table index. The EWA is thereby able to achieve high-quality filtering at substantially lower cost. After all the points in the ellipse have been scanned, the sum of the weighted values is divided by the sum of the weights (for normalization) and assigned to the output pixel.

All direct convolution methods have a computational cost proportional to the number of input pixels accessed. This cost is exacerbated in [Feibush 80] and [Gangnet 82] where the collected input samples must be mapped into screen space to be weighted with the kernel. By achieving identical results without this costly mapping, the EWA is the most cost-effective high-quality filtering method.

6.5. PREFILTERING

The direct convolution methods described above impose minimal constraints on the filter area (quadrilateral, ellipse) and filter kernel (precomputed lookup table entries). Additional speedups are possible if further constraints are imposed. Pyramids and preintegrated tables are introduced to approximate the convolution integral with a constant number of accesses. This compares favorably against direct convolution which requires a large number of samples that grow proportionately to preimage area. As we shall see, though, the filter area will be limited to squares or rectangles, and the kernel will consist of a box filter. Subsequent advances have extended their use to more general cases with only marginal increases in cost.

6.5.1. Pyramids

Pyramids are multi-resolution data structures commonly used in image processing and computer vision. They are generated by successively bandlimiting and subsampling the original image to form a hierarchy of images at ever decreasing resolutions. The original image serves as the base of the pyramid, and its coarsest version resides at the apex. Thus, in a lower resolution version of the input, each pixel represents the average of some number of pixels in the higher resolution version.

The resolution of successive levels typically differ by a power of two. This means that successively coarser versions each have one quarter of the total number of pixels as their adjacent predecessors. The memory cost of this organization is modest: $1 + 1/4 + 1/16 + \cdots = 4/3$ times that needed for the original input. This requires only 33% more memory.

To filter a preimage, one of the pyramid levels is selected based on the size of its bounding square box. That level is then point sampled and assigned to the respective output pixel. The primary benefit of this approach is that the cost of the filter is constant, requiring the same number of pixel accesses independent of the filter size. This performance gain is the result of the filtering that took place while creating the pyramid. Furthermore, if preimage areas are adequately approximated by squares, the direct convolution methods amount to point sampling a pyramid. This approach was first applied to texture mapping in [Catmull 74] and described in [Dungan 78].

There are several problems with the use of pyramids. First, the appropriate pyramid level must be selected. A coarse level may yield excessive blur while the adjacent finer level may be responsible for aliasing due to insufficient bandlimiting. Second, preimages are constrained to be squares. This proves to be a crude approximation for elongated preimages. For example, when a surface is viewed obliquely the texture may be compressed along one dimension. Using the largest bounding square will include the contributions of many extraneous samples and result in excessive blur. These two issues were addressed in [Williams 83] and [Crow 84], respectively, along with extensions proposed by other researchers.

Williams proposed a pyramid organization called *mip map* to store color images at multiple resolutions in a convenient memory organization [Williams 83]. The acronym "mip" stands for "multum in parvo," a Latin phrase meaning "many things in a small place." The scheme supports trilinear interpolation, where both intra- and inter-level interpolation can be computed using three normalized coordinates: u, v, and d. Both u and v are spatial coordinates used to access points within a pyramid level. The d coordinate is used to index, and interpolate between, different levels of the pyramid. This is depicted in Fig. 6.12.

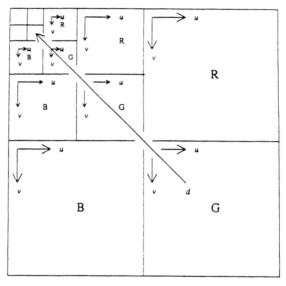

Figure 6.12: Mip Map memory organization.

The quadrants touching the east and south borders contain the original red, green, and blue (RGB) components of the color image. The remaining upper-left quadrant contains all the lower resolution copies of the original. The memory organization depicted in Fig. 6.12 clearly supports the earlier claim that memory cost is 4/3 times that required for the original input. Each level is shown indexed by the $[u,v,d]$ coordinate system, where d is shown slicing through the pyramid levels. Since corresponding points in different pyramid levels have indices which are related by some power of two, simple binary shifts can be used to access these points across the multi-resolution copies. This is a particularly attractive feature for hardware implementation.

The primary difference between mip maps and ordinary pyramids is the trilinear interpolation scheme possible with the $[u,v,d]$ coordinate system. By allowing a continuum of points to be accessed, mip maps are referred to as pyramidal parametric data structures. In Williams' implementation, a box filter (Fourier window) was used to create the mip maps, and a triangle filter (Bartlett window) was used to perform intra- and inter-level interpolation. The value of d must be chosen to balance the tradeoff between aliasing and blurring. Heckbert suggests

$$d^2 = \text{MAX}\left[\left[\frac{\partial u}{\partial x}\right]^2 + \left[\frac{\partial v}{\partial x}\right]^2 , \left[\frac{\partial u}{\partial y}\right]^2 + \left[\frac{\partial v}{\partial y}\right]^2 \right] \qquad (6.5.1)$$

where d is proportional to the span of the preimage area, and the partial derivatives can be computed from the surface projection [Heckbert 83].

6.5.2. Summed-Area Tables

An alternative to pyramidal filtering was proposed by Crow in [Crow 84]. It extends the filtering possible in pyramids by allowing rectangular areas, oriented parallel to the coordinate axes, to be filtered in constant time. The central data structure is a preintegrated buffer of intensities, known as the *summed-area table*. This table is generated by computing a running total of the input intensities as the image is scanned along successive scanlines. For every position P in the table, we compute the sum of intensities of pixels contained in the rectangle between the origin and P. The sum of all intensities in any rectangular area of the input may easily be recovered by computing a sum and two differences of values taken from the table. For example, consider the rectangles R_0, R_1, R_2, and R shown in Fig. 6.13. The sum of intensities in rectangle R can be computed by considering the sum at $[x1, y1]$, and discarding the sums of rectangles R_0, R_1, and R_2. This corresponds to removing all area lying below and to the left of R. The resulting area is rectangle R, and its sum S is given as

$$S = T[x1, y1] - T[x1, y0] - T[x0, y1] + T[x0, y0] \qquad (6.5.2)$$

where $T[x, y]$ is the value in the summed-area table indexed by coordinate pair $[x, y]$.

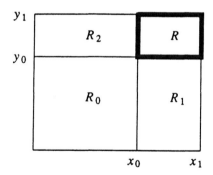

Figure 6.13: Summed-area table calculation.

Since $T[x1, y0]$ and $T[x0, y1]$ both contain R_0, the sum of R_0 was subtracted twice in Eq. (6.5.2). As a result, $T[x0, y0]$ was added back to restore the sum. Once S is determined it is divided by the area of the rectangle. This gives the average intensity over the rectangle, a process equivalent to filtering with a Fourier window (box filtering).

There are two problems with the use of summed-area tables. First, the filter area is restricted to rectangles. This is addressed in [Glassner 86], where an adaptive, iterative technique is proposed for obtaining arbitrary filter areas by removing extraneous regions from the rectangular bounding box. Second, the summed-area table is restricted to box filtering. This, of course, is attributed to the use of unweighted averages that keeps the algorithm simple. In [Perlin 85] and [Heckbert 86a], the summed-area table is generalized to support more sophisticated filtering by repeated integration.

It is shown that by repeatedly integrating the summed-area table n times, it is possible to convolve an orthogonally oriented rectangular region with an nth-order box filter (B-spline). Kernels for small n are shown in Fig. 5.10. The output value is computed by using $(n + 1)^2$ weighted samples from the preintegrated table. Since this result is independent of the size of the rectangular region, this method offers a great reduction in computation over that of direct convolution. Perlin called this a *selective image filter* because it allows each sample to be blurred by different amounts.

Repeated integration has rather high memory costs relative to pyramids. This is due to the number of bits necessary to retain accuracy in the large summations. Nevertheless, it allows us to filter rectangular or elliptical regions, rather than just squares as in pyramid techniques. Since pyramid and summed-area tables both require a setup time, they are best suited for input that is intended to be used repeatedly, i.e., a stationary background scene. In this manner, the initialization overhead can be amortized over each use. However, if the texture is only to be used once, the direct convolution methods raise a challenge to the cost-effectiveness offered by pyramids and summed-area tables.

6.6. FREQUENCY CLAMPING

The antialiasing methods described above all attempt to bandlimit the input by convolving input samples with a filter in the spatial domain. An alternative to this approach is to transform the input to the frequency domain, apply an appropriate low-pass filter to the spectrum, and then compute the inverse transform to display the bandlimited result. This was, in fact, already suggested as a viable technique for space-invariant filtering in which the low-pass filter can remain constant throughout the image. Norton, Rockwood, and Skolmoski explore this approach for space-variant filtering, where each pixel may require different bandlimiting to avoid aliasing [Norton 82].

The authors propose a simple technique for clamping, or suppressing, the offending high frequencies at each point in the image. This clamping function technique requires some a priori knowledge about the input image. In particular, the input should not be given as an array of samples but rather it should be represented by a Fourier series, i.e., a sum of bandlimited terms of increasing frequencies. When the frequency of a term exceeds the Nyquist rate at a given pixel, that term is forced to the local average value. This method has been successfully applied in a real-time visual system for flight simulators. It is used to solve the aliasing problem for textures of clouds and water, patterns which are convincingly generated using only a few low-order Fourier terms.

6.7. ANTIALIASED LINES AND TEXT

A large body of work has been directed towards efficient antialiasing methods for eliminating the jagged appearance of lines and text in raster images. These two applications have attracted a lot of attention due to their practical importance in the ever growing workstation and personal computer markets. While *images* of lines and text can be handled with the algorithms described above, antialiasing techniques have been developed which embed the filtering process directly within the drawing routines. Although a full treatment of this topic is outside the scope of this text, some pointers are provided below.

Shaded (gray) pixels for lines can be generated, for example, with the use of a lookup table indexed by the distance between each pixel center and the line (or curve). Since arbitrary kernels can be stored in the lookup table at no extra cost, this approach shares the same merits as [Feibush 80]. Conveniently, the point-line distance can be computed incrementally by the same Bresenham algorithm used to determine which pixels must be turned on. This algorithm is described in [Gupta 81].

In [Turkowski 82], the CORDIC rotation algorithm is used to calculate the point-line distance necessary for indexing into the kernel lookup table. Other related papers describing the use of lookup tables and bitmaps for efficient antialiasing of lines and polygons can be found in [Pitteway 80], [Fiume 83], and [Abram 85]. Recent work in this area is described in [Chen 88]. For a description of recent advances in antialiased text, the reader is referred to [Naiman 87].

6.8. DISCUSSION

This chapter has reviewed methods to combat the aliasing artifacts that may surface upon performing geometric transformations on digital images. Aliasing becomes apparent when the mapping of input pixels onto the output is many-to-one. Sampling theory suggests theoretical limitations and provides insight into the solution. In the majority of cases, increasing display resolution is not a parameter that the user is free to adjust. Consequently, the approaches have dealt with bandlimiting the input so that it may conform to the available output resolution.

All contributions in this area fall into one of two categories: direct convolution and prefiltering. Direct convolution calls for increased sampling to accurately resolve the input preimage that maps onto the current output pixel. A low-pass filter is applied to these samples, generating a single bandlimited output value. This approach raises two issues: sampling techniques and efficient convolution. The first issue has been addressed by the work on regular and irregular sampling, including the recent advances in stochastic sampling. The second issue has been treated by algorithms which embed the filter kernels in lookup tables and provide fast access to the appropriate weights. Despite all possible optimizations, the computational complexity of this approach is inherently coupled with the number of samples taken over the preimage. Thus, larger preimages will incur higher sampling and filtering costs.

A cheaper approach providing lower quality results is obtained through prefiltering. By precomputing pyramids and summed-area tables, filtering is possible with only a constant number of computations, independent of the preimage area. Combining the partially filtered results contained in these data structures produces large performance gains. The cost, however, is in terms of constraints on the filter kernel and approximations to the preimage area. Designing efficient filtering techniques that support arbitrary preimage areas and filter kernels remains a great challenge. It is a subject that will continue to receive much attention.

7

SCANLINE ALGORITHMS

Scanline algorithms comprise a special class of geometric transformation techniques that operate only along rows and columns. The purpose for using such algorithms is simplicity: resampling along a scanline is a straightforward 1-D problem that exploits simplifications in digital filtering and memory access. The geometric transformations that are best suited for this approach are those that can be shown to be *separable*, i.e., each dimension can be resampled independently of the other.

Separable algorithms spatially transform 2-D images by decomposing the mapping into a sequence of orthogonal 1-D transformations. For instance, 2-pass scanline algorithms typically apply the first pass to the image rows and the second pass to the columns. Although separable algorithms cannot handle all possible mapping functions, they can be shown to work particularly well for a wide class of common transformations, including affine and perspective mappings. Recent work in this area has shown how they may be extended to deal with arbitrary mapping functions. This is all part of an effort to cast image warping into a framework that is amenable to hardware implementation.

The flurry of activity now drawn to separable algorithms is a testimony to its practical importance. Growing interest in this area has gained impetus from the widespread proliferation of advanced workstations and digital signal processors. This has resulted in dramatic developments in both hardware and software systems. Examples include real-time hardware for video effects, texture mapping, and geometric correction. The speed offered by these products also suggests implications in new technologies that will exploit interactive image manipulation, of which image warping is an important component.

This chapter is devoted to geometric transformations that may be implemented with scanline algorithms. In general, this will imply that the mapping function is separable, although this need not always be the case. Consequently, space-variant digital filtering plays an increasingly important role in preventing aliasing artifacts. Despite the assumptions and errors that fall into this model of computation, separable algorithms perform surprisingly well.

7.1. INTRODUCTION

Geometric transformations have traditionally been formulated as either forward or inverse mappings operating entirely in 2-D. Their advantages and drawbacks have already been described in Chapter 3. We briefly restate these features in order to better motivate the case for scanline algorithms and separable geometric transformations. As we shall see, there are many compelling reasons for their use.

7.1.1. Forward Mapping

Forward mappings deposit input pixels into an output accumulator array. A distinction is made here based on the order in which pixels are fetched and stored. In forward mappings, the input arrives in scanline order (row by row) but the results are free to leave in any order, projecting into arbitrary areas in the output. In the general case, this means that no output pixel is guaranteed to be totally computed until the entire input has been scanned. Therefore, a full 2-D accumulator array must be retained throughout the duration of the mapping. Since the square input pixels project onto quadrilaterals at the output, costly intersection tests are needed to properly compute their overlap with the discrete output cells. Furthermore, an adaptive algorithm must be used to determine when supersampling is necessary in order to avoid blocky appearances upon one-to-many mappings.

7.1.2. Inverse Mapping

Inverse mappings are more commonly used to perform spatial transformations. By operating in scanline order at the output, square output pixels are projected onto arbitrary quadrilaterals. In this case, the projected areas lie in the input and are not generated in scanline order. Each preimage must be sampled and convolved with a low-pass filter to compute an intensity at the output. In Chapter 6, we reviewed clever approaches to efficiently approximate this computation. While either forward or inverse mappings can be used to realize *arbitrary* mapping functions, there are many transformations that are adequately approximated when using separable mappings. They exploit scanline algorithms to yield large computational savings.

7.1.3. Separable Mapping

There are several advantages to decomposing a mapping into a series of 1-D transforms. First, the resampling problem is made simpler since reconstruction, area sampling, and filtering can now be done entirely in 1-D. Second, this lends itself naturally to digital hardware implementation. Note that no sophisticated digital filters are necessary to deal explicitly with the 2-D case. Third, the mapping can be done in scanline order both in scanning the input image and in producing the projected image. In this manner, an image may be processed in the same format in which it is stored in the framebuffer: rows and columns. This leads to efficient data access and large savings in I/O time. The approach is amenable to stream-processing techniques such as pipelining and facilitates the design of hardware that works at real-time video rates.

7.2. INCREMENTAL ALGORITHMS

In this section, we examine the problem of image warping with several incremental algorithms that operate in scanline order. We begin by considering an incremental scanline technique for texture mapping. The ideas are derived from shading interpolation methods in computer graphics.

7.2.1. Texture Mapping

Texture mapping is a powerful technique used to add visual detail to synthetic images in computer graphics. It consists of a series of spatial transformations: a texture plane, $[u,v]$, is transformed onto a 3-D surface, $[x,y,z]$, and then projected onto the output screen, $[x,y]$. This sequence is shown in Fig. 7.1, where f is the transformation from $[u,v]$ to $[x,y,z]$ and p is the projection from $[x,y,z]$ onto $[x,y]$. For simplicity, we have assumed that p realizes an orthographic projection. The forward mapping functions X and Y represent the composite function $p(f(u,v))$. The inverse mapping functions are U and V.

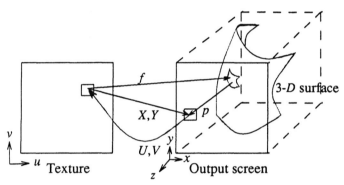

Figure 7.1: Texture mapping functions.

Texture mapping serves to create the appearance of complexity by simply applying image detail onto a surface, in much the same way as wallpaper. Textures are rather loosely defined. They are usually taken to be images used for mapping color onto the targeted surface. Textures are also used to perturb surface normals, thus allowing us to simulate bumps and wrinkles without the tedium of modeling them geometrically. Additional applications are included in [Heckbert 86b], a recent survey article on texture mapping.

The 3-D objects are usually modeled with planar polygons or bicubic patches. Patches are quite popular since they easily lend themselves for efficient rendering [Catmull 74, 80] and offer a natural parameterization that can be used as a curvilinear coordinate system. Polygons, on the other hand, are defined implicitly. Several parameterizations for planes and polygons are described in [Heckbert 89].

Once the surfaces are parameterized, the mapping between the input and output images is usually treated as a four-corner mapping. In inverse mapping, square output pixels must be projected back onto the input image for resampling purposes. In forward mapping, we project square texture pixels onto the output image via mapping functions X and Y. Below we describe an inverse mapping technique.

Consider an input square texture in the uv plane mapped onto a planar quadrilateral in the xyz coordinate system. The mapping can be specified by designating texture coordinates to the quadrilateral. For simplicity we select four corner mapping, as depicted in Fig. 7.2. In this manner, the four point correspondences are $(u_i, v_i) \rightarrow (x_i, y_i, z_i)$ for $0 \leq i < 4$. The problem now remains to determine the correspondence for all interior quadrilateral points. Careful readers will notice that this task is reminiscent of the surface interpolation paradigm already considered in Chapter 3. In the subsections that follow, we turn to a simplistic approach drawn from the computer graphics field.

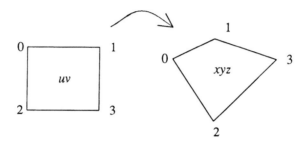

Figure 7.2: Four corner mapping.

7.2.2. Gouraud Shading

Gouraud shading is a popular intensity interpolation algorithm used to shade polygonal surfaces in computer graphics [Gouraud 71]. It serves to enhance realism in rendered scenes that approximate curved surfaces with planar polygons. Although we have no direct use for shading algorithms here, we use a variant of this approach to interpolate texture coordinates. We begin with a review of Gouraud shading in this section, followed by a description of its use in texture mapping in the next section.

Gouraud shading interpolates the intensities all along a polygon, given only the true values at the vertices. It does so while operating in scanline order. This means that the output screen is rendered in a raster fashion, (e.g., scanning the polygon from top-to-bottom, with each scan moving left-to-right). This spatial coherence lends itself to a fast incremental method for computing the interior intensity values. The basic approach is illustrated in Fig. 7.3.

For each scanline, the intensities at endpoints x_0 and x_1 are computed. This is achieved through linear interpolation between the intensities of the appropriate polygon vertices. This yields I_0 and I_1 in Fig. 7.3, where

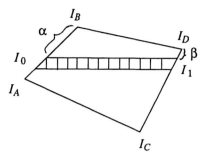

Figure 7.3: Incremental scanline interpolation.

$$I_0 = \alpha I_A + (1-\alpha) I_B, \qquad 0 \le \alpha \le 1 \qquad\qquad (7.2.1a)$$

$$I_1 = \beta I_C + (1-\beta) I_D, \qquad 0 \le \beta \le 1 \qquad\qquad (7.2.1b)$$

Then, beginning with I_0, the intensity values along successive scanline positions are computed incrementally. In this manner, I_{x+1} can be determined directly from I_x, where the subscripts refer to positions along the scanline. We thus have

$$I_{x+1} = I_x + dI \qquad\qquad (7.2.2)$$

where

$$dI = \frac{(I_1 - I_0)}{(x_1 - x_0)} \qquad\qquad (7.2.3)$$

Note that the scanline order allows us to exploit incremental computations. As a result, we are spared from having to evaluate two multiplications and two additions per pixel, as in Eq. (7.2.1). Additional savings are possible by computing I_0 and I_1 incrementally as well. This requires a different set of constant increments to be added along the polygon edges.

7.2.3. Incremental Texture Mapping

Although Gouraud shading has traditionally been used to interpolate intensity values, we now use it to interpolate texture coordinates. The computed (u,v) coordinates are used to index into the input texture. This permits us to obtain a color value that is then applied to the output pixel. The following segment of C code is offered as an example of how to process a single scanline.

```
dx = 1.0 / (x1 - x0);          /* normalization factor */
du = (u1 - u0) * dx;           /* constant increment for u */
dv = (v1 - v0) * dx;           /* constant increment for v */
dz = (z1 - z0) * dx;           /* constant increment for z */
for(x = x0; x < x1; x++) {     /* visit all scanline pixels */
    if(z < zbuf[x]) {          /* is new point closer? */
        zbuf[x] = z;           /* update z-buffer */
        scr[x] = tex(u,v);     /* write texture value to screen */
    }
    u += du;                   /* increment u */
    v += dv;                   /* increment v */
    z += dz;                   /* increment z */
}
```

The procedure given above assumes that the scanline begins at $(x0, y, z0)$ and ends at $(x1, y, z1)$. These two endpoints correspond to points $(u0, v0)$ and $(u1, v1)$, respectively, in the input texture. For every unit step in x, coordinates u and v are incremented by a constant amount, e.g., du and dv, respectively. This equates to an affine mapping between a horizontal scanline in screen space and an arbitrary line in texture space with slope dv/du (see Fig. 7.4).

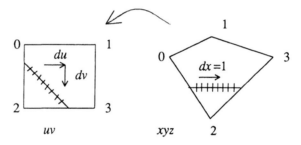

Figure 7.4: Incremental interpolation of texture coordinates.

Since the rendered surface may contain occluding polygons, the z-coordinates of visible pixels are stored in *zbuf*, the z-buffer for the current scanline. When a pixel is visited, its z-buffer entry is compared against the depth of the incoming pixel. If the incoming pixel is found to be closer, then we proceed with the computations involved in determining the output value and update the z-buffer with the depth of the closer point. Otherwise, the incoming point is occluded and no further action is taken on that pixel.

The function $tex(u,v)$ in the above code samples the texture at point (u,v). It returns an intensity value that is stored in *scr*, the screen buffer for the current scanline. For color images, RGB values would be returned by *tex* and written into three separate color channels. In the examples that follow, we let *tex* implement point sampling, e.g., no filtering. Although this introduces well-known artifacts, our goal here is to examine the geometrical properties of this simple approach. We will therefore tolerate artifacts, such as jagged edges, in the interest of simplicity.

Figure 7.5 shows the Checkerboard image mapped onto a quadrilateral using the approach described above. There are several problems that are readily noticeable. First, the textured polygon shows undesirable discontinuities along horizontal lines passing through the vertices. This is due to a sudden change in du and dv as we move past a vertex. It is an artifact of the linear interpolation of u and v. Second, the image does not exhibit the foreshortening that we would expect to see from perspective. This is due to the fact that this approach is consistent with the bilinear transformation scheme described in Chapter 3. As a result, it can be shown to be exact for affine mappings but it is inadequate to handle perspective mappings [Heckbert 89].

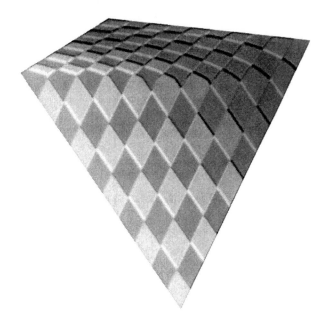

Figure 7.5: Naive approach applied to Checkerboard.

The constant increments used in the linear interpolation are directly related to the general transformation matrix elements. Referring to these terms, as defined in Chapter 3, we have

$$xw = a_{11}u + a_{21}v + a_{31}$$
$$yw = a_{12}u + a_{22}v + a_{32} \qquad (7.2.4)$$
$$w = a_{13}u + a_{23}v + a_{33}$$

For simplicity, we select $a_{33} = 1$ and leave eight degrees of freedom for the general transformation. Solving for u and v in terms of x, y, and w, we have

$$u = \frac{a_{22}xw - a_{21}yw + a_{21}a_{32} - a_{22}a_{31}}{a_{11}a_{22} - a_{12}a_{21}} \qquad (7.2.5a)$$

$$v = \frac{-a_{12}xw + a_{11}yw - a_{11}a_{32} + a_{12}a_{31}}{a_{11}a_{22} - a_{12}a_{21}} \qquad (7.2.5b)$$

This gives rise to expressions for du and dv. These terms represent the increment added to the interpolated coordinates at position x to yield a value for the next point at $x+1$. If we refer to these positions with subscripts 0 and 1, respectively, then we have

$$du = u_1 - u_0 = \frac{a_{22}(x_1w_1 - x_0w_0)}{a_{11}a_{22} - a_{12}a_{21}} \qquad (7.2.6a)$$

$$dv = v_1 - v_0 = \frac{-a_{12}(x_1w_1 - x_0w_0)}{a_{11}a_{22} - a_{12}a_{21}} \qquad (7.2.6b)$$

For affine transformations, $w_0 = w_1 = 1$ and Eqs. (7.2.6a) and (7.2.6b) simplify to

$$du = \frac{a_{22}}{a_{11}a_{22} - a_{12}a_{21}} \qquad (7.2.7a)$$

$$dv = \frac{-a_{12}}{a_{11}a_{22} - a_{12}a_{21}} \qquad (7.2.7b)$$

The expression for dw can be derived from du and dv as follows.

$$dw = a_{13}du + a_{23}dv \qquad (7.2.8)$$

$$= \frac{(a_{13}a_{22} - a_{23}a_{12})(x_1w_1 - x_0w_0)}{a_{11}a_{22} - a_{12}a_{21}}$$

The error of the linear interpolation method vanishes as $dw \rightarrow 0$. A simple ad hoc solution to achieve this goal is to continue with linear interpolation, but to finely subdivide the polygon. If the texture coordinates are correctly computed for the vertices of the new polygons, the resulting picture will exhibit less discontinuities near the vertices. The problem with this method is that costly computations must be made to correctly compute the texture coordinates at the new vertices, and it is difficult to determine how much subdivision is necessary. Clearly, the more parallel the polygon lies to the viewing plane, the less subdivision is warranted.

In order to provide some insight into the effect of subdivision, Fig. 7.6 illustrates the result of subdividing the polygon of Fig. 7.5 several times. In Fig. 7.6a, the edges of the polygon were subdivided into two equal parts, generating four smaller polygons. Their borders can be deduced in the figure by observing the persisting discontinuities. Due to the foreshortening effects of the perspective mapping, the placement of these borders are shifted from the apparent midpoints of the edges. Figures 7.6b, 7.6c, and 7.6d show the same polygon subdivided 2, 4, and 8 times, respectively. Notice that the artifacts diminish with each subdivision.

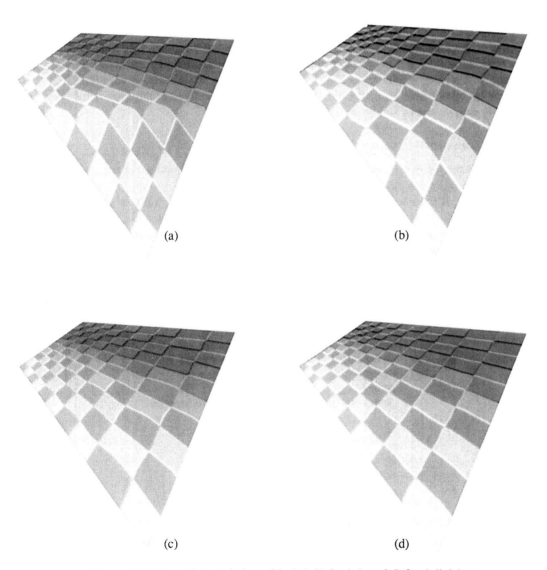

(a)

(b)

(c)

(d)

Figure 7.6: Linear interpolation with a) 1; b) 2; c) 4; and d) 8 subdivisions.

One physical interpretation of this problem can be given as follows. Let the planar polygon be bounded by a cube. We would like the depth of that cube to approach zero, leaving a plane parallel to the viewing screen where the transformation becomes an affine mapping. Some user-specified limit to the depth of the bounding cube and its displacement from the viewing plane must be given in order to determine how much polygon subdivision is necessary. Such computations are themselves costly and difficult to justify. As a result, *a priori* estimates to the number of subdivisions are usually made on the basis of the expected size and tilt of polygons.

At the time of this writing, this approach has been introduced into the most recent wave of graphics workstations that feature real-time texture mapping. One such method is reported in [Oka 87]. It is important to note that Gouraud shading has been used for years without major noticeable artifacts because shading is a slowly-varying function. However, applications such as texture mapping bring out the flaws of this approach more readily with the use of highly-varying texture patterns.

7.2.4. Incremental Perspective Transformations

A theoretically correct solution results by more closely examining the requirements of a perspective mapping. Since a perspective transformation is a ratio of two linear interpolants, it becomes possible to achieve theoretically correct results by introducing the divisor, i.e., homogeneous coordinate w. We thus interpolate w alongside u and v, and then perform two divisions per pixel. The following code contains the necessary adjustments to make the scanline approach work for perspective mappings.

```
dx = 1.0 / (x1 - x0);            /* normalization factor */
du = (u1 - u0) * dx;             /* constant increment for u */
dv = (v1 - v0) * dx;             /* constant increment for v */
dz = (z1 - z0) * dx;             /* constant increment for z */
dw = (w1 - w0) * dx;             /* constant increment for w */
for(x = x0; x < x1; x++) {       /* visit all scanline pixels */
        if(z < zbuf[x]) {        /* is new point closer? */
                zbuf[x] = z;     /* update z-buffer */
                scr[x] = tex(u/w,v/w); /* write texture value to screen */
        }
        u += du;                 /* increment u */
        v += dv;                 /* increment v */
        z += dz;                 /* increment z */
        w += dw;                 /* increment w */
}
```

Figure 7.7 shows the result of this method after it was applied to the Checkerboard texture. Notice the proper foreshortening and the continuity near the vertices.

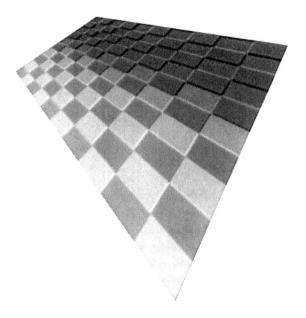

Figure 7.7: Perspective mapping using scanline algorithm.

7.2.5. Approximations

The main objective of the scanline algorithm described above is to exploit the use of incremental computation for fast texture mapping. However, the division operations needed for perspective mappings are expensive and undermine some of the computational gains. Although it can be argued that division requires only marginal cost relative to antialiasing, it is worthwhile to examine optimizations that can be used to approximate the correct solution. Before we do so, we review the geometric nature of the problem at hand.

Consider a planar polygon lying parallel to the viewing plane. All points on the polygon thereby lie equidistant from the viewing plane. This allows equal increments in screen space (the viewing plane) to correspond to equal, albeit not the same, increments on the polygon. As a result, linear interpolation of u and v is consistent with this spatial transformation, an affine mapping. However, if the polygon lies obliquely relative to the viewing plane, then foreshortening is introduced. This no longer preserves equispaced points along lines. Consequently, linear interpolation of u and v is inconsistent with the perspective mapping.

Although both mappings interpolate the same lines connecting $u0$ to $u1$ and $v0$ to $v1$, it is the *rates* at which these lines are sampled that is different. Affine mappings cause the line to be uniformly sampled, while perspective mappings sample the line more densely at distant points where foreshortening has a greater effect. This is depicted in Fig. 7.8 which shows a plot of the u-coordinates spanned using both affine and perspective mappings.

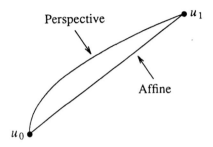

Figure 7.8: Interpolating texture coordinates.

We have already shown that division is necessary to achieve the correct results. Although the most advanced processors today can perform division at rates comparable to addition, there are many applications that look for cheaper approximations on more conventional hardware. Consequently, we examine how to approximate the nonuniform sampling depicted in Fig. 7.8. The most straightforward approach makes use of the Taylor series to approximate division. The Taylor series of a function $f(x)$ evaluated about the point x_0 is given as

$$f(x) = f(x_0) + f'(x_0)\delta + \frac{f''(x_0)}{2!}\delta^2 + \frac{f'''(x_0)}{3!}\delta^3 + \cdots \qquad (7.2.9)$$

where $\delta = x - x_0$. If we let f be the reciprocal function for w, i.e., $f(w) = 1/w$, then we may use the following first-order truncated Taylor series approximation [Lien 87].

$$\frac{1}{w} = \frac{1}{w_0} - \frac{\delta}{w_0^2} \qquad (7.2.10)$$

The authors of that paper suggest that w_0 be the most significant 8 bits of a 32-bit fixed point integer storing w. A lookup-table, indexed by an 8-bit w_0, contains the entries for $1/w_0$. That result may be combined with the lower 24-bit quantity δ to yield the approximated quotient. In particular, if we let a be the rough estimate $1/w_0$ that is retrieved from the lookup table, and b be the least significant 24-bit quantity of w, then from Eq. (7.2.10) we have $1/w = a - a*a*b$. In this manner, division has been replaced with addition and multiplication operations. The reader can verify that an 8-bit w_0 and 24-bit δ yields 18 bits of accuracy in the result. The full 32 bits of precision can be achieved with the use of the 16 higher-order bits for w_0 and the low-order 16 bits for δ.

7.2.6. Quadratic Interpolation

We continue to search for fast incremental methods to approximate the nonuniformity introduced by perspective. Instead of incremental linear interpolation, we examine higher-order interpolating functions. By inspection of Fig. 7.8, it appears that quadratic interpolation might suffice. A second-degree polynomial mapping function for u and v, has the form

$$u = a_2x^2 + a_1x + a_0 \qquad (7.2.11a)$$

$$v = b_2x^2 + b_1x + b_0 \qquad (7.2.11b)$$

where x is a normalized parameter in the range from 0 to 1, spanning the length of the scanline. Since we have three unknown coefficients for each mapping function, three (x_i,u_i) and (x_i,v_i) pairs must be supplied, for $0 \le i \le 2$. The three points we select are the two ends of the scanline and its midpoint. They shall be referred to with subscripts 0 and 2 for the left and right endpoints, and subscript 1 for the midpoint. At these points, the texture coordinates are computed exactly. The general solution for the polynomial coefficients is

$$a_2 = \frac{2(u_0 - 2u_1 + u_2)}{(x_0 - x_2)^2}$$

$$a_1 = \frac{-(x_0u_0 + 3x_2u_0 - 4x_0u_1 - 4x_2u_1 + 3x_0u_2 + x_2u_2)}{(x_0 - x_2)^2} \qquad (7.2.12)$$

$$a_0 = \frac{x_0x_2u_0 + x_2^2u_0 - 4x_0x_2u_1 + x_0^2u_2 + x_0x_2u_2}{(x_0 - x_2)^2}$$

By normalizing the span so that $x_0 = 0$ and $x_2 = 1$, we have the following coefficients for the quadratic polynomial.

$$a_2 = 2u_0 - 4u_1 + 2u_2$$

$$a_1 = -3u_0 + 4u_1 - u_2 \qquad (7.2.13)$$

$$a_0 = u_0$$

A similar result is obtained for b_i, except that v replaces u.

Now that we have the mapping function in a polynomial form, we may return to computing the texture coordinates incrementally. However, since higher-order polynomials are now used, the incremental computation makes use of *forward differencing*. This introduces two forward difference constants to be used in the approximation of the perspective mapping that is modeled with a quadratic polynomial. Expressed in terms of the polynomial coefficients, we have

$$UD1 = a_1 + a_2 \qquad\qquad (7.2.14)$$

$$UD2 = 2a_2$$

A full explanation of the method of forward differences is given in Appendix 3. The following segment of C code demonstrates its use in the quadratic interpolation of texture coordinates.

```
dx = 1.0 / (x2 - x0);              /* normalization factor */
dz = (z2 - z0) * dx;               /* constant increment for z */

/* evaluate texture coordinates at endpoints and midpoint of scanline */
u1 = (u0+u2) / (w0+w2);                    /* midpoint */
v1 = (v0+v2) / (w0+w2);                    /* midpoint */
u0 = u0 / w0;          v0 = v0 / w0;       /* left endpoint */
u2 = u2 / w2;          v2 = v2 / w2;       /* right endpoint */

/* compute quadratic polynomial coefficients: a2x^2 + a1x + a0 */
a0 = u0;                         b0 = v0;
a1 = (-3*u0 + 4*u1 - u2) * dx;   b1 = (-3*v0 + 4*v1 - v2) * dx;
a2 = 2*( u0 - 2*u1 + u2) * dx*dx; b2 = 2*( v0 - 2*v1 + v2) * dx*dx;

/* forward difference parameters for quadratic polynomial */
UD1 = a1 + a2;       VD1 = b1 + b2;       /* 1st forward difference */
UD2 = 2 * a2;        VD2 = 2 * b2;        /* 2nd forward difference */

/* init u,v with texture coordinates of left end of scanline */
u = u0;
v = v0;

for(x = x0; x < x2; x++) {       /* visit all scanline pixels */
    if(z < zbuf[x]) {            /* is new point closer? */
        zbuf[x] = z;             /* update z-buffer */
        scr[x] = tex(u,v);       /* write texture value to screen */
    }
    u += UD1;                    /* increment u with 1st fwd diff */
    v += VD1;                    /* increment v with 1st fwd diff */
    z += dz;                     /* increment z */
    UD1 += UD2;                  /* update 1st fwd diff */
    VD1 += VD2;                  /* update 1st fwd diff */
}
```

This method quickly converges to the correct solution, as demonstrated in Fig. 7.9. The same quadrilateral which had previously required several subdivisions to approach the correct solution is now directly transformed by quadratic interpolation. Introducing a single subdivision rectifies the slight distortion that appears near the rightmost corner of the figure. Since quadratic interpolation converges faster than linear interpolation, it is a superior cost-effective method for computing texture coordinates.

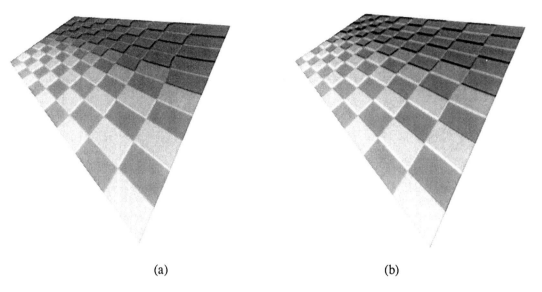

(a)   (b)

Figure 7.9: Quadratic interpolation with (a) 0 and (b) 1 subdivision.

7.2.7. Cubic Interpolation

Given the success of quadratic interpolation, it is natural to investigate how much better the results may be with cubic interpolation. A third-degree polynomial mapping function for u and v has the form

$$u = a_3 x^3 + a_2 x^2 + a_1 x + a_0 \qquad (7.2.15a)$$

$$v = b_3 x^3 + b_2 x^2 + b_1 x + b_0 \qquad (7.2.15b)$$

where x is a normalized parameter in the range from 0 to 1, spanning the length of the scanline. In the discussion that follows, we will restrict our attention to u. The same derivations apply to v.

Since we have four unknown coefficients for each mapping function, four constraints must be imposed. We choose to use the same constraints that apply to Hermite cubic interpolation: the polynomial must pass through the two endpoints of the span while satisfying imposed conditions on the first derivative. Therefore, given a span between x_0 and x_1, we must be given u_0, u_1, as well as derivatives u_0' and u_1' in order to solve for the polynomial coefficients. With these coefficients, the mapping function is defined across the entire scanline. The expressions for the four polynomial coefficients are derived in Appendix 2 (see Eq. A2.3.1) and will be restated later in this section. First, though, we discuss how the derivatives are computed.

Although u_0 and u_1 are readily available, the first derivatives u_0' and u_1' are generally not given directly. Instead, they must be determined indirectly from u_0 and u_1, the known texture coordinates at both ends of the scanline. We begin by rewriting the texture coordinates as a ratio of two linear interpolants. That is,

$$\frac{u}{w} = \frac{ax + b}{cx + d} \qquad (7.2.16)$$

The true function value at the endpoints are computed directly from Eq. (7.2.16) at x_0 and x_1. The first derivative of $f = u/w$ is computed as follows.

$$f' = \frac{a(cx + d) - c(ax + b)}{(cx + d)^2} \qquad (7.2.17)$$

$$= \frac{aw - cu}{w^2}$$

where the parameters a, b, c, and d are determined by using the boundary conditions for u and w. This yields

$$a = \frac{u_1 - u_0}{x_1 - x_0}$$

$$b = u_0 \qquad (7.2.18)$$

$$c = \frac{w_1 - w_0}{x_1 - x_0}$$

$$d = w_0$$

Substituting these values into Eq. (7.2.17) gives us

$$f' = \frac{(u_1 - u_0)(w_0) - (u_0)(w_1 - w_0)}{(x_1 - x_0)(w_1 - w_0)} \qquad (7.2.19)$$

$$= \frac{u_1 w_0 - u_0 w_1}{(x_1 - x_0)(w_1 - w_0)}$$

This serves to express the first derivatives in terms of the known values. Now having data in the form of both function values and first derivatives, the coefficients of the cubic polynomial are given as

$$a_0 = \frac{ax_0 + b}{cx_0 + d} = \frac{u_0}{w_0}$$

$$a_1 = \frac{ad - bc}{(cx_0 + d)^2} = \frac{u_1 w_0 - u_0 w_1}{(x_1 - x_0)(w_1 - w_0)} \qquad (7.2.20)$$

$$a_2 = \left[\frac{1}{x_1 - x_0} \right] \left[3\frac{u_1 - u_0}{x_1 - x_0} - 2\frac{ad - bc}{(cx_0 + d)^2} - \frac{ad - bc}{(cx_1 + d)^2} \right]$$

$$a_3 = \left[\frac{1}{(x_1 - x_0)^2} \right] \left[-2\frac{u_1 - u_0}{x_1 - x_0} + \frac{ad - bc}{(cx_0 + d)^2} + \frac{ad - bc}{(cx_1 + d)^2} \right]$$

Again, forward differences are used to evaluate the cubic polynomial. Expressed in terms of the polynomial coefficients, the three forward difference constants are

$$UD1 = a_1 + a_2 + a_3$$

$$UD2 = 6a_3 + 2a_2 \qquad (7.2.21)$$

$$UD3 = 6a_3$$

These terms are derived in Appendix 3. The following segment of C code demonstrates its use in the cubic interpolation of texture coordinates.

```
dx = 1.0 / (x1 - x0);          /* normalization factor */
dz = (z1 - z0) * dx;           /* constant increment for z */

/* evaluate some intermediate products */
t1 = 1.0 / (w1*w1);            t2 = 1.0 / (w2*w2);
t3 = (u2*w1 - u1*w2) * dx;     t4 = (v2*w1 - v1*w2) * dx;
du = (u2 - u1) * dx;           dv = (v2 - v1) * dx;

/* compute cubic polynomial coefficients: a3x^3 + a2x^2 + a1x + a0 */
a0 = u1 / w1;                  b0 = v1 / w1;
a1 = t1 * t3;                  b1 = t1 * t4;
a2 = (3*du - 2*a1 - t2*t3) * dx;    b2 = (3*dv - 2*b1 - t2*t4) * dx;
a3 = (-2*du + a1 + t2*t3) * dx*dx;  b3 = (-2*dv + b1 + t2*t4) * dx*dx;

/* forward difference parameters for cubic polynomial */
UD1 = a1 + a2 + a3;    VD1 = b1 + b2 + b3;    /* 1st forward difference */
UD2 = 6*a3 + 2*a2;     VD2 = 6*b3 + 2*b2;     /* 2nd forward difference */
UD3 = 6*a3;            VD2 = 6*b3;            /* 3rd forward difference */

/* init u,v with texture coordinates of left end of scanline */
u = a0;
v = b0;

for(x = x1; x < x2; x++) {     /* visit all scanline pixels */
    if(z < zbuf[x]) {          /* is new point closer? */
```

```
        zbuf[x] = z;              /* update z-buffer */
        scr[x] = tex(u,v);        /* write texture value to screen */
    }
    u += UD1;                     /* increment u with 1st fwd diff */
    v += VD1;                     /* increment v with 1st fwd diff */
    z += dz;                      /* increment z */
    UD1 += UD2;                   /* update 1st  fwd diff */
    VD1 += VD2;                   /* update 1st  fwd diff */
    UD2 += UD3;                   /* update 2nd fwd diff */
    VD2 += VD3;                   /* update 2nd fwd diff */
```

Although intuition would lead one to believe that this method should be superior to quadratic interpolation, it does not generally converge significantly faster to warrant its additional cost. Figure 7.10 shows the results of cubic interpolation with 0 and 1 subdivision. In practice, this approach requires the same number of subdivisions to achieve equivalent results. Readers are encouraged to compare these results for themselves.

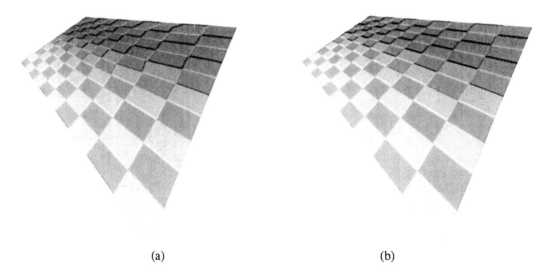

(a)   (b)

Figure 7.10: Cubic interpolation with (a) 0 and (b) 1 subdivision.

7.3. ROTATION

The incremental scanline algorithms described above all exploit the computational savings made possible by forward differences. While they may be fast at computing the transformation, they neglect filtering issues *between* scanlines. Rather than attempt to approximate the transformation along only one direction, separable algorithms decompose their mapping functions along orthogonal directions, i.e., rows and columns. In this manner, the computation of the transformation is more precise, and the associated resampling remains a straightforward 1-D filtering operation. The earliest separable geometric techniques can be traced back to the application of image rotation. Several of these algorithms are reviewed below.

7.3.1. Braccini and Marino, 1980

Braccini and Marino use a variant of the Bresenham line-drawing algorithm to rotate and shear images [Braccini 80]. While this does not qualify as a separable technique, it is included here because it is similar in spirit. In particular, the algorithm demonstrates the decomposition of the rotation matrix into simpler operations which can be efficiently computed.

Consider a straight line with slope n/m, where n and m are both integers. The line is rotated by an angle θ from the horizontal. The expressions for $\cos\theta$ and $\sin\theta$ can be given in terms of n and m as follows:

$$\cos\theta = \frac{m}{\sqrt{(n^2 + m^2)}} \tag{7.3.1}$$

$$\sin\theta = \frac{n}{\sqrt{(n^2 + m^2)}}$$

These terms can be substituted into the rotation matrix R to yield

$$R = \begin{bmatrix} \cos\theta & \sin\theta \\ -\sin\theta & \cos\theta \end{bmatrix} \tag{7.3.2}$$

$$= \frac{m}{\sqrt{(n^2 + m^2)}} \begin{bmatrix} 1 & n/m \\ -(n/m) & 1 \end{bmatrix}$$

The matrix in Eq. (7.3.2) is equivalent to generating a digital line with slope n/m, an operation conveniently implemented by the Bresenham line-drawing algorithm [Foley 90]. The scale factor that is applied to the matrix amounts to resampling the input pixels, an operation which can be formulated in terms of the Bresenham algorithm as well. This is evident by noting that the distribution of n input pixels onto m output pixels is equivalent to drawing a line with slope n/m. The primary advantage of this formulation is that it exploits the computational benefits of the Bresenham algorithm: an incremental technique using only simple integer arithmetic computations.

The rotation algorithm is thereby implemented by depositing the input pixels along a digital line. Both the position of points along the line and the resampling of the input array are determined using the Bresenham algorithm. Due to the inherent jaggedness of digital lines, holes may appear between adjacent lines. Therefore, an extra pixel is drawn at each bend in the line to fill any gap that may otherwise be present. Clearly, this is a crude attempt to avoid holes, a problem inherent in this forward mapping approach.

The above procedure has been used for rotation and scale changes. It has been generalized into a 2-pass technique to realize all affine transformations. This is achieved by using different angles and scale factors along each of the two image axes. Further nonlinear extensions are possible if the parameters are allowed to vary depending upon spatial position, e.g., space-variant mapping.

7.3.2. Weiman, 1980

Weiman describes a rotation algorithm based on cascading simpler 1-D scale and shear operations [Weiman 80]. These transformations are determined by decomposing the rotation matrix R into four submatrices.

$$R = \begin{bmatrix} \cos\theta & \sin\theta \\ -\sin\theta & \cos\theta \end{bmatrix}$$

$$= \begin{bmatrix} 1 & \tan\theta \\ 0 & 1 \end{bmatrix} \begin{bmatrix} 1 & 0 \\ -\sin\theta\cos\theta & 1 \end{bmatrix} \begin{bmatrix} 1 & 0 \\ 0 & \cos\theta \end{bmatrix} \begin{bmatrix} 1/\cos\theta & 0 \\ 0 & 1 \end{bmatrix} \tag{7.3.3}$$

This formulation represents a separable algorithm in which 1-D scaling and shearing are performed along both image axes. As in the Braccini-Marino algorithm, an efficient line-drawing algorithm is used to resample the input pixels and perform shearing. Instead of using the incremental Bresenham algorithm, Weiman uses a periodic code algorithm devised by Rothstein. By averaging over all possible cyclic shifts in the code, the transformed image is shown to be properly filtered. In this respect, the Weiman algorithm is superior to that in [Braccini 80]. An earlier incarnation of this 4-pass approach can be traced back to [Casey 71].

7.3.3. Catmull and Smith, 1980

Catmull and Smith describe a 2-pass solution to a wide class of spatial transformations in [Catmull 80]. Their work is quite general, including affine and perspective transformations onto planar surfaces, biquadratic patches, bicubic patches, and superquadrics. Image rotation, being an affine transformation, is of course treated in their work. The resulting 2-pass transform decomposes the rotation matrix R into two submatrices, each producing a scale/shear transformation.

$$R = \begin{bmatrix} \cos\theta & \sin\theta \\ -\sin\theta & \cos\theta \end{bmatrix}$$

$$= \begin{bmatrix} \cos\theta & 0 \\ -\sin\theta & 1 \end{bmatrix} \begin{bmatrix} 1 & \tan\theta \\ 0 & 1/\cos\theta \end{bmatrix} \tag{7.3.4}$$

The algorithm first skews and scales the image along the horizontal direction. The result then undergoes a similar process in the vertical direction. This 2-pass approach is illustrated in Fig. 7.11. A description of a hardware system to implement this process is found in [Tabata 86].

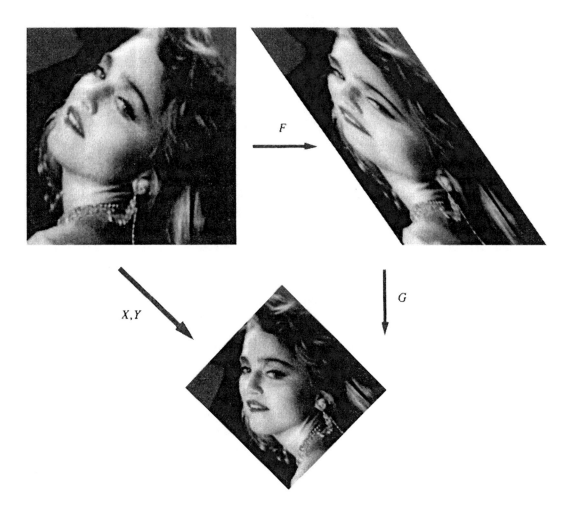

Figure 7.11: 2-pass scale/shear rotation algorithm.

7.3.4. Paeth, 1986 / Tanaka, et. al., 1986

The most significant algorithm to be developed for image rotation was proposed independently in [Paeth 86] and [Tanaka 86, 88]. They demonstrate that rotation can be implemented by cascading three shear transformations.

$$R = \begin{bmatrix} \cos\theta & \sin\theta \\ -\sin\theta & \cos\theta \end{bmatrix}$$
$$= \begin{bmatrix} 1 & 0 \\ -\tan(\theta/2) & 1 \end{bmatrix} \begin{bmatrix} 1 & \sin\theta \\ 0 & 1 \end{bmatrix} \begin{bmatrix} 1 & 0 \\ -\tan(\theta/2) & 1 \end{bmatrix} \quad (7.3.5)$$

The algorithm first skews the image along the horizontal direction by displacing each row. The result is then skewed along the vertical direction. Finally, an additional skew in the horizontal direction yields the rotated image. This sequence is illustrated in Fig. 7.12.

The primary advantage to the 3-pass shear transformation algorithm is that it avoids a costly scale operation. In this manner, it differs significantly from the 2-pass Catmull-Smith algorithm which combined scaling and shearing in each pass, and the 4-pass Weiman algorithm which further decomposed the scale/shear sequence. By not introducing a scale operation, the algorithm avoids complications in sampling, filtering, and the associated degradations. Note, for instance, that this method is not susceptible to the bottleneck problem.

Simplifications are based in the particularly efficient means available to realize a shear transformation. The skewed output is the result of displacing each scanline differently. The displacement is generally not integral, but remains constant for all pixels on a given scanline. This allows intersection testing to be computed once for each scanline, noting that each input pixel can overlap at most two output pixels in the skewed image. The result is used to weigh each input intensity as it contributes to the output. Since the filter support is limited to two pixels, a simple triangle filter (linear interpolation) is adequate. Furthermore, the sum of the pixel intensities along any scanline can be shown to remain unchanged after the shear operation. Thus, the algorithm produces no visible spatial-variant artifacts or holes. Finally, images on bitmap displays can be rotated using conventional hardware supporting *bitblt*, the bit block transfer operation useful for translations. A C program to implement this algorithm is given below.

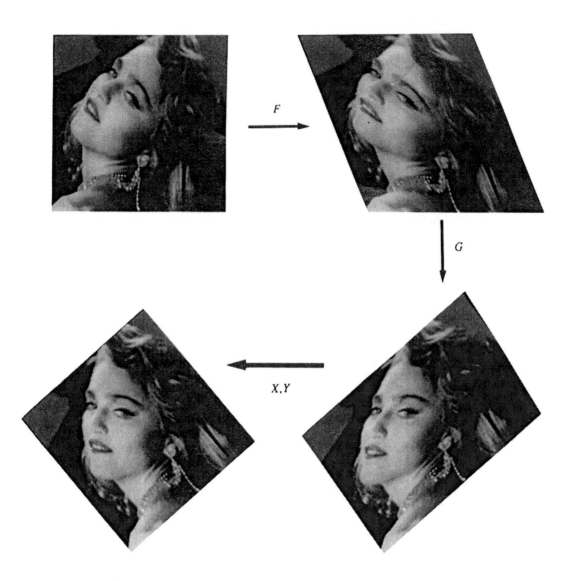

Figure 7.12: 3-pass shear rotation algorithm.

```
/*********************************************************************
        Rotate image IN about its center by angle ang (in radians)
        IN has height h and width w. The output is stored in OUT
        We assume that -π < ang < π
 *********************************************************************/

rotate(IN, h, w, ang, OUT)
unsigned char *IN, *OUT;
int h, w;
double ang;
{
        int x, y, wmax, newh, neww;
        double sine, tangent, offst;

        /* the dimensions of the rotated image as it is processed are:
         * (h)(w) -> (h)(wmax) -> (newh)(wmax) -> (newh)(neww).
         * +1 will be added to dimensions due to last fractional pixel */
         * Temporary buffer TMP is used to hold intermediate image. */
        sine = sin(ang);
        tangent = tan(ang / 2.0);
        wmax = w + h*tangent + 1;                    /* width of intermediate image */
        newh = w*sine + h*cos(ang) + 1;              /* final image height */
        neww = h*sine + w*cos(ang) + 1;              /* final image width */

        /* 1st pass: skew x (horizontal scanlines) */
        for(y = 0; y < h; y++) {                      /* visit each row in IN */
                src = &IN[y * w];                     /* input scanline pointer */
                dst = &OUT[y * wmax];                 /* output scanline pointer */
                skew(src, w, wmax, y*tangent, 1, dst);   /* skew row */
        }

        /* 2nd pass: skew y (vertical scanlines). Use TMP for intermediate image */
        offst = (w-1) * sine;                         /* offset from top of image */
        for(x = 0; x < wmax; x++) {                   /* visit each column in OUT */
                src = &OUT[x];                        /* input scanline pointer */
                dst = &TMP[x];                        /* output scanline pointer */
                skew(src, h, newh, offst - x*sine, wmax, dst);   /* skew column */
        }

        /* 3rd pass: skew x (horizontal scanlines) */
        for(y = 0; y < newh; y++) {                   /* visit each row in TMP */
                src = &TMP[y * wmax];                 /* input scanline pointer */
                dst = &OUT[y * neww];                 /* output scanline pointer */
                skew(src, wmax, neww, (y-offst)*tangent, 1, dst);   /* skew row */
        }
}
```

```
/*******************************************************************
        Skew scanline in src (length len) into dst (length nlen)
        starting at position strt. The offset between each scanline
        pixel is offst. offst=1 for rows; offst=width for columns
********************************************************************/

skew(src, len, nlen, strt, offst, dst)
unsigned char *src, *dst;
int len, nlen, offst;
double strt;
{
        int i, istrt, lim;
        double f, g, w1, w2;

        /* process left end of output: either prepare for clipping or add padding */
        istrt = (int) strt;                     /* integer index */
        if(istrt < 0) src -= istrt;             /* advance input pointer for clipping */
        lim = MIN(len+istrt, nlen);             /* find index for right edge (valid range) */
        for(i = 0; i < istrt; i++) {            /* visit all null output pixels at left edge */
                *dst = 0;                       /* pad with 0 */
                dst += offst;                   /* advance output pointer */
        }

        f = ABS(strt - istrt);                  /* weight for right straddle */
        g = 1. - f;                             /* weight for left straddle */
        if(f == 0.) {                           /* simple integer shift: no interpolation */
                for(; i < lim; i++) {           /* visit all pixels in valid range */
                        *dst = *src;            /* copy input to output */
                        src += offst;           /* advance input pointer */
                        dst += offst;           /* advance output pointer */
                }
        } else {                                /* fractional shift: interpolate */
                if(strt > 0.) {
                        w1 = f;                 /* weight for left pixel */
                        w2 = g;                 /* weight for right pixel */
                        *dst = g * srcp[0];     /* first pixel */
                        dst += offst;           /* advance output pointer */
                        i++;                    /* increment index */
                } else {
                        w1 = g;                 /* weight for left pixel */
                        w2 = f;                 /* weight for right pixel */
                        if(lim < nlen) lim--;
                }
                for(; i < lim; i++) {                           /* visit all pixels in valid range */
                        /* src[0] is left (top) pixel, and src[offst] is right (bottom) pixel */
                        *dst = w1*src[0] + w2*src[offst];   /* linear interpolation */
                        dst += offst;                       /* advance output pointer */
                        src += offst;                       /* advance input pointer */
                }
                if(i < nlen) {
```

```
                *dst = w1 * src[0];          /* src[0] is last pixel */
                dst += offst;                /* advance output pointer */
                i++;                         /* increment output index */
            }
        }
        for(; i < nlen; i++) {               /* visit all remaining pixels at right edge */
            *dst = 0;                        /* pad with 0 */
            dst += offst;                    /* advance output pointer */
        }
    }
}
```

7.3.5. Cordic Algorithm

Another rotation algorithm worth mentioning is the CORDIC algorithm. CORDIC is an acronym for *co*ordinate *r*otation *di*gital *c*omputer. It was originally introduced in [Volder 59], and has since been applied to calculating Discrete Fourier Transforms, exponentials, logarithms, square roots, and other trigonometric functions. It has also been applied to antialiasing calculations for lines and polygons [Turkowski 82]. Although this is an iterative technique, and not a scanline algorithm, it is nevertheless a fast rotation method for points that exploits fast shift and add operations.

The CORDIC algorithm is based on cascading several rotations that are each smaller and easier to compute. The rotation matrix is decomposed into the following form.

$$R = \begin{bmatrix} \cos\theta & \sin\theta \\ -\sin\theta & \cos\theta \end{bmatrix} \tag{7.3.6}$$
$$= \cos\theta \begin{bmatrix} 1 & -\tan\theta \\ -\tan\theta & 1 \end{bmatrix}$$

The composite rotation θ is realized with a series of smaller rotations θ_i such that

$$\theta = \sum_{i=0}^{N-1} \theta_i \tag{7.3.7}$$

where N is the number of iterations in the computation. This method increasingly refines the accuracy of the rotated vector with each iteration. Rotation is thus formulated as a product of smaller rotations, giving us

$$R = \prod_{i=0}^{N-1} \begin{bmatrix} \cos\theta_i & \sin\theta_i \\ -\sin\theta_i & \cos\theta_i \end{bmatrix} \tag{7.3.8}$$

$$= \prod_{i=0}^{N-1} \cos\theta_i \begin{bmatrix} 1 & \tan\theta_i \\ -\tan\theta_i & 1 \end{bmatrix}$$

$$= \left\{ \prod_{i=0}^{N-1} \cos\theta_i \right\} \prod_{i=0}^{N-1} \begin{bmatrix} 1 & \tan\theta_i \\ -\tan\theta_i & 1 \end{bmatrix}$$

The underlying rationale for this decomposition is that large computational savings are gained if the θ_i's are constrained such that

$$\tan\theta_i = \pm_i 2^{-i} \tag{7.3.9}$$

where the sign is chosen to converge to θ in Eq. (7.3.7). This permits the series of matrix multiplications to be implemented by simply shifting and adding intermediate results. The convergence of this series is guaranteed with θ in the range from $-90°$ to $90°$ when i starts out at 0, although convergence is faster when i begins at -1. With this constraint, we have

$$R = \left\{ \prod_{i=0}^{N-1} \cos\left(\tan^{-1} 2^{-i}\right) \right\} \prod_{i=0}^{N-1} \begin{bmatrix} 1 & \pm 2^{-i} \\ \mp 2^{-i} & 1 \end{bmatrix} \tag{7.3.10}$$

$$= \left\{ \prod_{i=0}^{N-1} \frac{1}{\sqrt{1+2^{-2i}}} \right\} \prod_{i=0}^{N-1} \begin{bmatrix} 1 & \pm 2^{-i} \\ \mp 2^{-i} & 1 \end{bmatrix}$$

The reader should note several important properties of the matrices in Eq. (7.3.10). First, the matrices are not orthogonal, i.e., the determinant $1^2 + 2^{-2i} \neq 1$. As a result, the matrix multiplication is called a *pseudorotation* because it enlarges the vector in addition to rotating it. Second, the terms 2^{-i} refer to binary shift operations which are easily realized in fast hardware. Third, the term in braces is a constant for a fixed number of rotation iterations, and converges quickly to 0.27157177. Consequently, it can be precomputed once before processing. Finally, the CORDIC algorithm improves the precision of the results by approximately one bit for each iteration. Such linear convergence can be faster than other methods if multiplications are slower than addition, which is less true of modern signal processors.

The main body of the CORDIC rotation algorithm is presented in the C program given below. Preprocessing is necessary to get the angle between the $-90°$ and $90°$ range, while postscaling is necessary to keep the magnitude of the vector the same.

```
for(i = 0; i < N; i++) {          /* iterate N times */
    if(theta > 0) {               /* positive pseudorotation */
        tmp = x - (y >> i);
        y = y + (x >> i);         /* y = y + x*tan(theta) */
        x = tmp;                  /* x = x - y*tan(theta) */
```

```
            theta -= atantab[i];        /* arctan table of 2⁻ⁱ */
        } else {                        /* negative pseudorotation */
            tmp = x + (y >> i);
            y = y - (x >> i);           /* y = y - x*tan(theta) */
            x = tmp;                    /* x = x + y*tan(theta) */
            theta += atantab[i];        /* arctan table of 2⁻ⁱ */
        }
    }
}
```

where $(a >> b)$ means that a is shifted right by b bits.

The algorithm first checks to see whether the angle *theta* is positive. If so, a pseudorotation is done by an angle of $\tan^{-1} 2^{-i}$. Otherwise, a pseudorotation is done by an angle of $-\tan^{-1} 2^{-i}$. In either case, that angle is subtracted from *theta*. The check for the sign of the angle is done again, and a sequence of pseudorotations iterate until the loop has been executed N times. At each step of the iteration, the angle *theta* fluctuates about zero during the course of the iterative refinement.

Although the CORDIC algorithm is a fast rotation algorithm for *points*, it is presented here largely for the sake of completeness. It is not particularly useful for image rotation because it does not resolve filtering issues. Unless priority is given to filtering, the benefits of a fast algorithm to compute the coordinate transformation of each point is quickly diluted. As we have seen earlier, the 3-pass technique resolves the coordinate transformation *and* filtering problems simultaneously. As a result, that approach is taken to be the method of choice for the special case of rotation. It must be noted that these comments apply for software implementation. Of course if enough hardware is thrown at the problem, then the relative costs and merits change based on what is now considered to be computationally cheap.

7.4. 2-PASS TRANSFORMS

Consider a spatial transformation specified by forward mapping functions X and Y such that

$$[x, y] = T(u,v) = [X(u,v), Y(u,v)] \tag{7.4.1}$$

The transformation T is said to be *separable* if $T(u,v) = F(u) G(v)$. Since it is understood that G is applied only after F, the mapping $T(u,v)$ is said to be *2-pass transformable*, or simply *2-passable*. Functions F and G are called the *2-pass functions*, each operating along different axes. Consequently, the forward mapping in Eq. (7.4.1) can be rewritten as a succession of two 1-D mappings F and G, the horizontal and vertical transformations, respectively.

It is important to elaborate on our use of the term separable. As mentioned above, the signal processing literature refers to a filter T as separable if $T(u,v) = F(u) G(v)$. This certainly applied to the rotation algorithms described earlier. We extend this definition by defining T to be separable if $T(u,v) = F(u) \circ G(v)$. This simply replaces multiplication with the composition operator in combining both 1-D functions. The definition we offer for separablity in this book is consistent with standard implementation practices. For instance, the 2-D Fourier transform, separable in the classic sense, is generally implemented by a 2-pass algorithm. The first pass applies a 1-D Fourier transform to each row, and the second applies a 1-D Fourier transform along each column of the intermediate result. Multi-pass scanline algorithms that operate in this sequential row-column manner will be referred to as separable. The underlying theme is that processing is decomposed into a series of 1-D stages that each operate along orthogonal axes.

7.4.1. Catmull and Smith, 1980

The most general presentation of the 2-pass technique appears in the seminal work described by Catmull and Smith in [Catmull 80]. This paper tackles the problem of mapping a 2-D image onto a 3-D surface and then projecting the result onto the 2-D screen for viewing. The contribution of this work lies in the decomposition of these steps into a sequence of computationally cheaper mapping operations. In particular, it is shown that a 2-D resampling problem can be replaced with two orthogonal 1-D resampling stages. This is depicted in Fig. 7.13.

7.4.1.1. First Pass

In the first pass, each horizontal scanline (row) is resampled according to spatial transformation $F(u)$, generating an intermediate image I in scanline order. All pixels in I have the same x-coordinates that they will assume in the final output; only their y-coordinates now remain to be computed. Since each scanline will generally have a different transformation, function $F(u)$ will usually differ from row to row. Consequently, F can be considered to be a function of both u and v. In fact, it is clear that mapping function F is identical to X, generating x-coordinates from points in the $[u,v]$ plane. To remain consistent with earlier notation, we rewrite $F(u,v)$ as $F_v(u)$ to denote that F is applied to horizontal scanlines, each having constant v. Therefore, the first pass is expressed as

$$[x, v] = [F_v(u), v] \tag{7.4.2}$$

where $F_v(u) = X(u,v)$. This relation maps all $[u,v]$ points onto the $[x,v]$ plane.

7.4.1.2. Second Pass

In the second pass, each vertical scanline (column) in I is resampled according to spatial transformation $G(v)$, generating the final image in scanline order. The second pass is more complicated than the first pass because the expression for G is often difficult to derive. This is due to the fact that we must invert $[x,v]$ to get $[u,v]$ so that G can directly access $Y(u,v)$. In doing so, new y-coordinates can be computed for each point in I.

Inverting f requires us to solve the equation $X(u,v) - \tilde{x} = 0$ for u to obtain $u = H_x(v)$ for vertical scanline (column) $\tilde{x}$. Note that $\tilde{x}$ contains all the pixels along the column at x. Function H, known as the *auxiliary function*, represents the u-coordinates of the inverse projection of $\tilde{x}$, the column we wish to resample. Thus, for every column in I, we

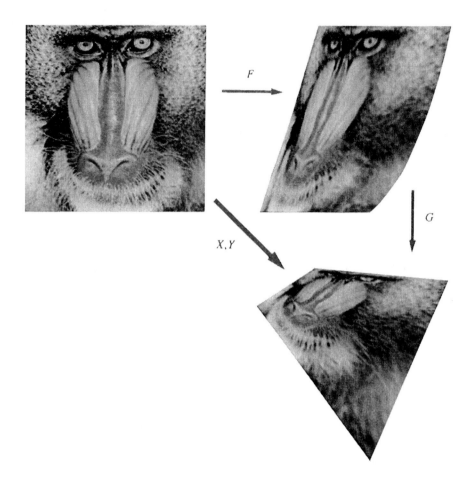

Figure 7.13: 2-pass geometric transformation.

compute $H_x(v)$ and use it together with the available v-coordinates to index into mapping function Y. This specifies the vertical spatial transformation necessary for resampling the column. The second pass is therefore expressed as

$$[x, y] = [x, G_x(v)]$$ (7.4.3)

where $G_x(v)$ refers to the evaluation of $G(x,v)$ along vertical scanlines with constant x. It is given by

$$G_x(v) = Y(H_x(v), v)$$ (7.4.4)

The relation in Eq. (7.4.3) maps all points in I from the $[x, v]$ plane onto the $[x, y]$ plane, the coordinate system of the final image.

7.4.1.3. 2-Pass Algorithm

In summary, the 2-pass algorithm has three steps. They correspond directly to the evaluation of scanline functions F and G, as well as the auxiliary function H.

1. The horizontal scanline function is defined as $F_v(u) = X(u,v)$. Each row is resampled according to this spatial transformation, yielding intermediate image I.

2. The auxiliary function $H_x(v)$ is derived for each vertical scanline $\tilde{x}$ in I. It is defined as the solution to $\tilde{x} = X(u,v)$ for u, if such a solution can be derived. Sometimes a closed form solution for H is not possible and numerical techniques such as the Newton-Raphson iteration method must be used. As we shall see later, computing H is the principal difficulty with the 2-pass algorithm.

3. Once $H_x(v)$ is determined, the second pass plugs it into the expression for $Y(u,v)$ to evaluate the target y-coordinates of all pixels in column x in image I. The vertical scanline function is defined as $G_x(v) = Y(H_x(v),v)$. Each column in I is resampled according to this spatial transformation, yielding the final image.

7.4.1.4. An Example: Rotation

The above procedure is demonstrated on the simple case of rotation. The rotation matrix is given as

$$[x, y] \;=\; [u, v] \begin{bmatrix} \cos\theta & \sin\theta \\ -\sin\theta & \cos\theta \end{bmatrix} \tag{7.4.5}$$

We want to transform every pixel in the original image in scanline order. If we scan a row by varying u and holding v constant, we immediately notice that the transformed points are not being generated in scanline order. This presents difficulties in antialiasing filtering and fails to achieve our goals of scanline input and output.

Alternatively, we may evaluate the scanline by holding v constant in the output as well, and only evaluating the new x values. This is given as

$$[x, v] \;=\; [u\cos\theta - v\sin\theta, \, v] \tag{7.4.6}$$

This results in a picture that is skewed and scaled along the horizontal scanlines.

The next step is to transform this intermediate result by holding x constant and computing y. However, the equation $y = u\sin\theta + v\cos\theta$ cannot be applied since the variable u is referenced instead of the available x. Therefore, it is first necessary to express u in terms of x. Recall that $x = u\cos\theta - v\sin\theta$, so

$$u = \frac{x + v\sin\theta}{\cos\theta} \tag{7.4.7}$$

Substituting this into $y = u\sin\theta + v\cos\theta$ yields

$$y = \frac{x\sin\theta + v}{\cos\theta} \tag{7.4.8}$$

The output picture is now generated by computing the y-coordinates of the pixels in the intermediate image, and resampling in vertical scanline order. This completes the 2-pass rotation. Note that the transformations specified by Eqs. (7.4.6) and (7.4.8) are embedded in Eq. (7.3.4). An example of this procedure for a 45° clockwise rotation has been shown in Fig. 7.11.

The stages derived above are directly related to the general procedure described earlier. The three expressions for F, G, and H are explicitly listed below.

1.   The first pass is defined by Eq. (7.4.6). In this case, $F_v(u) = u\cos\theta - v\sin\theta$.

2.   The auxiliary function H is given in Eq. (7.4.7). It is the result of isolating u from the expression for x in mapping function $X(u,v)$. In this case, $H_x(v) = (x + v\sin\theta) / \cos\theta$.

3.   The second pass then plugs $H_x(v)$ into the expression for $Y(u,v)$, yielding Eq. (7.4.8). In this case, $G_x(v) = (x\sin\theta + v) / \cos\theta$.

7.4.1.5. Another Example: Perspective

Another typical use for the 2-pass method is to transform images onto planar surfaces in perspective. In this case, the spatial transformation is defined as

$$[x', y', w'] = [u, v, 1] \begin{bmatrix} a_{11} & a_{12} & a_{13} \\ a_{21} & a_{22} & a_{23} \\ a_{31} & a_{32} & a_{33} \end{bmatrix} \tag{7.4.9}$$

where $x = x'/w'$ and $y = y'/w'$ are the final coordinates in the output image. In the first pass, we evaluate the new x values, giving us

$$[x, v] = \left[\frac{a_{11}u + a_{21}v + a_{31}}{a_{13}u + a_{23}v + a_{33}}, v \right] \tag{7.4.10}$$

Before the second pass can begin, we use Eq. (7.4.10) to find u in terms of x and v:

$$(a_{13}u + a_{23}v + a_{33})x = a_{11}u + a_{21}v + a_{31} \tag{7.4.11}$$

$$(a_{13}x - a_{11})u = -(a_{23}v + a_{33})x + a_{21}v + a_{31}$$

$$u = \frac{-(a_{23}v + a_{33})x + a_{21}v + a_{31}}{a_{13}x - a_{11}}$$

Substituting this into our expression for y yields

$$y = \frac{a_{12}u + a_{22}v + a_{32}}{a_{13}u + a_{23}v + a_{33}} \tag{7.4.12}$$

$$= \frac{[-a_{12}(a_{23}v + a_{33})x + a_{12}a_{21}v + a_{12}a_{31}] + [(a_{13}x - a_{11})(a_{22}v + a_{32})]}{[-a_{13}(a_{23}v + a_{33})x + a_{13}a_{21}v + a_{13}a_{31}] + [(a_{13}x - a_{11})(a_{23}v + a_{33})]}$$

$$= \frac{[(a_{13}a_{22} - a_{12}a_{23})x + a_{12}a_{21} - a_{11}a_{22}]v + (a_{13}a_{32} - a_{12}a_{33})x + (a_{12}a_{31} - a_{11}a_{32})}{(a_{13}a_{21} - a_{11}a_{23})v + (a_{13}a_{31} - a_{11}a_{33})}$$

For a given column, x is constant and Eq. (7.4.12) is a ratio of two linear interpolants that are functions of v. As we make our way across the image, the coefficients of the interpolants change (being functions of x as well), and we get the spatially-varying results shown in Fig. 7.13.

7.4.1.6. Bottleneck Problem

After completing the first pass, it is sometimes possible for the intermediate image to collapse into a narrow area. If this area is much less than that of the final image, then there is insufficient data left to accurately generate the final image in the second pass. This phenomenon, referred to as the *bottleneck problem* in [Catmull 80], is the result of a many-to-one mapping in the first pass followed by a one-to-many mapping in the second pass.

The bottleneck problem occurs, for instance, upon rotating an image clockwise by 90°. Since the top row will map to the rightmost column, all of the points in the scanline will collapse onto the rightmost point. Similar operations on all the other rows will yield a diagonal line as the intermediate image. No possible separable solution exists for this case when implemented in this order. This unfortunate result can be readily observed by noting that the $\cos\theta$ term in the denominator of Eq. (7.4.7) approaches zero as θ approaches 90°, thereby giving rise to an undeterminable inverse.

The solution to this problem lies in considering all the possible orders in which a separable algorithm can be implemented. Four variations are possible to generate the intermediate image:

1. Transform u first.
2. Transform v first.
3. Rotate the input image by 90° and transform u first.
4. Rotate the input image by 90° and transform v first.

In each case, the area of the intermediate image can be calculated. The method that produces the largest intermediate area is used to implement the transformation. If a 90° rotation is required, it is conveniently implemented by reading horizontal scanlines and writing them in vertical scanline order.

In our example, methods (3) and (4) will yield the correct result. This applies equally to rotation angles near 90°. For instance, an 87° rotation is best implemented by first rotating the image by 90° as noted above and then applying a −3° rotation by using the 2-pass technique. These difficulties are resolved more naturally in a recent paper, described later, that demonstrates a separable technique for implementing arbitrary spatial lookup tables [Wolberg 89b].

7.4.1.7. Foldover Problem

The 2-pass algorithm is particularly well-suited for mapping images onto surfaces with closed form solutions to auxiliary function H. For instance, texture mapping onto rectangles that undergo perspective projection was first shown to be 2-passable in [Catmull 80]. This was independently discovered by Evans and Gabriel at Ampex Corporation where the result was implemented in hardware. The product was a real-time video effects generator called ADO (Ampex Digital Optics). It has met with great success in the television broadcasting industry where it is routinely used to map images onto rectangles in 3-space and move them around fluidly. Although the details of their design are not readily available, there are several patents documenting their invention [Bennett 84a, 84b, Gabriel 84].

The process is more complicated for surfaces of higher order, e.g., bilinear, biquadratic, and bicubic patches. Since these surfaces are often nonplanar, they may be self-occluding. This has the effect of making F or G become multi-valued at points where the image folds upon itself, a problem known as *foldover*.

Foldover can occur in either of the two passes. In the vertical pass, the solution for *single* folds in G is to compute the depth of the vertical scanline endpoints. At each column, the endpoint which is furthest from the viewer is transformed first. The subsequent closer points along the vertical scanline will obscure the distant points and remain visible. Generating the image in this back-to-front order becomes more complicated for surfaces with more than one fold. In the general case, this becomes a hidden surface problem.

This problem can be avoided by restricting the mappings to be nonfolded, or single-valued. This simplification reduces the warp to one that resembles those used in remote sensing. In particular, it is akin to mapping images onto distorted planar grids where the spatial transformation is specified by a polynomial transformation. For instance, the nonfolded biquadratic patch can be shown to correct common lens aberrations such as the barrel and pincushion distortions depicted in Fig. 3.12.

Once we restrict patches to be nonfolded, only one solution is valid. This means that only one u on each horizontal scanline can map to the current vertical scanline. We cannot attempt to use classic techniques to solve for H because n solutions may be obtained for an nth-order surface patch. Instead, we find a solution $u = H_x(0)$ for the first horizontal scanline. Since we are assuming smooth surface patches, the next adjacent scanline can be expected to lie in the vicinity. The Newton-Raphson iteration method can be used to solve for $H_x(1)$ using the solution from $H_x(0)$ as a first approximation (starting value). This exploits the spatial coherence of surface elements to solve the inverse problem at hand.

The complexity of this problem can be reduced at the expense of additional memory. The need to evaluate H can be avoided altogether if we make use of earlier computations. Recall that the values of u that we now need in the second pass were already computed in the first pass. Thus, by introducing an auxiliary framebuffer to store these u's, H becomes available by trivial lookup table access.

In practice, there may be many u's mapping onto the unit interval between x and $x+1$. Since we are only interested in the inverse projection of integer values of x, we compute x for a dense set of equally spaced u's. When the integer values of two successive x's differ, we take one of the two following approaches.

1. Iterate on the interval of their projections u_i and u_{i+1}, until the computed x is an integer.

2. Approximate u by $u = u_i + a(u_{i+1} - u_i)$ where $a = x - x_i$.

The computed u is then stored in the auxiliary framebuffer at location x.

7.4.2. Fraser, Schowengerdt, and Briggs, 1985

Fraser, Schowengerdt, and Briggs demonstrate the 2-pass approach for geometric correction applications [Fraser 85]. They address the problem of accessing data along vertical scanlines. This issue becomes significant when processing large multichannel images such as Landsat multispectral data. Accessing pixels along columns can be inefficient and can lead to major performance degradation if the image cannot be entirely stored in main memory. Note that paging will also contribute to excessive time delays. Consequently, the intermediate image should be transposed, making rows become columns and columns become rows. This allows the second pass to operate along easily accessible rows.

A fast transposition algorithm is introduced that operates directly on a multichannel image, manipulating the data by a general 3-D permutation. The three dimensions include the row, column, and channel indices. The transposition algorithm uses a bit-reversed indexing scheme akin to that used in the Fast Fourier Transform (FFT) algorithm. Transposition is executed "in place," with no temporary buffers, by interchanging all elements having corresponding bit-reversed index pairs.

7.4.3. Smith, 1987

The 2-pass algorithm has been shown to apply to a wide class of transformations of general interest. These mappings include the perspective projection of rectangles, bivariate patches, and superquadrics. Smith has discussed them in detail in [Smith 87].

The paper emphasizes the mathematical consequence of decomposing mapping functions X and Y into a sequence of F followed by G. Smith distinguishes X and Y as the *parallel warp*, and F and G as the *serial warp*, where *warp* refers to resampling. He shows that an nth-order serial warp is equivalent to an $(n^2 + n)^{th}$-order parallel warp. This higher-order polynomial mapping is quite different in form from the parallel polynomial warp. Smith also proves that the serial equivalent of a parallel warp is generally more complicated than a polynomial warp. This is due to the fact that the solution to H is typically not a polynomial.

7.5. 2-PASS MESH WARPING

The 2-pass algorithm formulated in [Catmull 80] has been demonstrated for warps specified by closed-form mapping functions. Another equally important class of warps are defined in terms of piecewise continuous mapping functions. In these instances, the input and output images can each be partitioned into a mesh of patches. Each patch delimits an image region over which a continuous mapping function applies. Mapping between both images now becomes a matter of transforming each patch onto its counterpart in the second image, i.e., mesh warping. This approach, typical in remote sensing, is appropriate for applications requiring a high degree of user interaction. By moving vertices in a mesh, it is possible to define arbitrary mapping functions with local control. In this section, we will investigate the use of the 2-pass technique for mesh warping. We begin with a motivation for mesh warping and then proceed to describe an algorithm that has been used to achieve fascinating special effects.

7.5.1. Special Effects

The 2-pass mesh warping algorithm described in this section was developed by Douglas Smythe at Industrial Light and Magic (ILM), the special effects division of Lucasfilm Ltd. This algorithm has been successfully used at ILM to generate special effects for the motion pictures *Willow, Indiana Jones and the Last Crusade*, and *The Abyss*.[†]. The algorithm was originally conceived to create a sequence of transformations: goat → ostrich → turtle → tiger → woman. In this context, a transformation refers to the geometric metamorphosis of one shape into another. It should not be confused with a *cross-dissolve* operation which simply blends one image into the next via point-to-point color interpolation. Although a cross-dissolve is one element of the effect, it is only invoked once the shapes are geometrically aligned to each other.

In the world of special effects, there are basically three approaches that may be taken to achieve such a cinematic illusion. The conventional approach makes use of physical and optical techniques, including air bladders, vacuum pumps, motion-control rigs, and optical printing. The next two approaches make use of computer processing. In particular, they refer to computer graphics and image processing, respectively.

In computer graphics, each of the animals would have to be modeled as 3-D objects and then be accurately rendered. The transformation would be the result of smoothly animating the interpolation between the models of the animals. There are several problems with this approach. First, computer-generated models that accurately resemble the animals are difficult to produce. Second, any technique to accurately render fur, feathers, and skin would be prohibitively expensive. On the other hand, the benefit of computer graphics in this application is the complete control that the director may have over each

---

† Winner of the 1990 Academy Award for special effects.

possible aspect of the illusion.

Image processing proves to be the best alternative. It avoids the problem of modeling the animals by starting directly from images of real animals. The transformation is now achieved by means of digital image warping. Whereas computer graphics renders a set of deforming 3-D models, image processing deforms the images themselves. This conforms with the notion that it is easier to create an effective illusion by distorting reality rather than synthesizing it from nothing. The roles of the two computer processing approaches in creating illusions are depicted in Fig. 7.14.

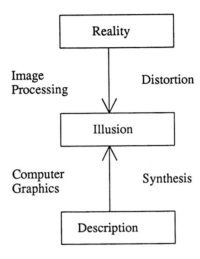

Figure 7.14: Two approaches to computer-generated special effects.

The drawback with the image processing approach is the lack of control. Since the distortions act upon what is already present in the image, the input scenes must be carefully selected and choreographed. For instance, movement of an animal may cause difficulties in alignment with the next animal in the sequence, or present problems with occlusion and shadows. Nevertheless, the benefits of the image processing approach to special effects greatly outweigh its drawbacks.

Special effects is one of many applications in which the mapping functions are conveniently specified by laying down two sets of control points: one set to select points from the input image, and a second set to specify their correspondence in the output image. Since the mapping function is defined only at these discrete points, it becomes necessary for us to determine the mapping function over all points in order to perform the warp. That is, given $X(u_i,v_i)$ and $Y(u_i,v_i)$ for $1 \leq i \leq N$, we must derive X and Y for all the (u,v) points. This is reminiscent of the surface interpolation paradigm presented in Chapter 3, where we formulated this problem as an interpolation of two surfaces X and Y given an arbitrary set of points (u_i,v_i,x_i) and (u_i,v_i,y_i) along them.

In that chapter, we considered various surface interpolation methods, including piecewise polynomials defined over triangulated regions, and global splines. The primary complication lied in the irregular distribution of points. A great deal of simplification is possible when a regular structure is imposed on the points. A rectilinear grid of (u,v) lines, for instance, facilitates mapping functions comprised of rectangular patches. Since many points of interest do not necessarily lie on a rectilinear grid, we allow the placement of control points to coincide with the vertices of a nonuniform mesh. This extension is particularly straightforward since we can consider a mesh to be a parametric grid. In this manner, the control points are indexed by integer (u,v) coordinates that now serve as pointers to the true position, i.e., there is an added level of indirection. The parametric grid partitions the image into a contiguous set of patches, as shown in Fig. 7.15. These patches can now be fitted with a bivariate function to realize a (piecewise) continuous mapping function.

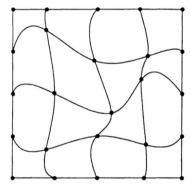

Figure 7.15: Mesh of patches.

7.5.2. Description of the Algorithm

The algorithm in [Smythe 90] accepts a source image and two 2-D arrays of coordinates. The first array, S, specifies the coordinates of control points in the source image. The second array, D, specifies their corresponding positions in the destination image. Both S and D must necessarily have the same dimensions in order to establish a one-to-one correspondence. Since the points are free to lie anywhere in the image plane, the coordinates in S and D are real-valued numbers.

The 2-D arrays in which the control points are stored impose a rectangular topology to the mesh. Each control point, no matter where it lies, is referenced by integer indices. This permits us to fit any bivariate function to them in order to produce a continuous mapping from the discrete set of correspondence points given in S and D. The only constraint is that the meshes defined by both arrays be topologically equivalent, i.e., no folding or discontinuities. Therefore, the entries in D are coordinates that may wander as far from S as necessary, as long as they do not cause self-intersection. Figure 7.16 shows vertices of overlaid meshes S and D.

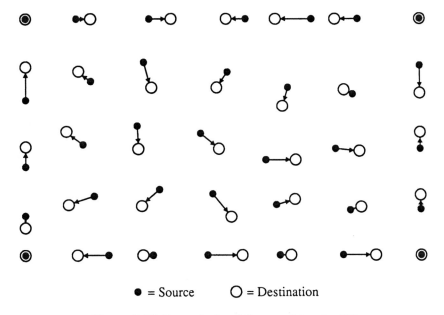

● = Source   ○ = Destination

Figure 7.16: Example S and D arrays [Smythe 90].

The 2-pass mesh warping algorithm is similar in spirit to the 2-pass Catmull-Smith algorithm described earlier. The first pass is responsible for resampling each row independently. It maps all (u,v) points to their (x,v) coordinates in the intermediate image I, thereby positioning each input point into its proper output column. In this manner, the intermediate image I is defined whose x-coordinates are the same as those in D and whose y-coordinates are taken from S (see Fig. 7.17). The second pass then resamples each column in I, mapping every (x,v) point to its final (x,y) position. In this manner, each point can now lie in its proper row, as well as column. We now describe both passes in more detail.

7.5.2.1. First Pass

The first pass requires the output x-coordinates of all pixels along each row. This information is derived directly from S and I in a two-phase process. We let S and I each have h rows and w columns. In practice, these dimensions are much smaller than those of the source image. For reasons described later, the source, intermediate, and destination images all share the same dimensions, $h_{in} \times w_{in}$. Since the control point coordinates are only available at sparse positions, the role of the two-phase process is to spread this data throughout the source image. This makes it possible for all pixels to have the x-coordinate data necessary for resampling along the horizontal direction.

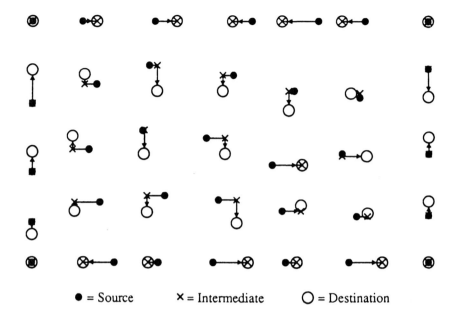

● = Source   ✕ = Intermediate   ○ = Destination

Figure 7.17: Intermediate grid I for S and D [Smythe 90].

In the first phase, each column in S and I is fitted with an interpolating spline through the x-coordinates of the control points. A Catmull-Rom spline was used in [Smythe 90] because it offers local control, although any spline would suffice. These vertical splines are then sampled as they cross each row, creating tables T_S and T_I of dimension $h_{in} \times w$ (see Fig. 7.18). This effectively scan converts each patch boundary in the vertical direction, spreading sparse coordinate data across all rows.

The second phase must now interpolate this data *along* each row. In this manner, each row of width w is resampled to w_{in}, the width of the input image. Since T_S and T_I have the same number of columns, every row in S and I has the same number of vertical patch boundaries; only their particular x-intercepts are different. For each patch interval that spans horizontally from one x-intercept to the next, a normalized index is defined. As we traverse each row in the second phase, we determine the index at every integer pixel boundary in I and we use that index to sample the corresponding spline segment in S. In this manner, the second phase has effectively scan converted T_s and T_I in the horizontal direction, while identifying corresponding intervals in S and I along each row. This form of inverse point sampling, used together with box filtering, achieved the high-quality warps in the feature films cited earlier.

For each pixel P in intermediate image I, box filtering amounts to weighting all input contributions from S by their fractional coverage to P. For minification, the value P is evaluted as a weighted sum from x_0 to x_1, the leftmost and rightmost positions in S that are the projections (inverse mappings) of the left and right integer-valued boundaries of P :

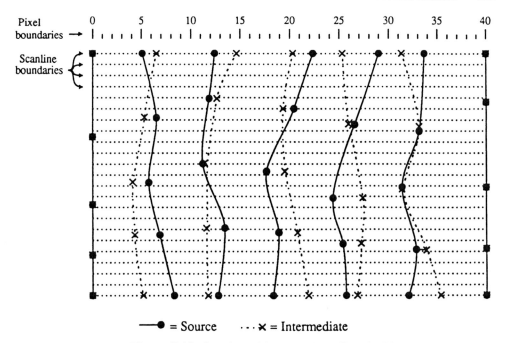

Pixel boundaries →

Scanline boundaries

——• = Source   · · ✗ = Intermediate

Figure 7.18: Creating tables T_S and T_I [Smythe 90].

$$P = \frac{\sum_{x=x_0}^{x_1} k_x S_x}{x_1 - x_0} \tag{7.5.1}$$

where k_x is the scale factor of source pixel S_x, and the subscript x denotes the integer-valued index that lies in the range $\text{floor}(x_0) \leq x < \text{ceil}(x_1)$. The scale factor k_x is defined to be

$$k_x = \begin{cases} \text{ceil}(x) - x_0 & \text{floor}(x) < x_0 \\ 1 & x_0 \leq x < x_1 \\ x_1 - \text{floor}(x) & \text{ceil}(x) > x_1 \end{cases} \tag{7.5.2}$$

The first condition in Eq. (7.5.2) deals with the partial contribution of source pixel S_x when it is clipped on the left edge of the input interval. The second condition applies when S_x lies totally embedded between x_0 and x_1. The final condition deals with the rightmost pixel in the interval in S that may be clipped.

The summation in Eq. (7.5.1) is avoided upon magnification. Instead, some interpolation scheme is applied. Linear interpolation is a popular choice due to its simplicity and effectiveness over a reasonable range of magnification factors.

Figure 7.19 shows the effect of applying the mesh in Fig. 7.15 to the Checkerboard image. In this case, S contains coordinates that lie on the rectilinear grid, and D contains the mesh vertices of Fig. 7.15. Notice that resampling is restricted to the horizontal direction. The second pass will now complete the warp by resampling in the vertical direction.

Figure 7.19: Warped Checkerboard image after first pass.

7.5.2.2. Second Pass

The second pass is virtually identical to that of the first pass. This time, however, we begin by fitting an interpolating spline through the y-coordinates of the control points in each row of I and D. These horizontal splines are then sampled as they cross each column, creating tables T_I and T_D of height h and width w_{in}. Interpolating splines are then fitted to each column in these tables. This facilitates vertical resampling to occur in much the same way as horizontal resampling was performed in the first pass. The collection of vertical splines fitted through S and I in the first pass, together with the horizontal splines fitted through I and D in the second pass, are shown in Fig. 7.20. The warped Checkerboard image, after it comes out of the second pass, is shown in Fig. 7.21.

7.5.2.3. Discussion

The algorithm as presented above requires that all four edges of S and D be frozen. This means that the first and last rows and columns all remain intact throughout the warp. As we shall discover shortly, this seemingly limiting constraint has important implications in the simplicity of the algorithm. Furthermore, if we consider the border to lie far beyond the region of interest in the image, then the frozen edge constraint proves to have little consequence on the class of warps that can be achieved.

In examining this 2-pass mesh warping algorithm more closely, it is worthwhile to compare it to the 2-pass Catmull-Smith transform. In the latter case, the forward map was given only in terms of the input coordinates u and v. Although nonfrozen edges were allowed, this formulation placed a heavy burden in computing an inverse function after the first pass. Afterall, after the first pass warps the (u,v) data into the (x,v)

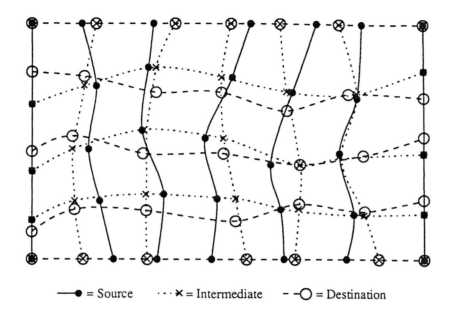

—●— = Source   ⋯✗ = Intermediate   --○ = Destination

Figure 7.20: Splines fitted through S, I, and D [Smythe 90].

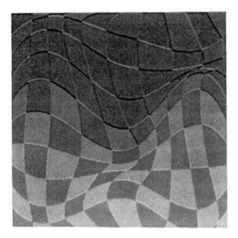

Figure 7.21: Warped Checkerboard image after second pass.

coordinate system, direct access into mapping function $Y(u,v)$ is no longer possible without the existence of an inverse. The 2-pass mesh warping algorithm, on the other hand, defines the forward mapping function in terms of *two* tables of control point coordinates. This formulation permits a straightforward use of interpolating splines, as described for the two-phase first pass.

Although the first pass could have permitted the image boundaries to be nonfrozen, difficulties would have surfaced for an equally simple second pass. In particular, each column in I and D would no longer be guaranteed of sharing the same number of horizontal splines that can be fitted in the vertical direction by just *one* spline. A single vertical spline in the second phase of the second pass proves most useful. It avoids boundary effects around discontinuities that would otherwise arise as a nonfrozen, possibly wiggly, edge is scan converted in the vertical direction. Clearly, slicing such an edge in the vertical direction would produce alternating intervals that lie inside and outside the mesh. Therefore, the frozen edge constraint is placed in order to make the process symmetric among the two passes, and simplify filtering problems in the second pass.

Like the Catmull-Smith algorithm, there is no graceful solution presented to the foldover problem. In fact, the user is refrained from creating such warps. Furthermore, there is no provision for handling the bottleneck problem. As a result, it is possible for distortion to arise when the warps contain large rotational components. This places additional constraints on the user. A 2-pass algorithm that treats the general case with attention to the bottleneck and foldover problems is described in Section 7.7.

7.5.3. Examples

The 2-pass mesh warping algorithm described in this section has been used to produce many fascinating warps. The primary application has been in the transformation between objects. Consider two image sequences of equal length, $F_1(t)$ and $F_2(t)$, where t varies from 0 to N. They are each moving images depicting two creatures, say an ostrich and a turtle. The original state of the metamorphosis begins at $F_1(0)$, with the first image of the ostrich. As t approaches N, the output $H(t)$ progresses towards $F_2(N)$, an uncorrupted image of the turtle at the end of the sequence. Along the way, the output is produced by warping corresponding images of $F_1(t)$ and $F_2(t)$ in some desired way, as specified by their respective control points grids. As a matter of convenience, we shall drop the argument t from the notation in the remaining discussion. It should be understood that when we speak of the image or grid sequences, we refer to one instance at a time.

For each image in the two sequences, grids G_1 and G_2 are defined such that each point in G_1 lies over the same feature in F_1 as the corresponding point in G_2 lies over F_2. F_1 is then warped into a new picture F_{1w} by using source grid G_1 and destination grid G_I, a grid whose coordinates are at some intermediate stage between G_1 and G_2. Similarly, F_2 is warped into a new image F_{2w} using source grid G_2 and destination grid G_I, the same grid that was used in making F_{1w}. In this manner, F_{1w} and F_{2w} are different creatures stretched into geometric alignment. A cross-dissolve between them now yields a frame in the transformation between the two creatures. This process is depicted in Fig. 7.22, where boldface is used to depict the keyframes. These are frames that the user determines to be important in the image sequence. Control grids G_1 and G_2 are precisely established for these keyframes. All intermediate images then get their grid assignments via interpolation.

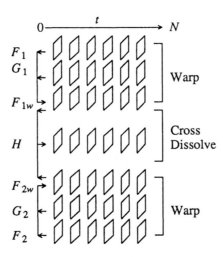

Figure 7.22: Transformation process: warp and cross-dissolve.

One key to making the transformations interesting is to apply a different rate of transition between F_1 and F_2 when creating G_I, so different parts of the creature can move and change at different rates. Figure 7.23 shows one such plot of point movement versus time. The curve moves from the position of the first creature (curve at the bottom early in time) toward the position of the second creature (curve moves to the top later in time). A similar approach is used to vary the rate of color blending from pixel to pixel. The user specifies this information via a digitizing tablet and mouse (Fig. 7.24).

Figure 7.25 shows four frames of Raziel's transformation sequence from *Willow* that warps an ostrich into a turtle. The more complete transformation process, including warps between a tiger and a woman, is depicted in the image on the front cover. The reader should note that the warping program is applied only to the transforming creatures. They are computed separately with a black background. The warped results are then optically composited with the background, the magic wand, and some smoke.

The same algorithm was also used as an integral element in other special effects where geometric alignment was a critical task. This appeared in the movie *Indiana Jones and the Last Crusade* in the scene where an actor underwent physical decomposition, as shown in Fig. 7.26. In order to create this illusion, the ILM creature shop constructed three motion-controlled puppet heads. Each was in a progressively more advanced stage of decomposition. Mechanical systems were used to achieve particular effects such as receding eyeballs and shriveling skin. Each of these was filmed separately, going through identical computer-controlled moves. The warping process was used to ensure a smooth and undetectable transition between the different sized puppet heads and their changing facial features and receding hair (and you thought you had problems!). This appears to be the first time that a feature film sequence was entirely digitally composited from film elements, without the use of an optical printer [Hu 90].

In *The Abyss*, warping was used for facial animation. Several frames of a face were scanned into the computer by using a Cyberware 3D video laser input system. The

Figure 7.23: User interface.
Courtesy of Industrial Light & Magic, a Division of Lucasfilm Ltd.
Copyright © 1990 Lucasfilm Ltd. All Rights Reserved.

resulting images consist of range data denoting the distance of each point from the sensor. Although this data can be used to directly generate 3D models of a human face, such models prove cumbersome for creating realistic facial animations with effective facial expressions. As a result, the range data is left in its 2D form and manipulated with image processing tools, including the 2-pass mesh warping algorithm. Each of the facial images is used as a keyframe in the animation process. Meshes are used to define and control a complex warp in each successive keyframe. In this manner, an animation is created in which one facial expression naturally moves into another. After the frames have been warped in 2D, they are rendered as 3D surfaces for viewing [Anderson 90].

Two additional examples of mesh warping are shown in Figs. 7.28 and 7.29. They serve to further highlight the wide range of transformations possible with this approach.

Figure 7.24: User inputs mesh grid via digitizing tablet.
Courtesy of Industrial Light & Magic, a Division of Lucasfilm Ltd.
Copyright © 1990 Lucasfilm Ltd. All Rights Reserved.

7.5.4. Source Code

A C program that implements the 2-pass mesh warping algorithm is given below. It warps input image *IN* into the output image *OUT*. Both *IN* and *OUT* have the same dimensions: height *IN_h* (rows) and width *IN_w* (columns). The images are assumed to have a single channel consisting of byte-sized pixels, as denoted by the *unsigned char* data type. Multi-channel images (e.g., color) can be handled by sending each channel through the program independently.

The source mesh is supplied through the 2-D arrays *Xs* and *Ys*. Similarly, the destination mesh coordinates are contained in *Xd* and *Yd*. Both mesh tables accommodate double-precision numbers and share the same dimensions: height *T_h* and width *T_w*.

Figure 7.25: Raziel's transformation sequence from *Willow*.
Courtesy of Industrial Light & Magic, a Division of Lucasfilm Ltd.
Copyright © 1988 Lucasfilm Ltd. All Rights Reserved.

The program makes use of *ispline_gen* and *resample_gen*, two functions defined elsewhere in this book. Function *ispline_gen* is used here to fit an interpolating cubic spline through the mesh coordinates. Since it can fit a spline through the data and resample it at arbitrary positions, *ispline_gen* is also used for scan conversion. This is simply achieved by resampling the spline at all integer coordinate values along a row or column. The program listing for *ispline_gen* can be found in Appendix 2. The function takes six arguments, i.e., *ispline_gen*(A,B,C,D,E,F). Arguments A and B are pointers to a list of (x,y) data points whose length is C. The spline is resampled at F positions whose coordinates are contained in D. The results are stored in E.

Figure 7.26: Donovan's destruction sequence from *Indiana Jones and the Last Crusade.*
Courtesy of Industrial Light & Magic, a Division of Lucasfilm Ltd.
Copyright © 1989 Lucasfilm Ltd. All rights reserved.

Once the forward mapping function is defined, function *resample_gen* is used to warp the data. Although an inverse mapping scheme was used in [Smythe 90], we choose a forward mapping formulation because it conveniently allows us to demonstrate algorithms derived earlier. This particular version of *resample_gen* is a variation to the spatially-varying version of Fant's algorithm given in Section 5.6. Although the segment of code given there is limited to processing horizontal scanlines, we now treat the more general case that includes vertical scanlines as well. This is accommodated with the use of an additional parameter that specifies the offset from one pixel to the next. Horizontal scanlines have a pixel-to-pixel offset of one, while vertical scanlines have an offset equal to the width of a row. The function *resample_gen* (A,B,C,D,E) applies the mapping function A to input scanline B, generating output C. The input (and output) dimension is

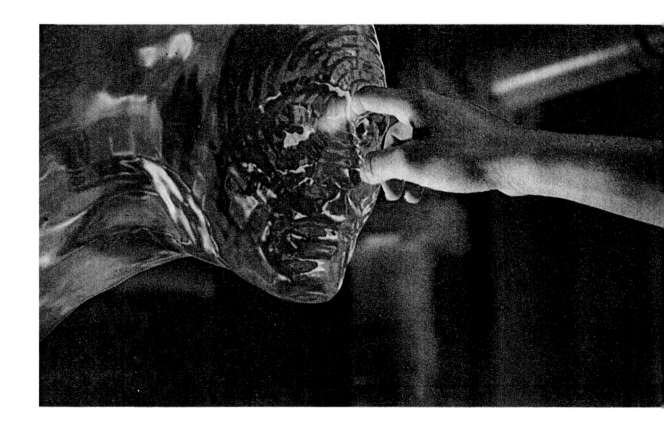

Figure 7.27: Facial Animation from the Pseudopod sequence in *The Abyss*.
Courtesy of Industrial Light & Magic, a Division of Lucasfilm Ltd.
Copyright © 1989 Twentieth Century Fox. All rights reserved.

D and the inter-pixel offset is C. The function performs linear interpolation for magnification and box filtering for minification. This is equivalent to the reconstruction and antialiasing methods used in [Smythe 90]. Superior filters can be added within this framework by incorporating the results of Chapters 5 and 6.

Figure 7.28: A warped image of Piazza San Marco.
Copyright © 1989 Pixar. All rights reserved.

```
/***************************************************************
    Two-pass mesh warping based on algorithm in [Smythe 90].

    Input image IN has height IN_h and width IN_w.
    Xs,Ys contain the x,y coordinates of the source mesh.
    Xd,Yd contain the x,y coordinates of the destination mesh.
    Their height and width dimensions are T_h and T_w.
    The output is stored in OUT. Due to the frozen edge
    assumption, OUT has same dimensions as IN.
 ***************************************************************/

warp_mesh(IN, OUT, Xs, Ys, Xd, Yd, IN_h, IN_w, T_h, T_w)
unsigned char *IN, *OUT;
double *Xs, *Ys, *Xd, *Yd;
```

Figure 7.29: A caricature of Albert Einstein.
Copyright © 1989 Pixar. All rights reserved.

```
int IN_h, IN_w, T_h, T_w;
{
      int a, b, x, y;
      unsigned char *src, *dst;
      double *x1, *y1, *x2, *y2 *xrow1, *yrow1, *xrow2, *yrow2, *map1, *map2, *indx, *Ts, *Ti, *Td;

      /*
       * allocate memory for buffers: indx stores indices used to sample splines;
       * xrow1, xrow2, yrow1, yrow2 store column data in row order for  ispline_gen();
       * map1, map2 store mapping functions computed in row order in ispline_gen()
       */
      a = MAX(IN_h, IN_w) + 1;
      b = sizeof(double);
      indx   = (double *) calloc(a, b);
```

```
xrow1 = (double *) calloc(a, b);          yrow1 = (double *) calloc(a, b);
xrow2 = (double *) calloc(a, b);          yrow2 = (double *) calloc(a, b);
map1  = (double *) calloc(a, b);          map2  = (double *) calloc(a, b);


/*
 * First pass (phase one): create tables Ts and Ti for x-intercepts of
 * vertical splines in S and I. Tables have T_w columns of height IN_h
 */
Ts = (double *) calloc(T_w * IN_h, sizeof(double));
Ti = (double *) calloc(T_w * IN_h, sizeof(double));
for(y=0; y<IN_h; y++) indx[y] = y;          /* indices used to sample vertical splines */
for(x=0; x<T_w; x++) {                       /* visit each vertical spline   */
        /* store columns as rows for ispline_gen */
        for(y=0; y<T_h; y++) {
                xrow1[y] = Xs[y*T_w + x];          yrow1[y] = Ys[y*T_w + x];
                xrow2[y] = Xd[y*T_w + x];          yrow2[y] = Yd[y*T_w + x];
        }

        /* scan convert vertical splines of S and I */
        ispline_gen(yrow1, xrow1, T_h, indx, map1, IN_h);
        ispline_gen(yrow2, xrow2, T_h, indx, map2, IN_h);

        /* store resampled rows back into columns */
        for(y=0; y<IN_h; y++) {
                Ts[y*T_w + x] = map1[y];
                Ti [y*T_w + x] = map2[y];
        }
}

/* First pass (phase two): warp x using Ts and Ti. TMP holds intermediate image. */
TMP = (unsigned char *) calloc(IN_h, IN_w);
for(x=0; x<IN_w; x++) indx[x] = x;          /* indices used to sample horizontal spline */
for(y=0; y<IN_h; y++) {                       /* visit each row */
        /* fit spline to x-intercepts; resample over all columns */
        x1 = &Ts[y * T_w];
        x2 = &Ti [y * T_w];
        ispline_gen(x1, x2, T_w, indx, map1, IN_w);

        /* resample source row based on map1 */
        src = &IN[y * IN_w];
        dst = &TMP[y * IN_w];
        resample_gen(map1, src, dst, w, 1);
}

/* free buffers */
cfree((char *) Ts);
cfree((char *) Ti );


/*
 * Second pass (phase one): create tables Ti and Td for y-intercepts of
```

```
 * horizontal splines in I and D. Tables have T_h rows of width IN_w
 */
Ti  = (double *) calloc(T_h * IN_w, sizeof(double));
Td = (double *) calloc(T_h * IN_w, sizeof(double));
for(x=0; x<IN_w; x++) indx[x] = x;        /* indices used to sample horizontal splines */
for(y=0; y<T_h; y++) {                     /* visit each horizontal spline   */
        /* scan convert horizontal splines of I and D */
        x1 = &X1[y * T_w];        y1 = &Y1[y * T_w];
        x2 = &X2[y * T_w];        y2 = &Y2[y * T_w];
        ispline_gen(x1, y1, T_w, indx, &Ti [y*IN_w], IN_w);
        ispline_gen(x2, y2, T_w, indx, &Td[y*IN_w], IN_w);
}

/* Second pass (phase two): warp y using Ti and Td */
for(y=0; y<IN_h; y++) indx[y] = y;
for(x=0; x<T_w; x++) {
        /* store column as row for ispline_gen */
        for(y=0; y<T_h; y++) {
                xrow1[y] = Ti [y*IN_w + x];
                yrow1[y] = Td[y*IN_w + x];
        }

        /* fit spline to y-intercepts; resample over all rows */
        ispline_gen(xrow1, yrow1, T_h, indx, map1, IN_h);

        /* resample intermediate image column based on map1 */
        src = &TMP[x];
        dst = &OUT[x];
        resample_gen(map1, src, dst, IN_h, IN_w);
}
cfree((char *) TMP);            cfree((char *) indx);
cfree((char *) Ti);            cfree((char *) Td);
cfree((char *) xrow1);        cfree((char *) yrow1);
cfree((char *) xrow2);        cfree((char *) yrow2);
cfree((char *) map1);        cfree((char *) map2);
}
```

7.6. MORE SEPARABLE MAPPINGS

Additional separable geometric transformations are described in this section. They rely on the simplifications of 1-D processing to perform perspective projections, mappings among arbitrary planar shapes, and spatial lookup tables.

7.6.1. Perspective Projection: Robertson, 1987

The perspective projection of 3-D surfaces has been shown to be reducible into a series of fast 1-D resampling operations [Robertson 87, 89]. In the traditional approach, this task has proved to be computationally expensive due to the problems in determining visibility and performing hidden-point removal. With the introduction of this algorithm, the problem can be decomposed into efficient separable components that can each be implemented at rates approaching real-time.

The procedure begins by rotating the image into alignment with the frontal (nearest) edge of the viewing window. Each horizontal scanline is then compressed so that all pixels which lie in a line of sight from the viewpoint are aligned into columns in the intermediate image. That is, each resulting column comprises a line of sight between the viewpoint and the surface.

Occlusion of a pixel can now only be due to another pixel in that column that lies closer to the viewer. This simplifies the perspective projection and hidden-pixel removal stages. These operations are performed along the vertical scanlines. By processing each column in back-to-front order, hidden-pixel removal is executed trivially.

Finally, the intermediate image undergoes a horizontal pass to apply the horizontal projection. This pass is complicated by the need to invert the previously applied horizontal compression. The difficulty arises since the image has already undergone hidden-pixel removal. Consequently, it is not directly known which surface point has been mapped to the current projected point. This can be uniquely determined only after additional calculations. The resulting image is the perspective transformation of the input, performed at rates which make real-time interactive manipulation possible.

7.6.2. Warping Among Arbitrary Planar Shapes: Wolberg, 1988

The advantages of 1-D resampling have been exploited for use in warping images among arbitrary planar shapes [Wolberg 88, 89a]. The algorithm addresses the following inadequately solved problem: mapping between two images that are delimited by arbitrary, closed, planar, curves, e.g., hand-drawn curves.

Unlike many other problems treated in image processing or computer graphics, the stretching of an arbitrary shape onto another, and the associated mapping, is a problem not addressed in a tractable fashion in the literature. The lack of attention to this class of problems can be easily explained. In image processing, there is a well-defined 2-D rectilinear coordinate system. Correcting for distortions amounts to mapping the four corners of a nonrectangular patch onto the four corners of a rectangular patch. In computer graphics, a parameterization exists for the 2-D image, the 3-D object, and the 2-D screen. Consequently, warping amounts to a change of coordinate system (2-D to 3-D) followed by a projection onto the 2-D screen. The problems considered in this work fail to meet the above properties. They are neither parameterized nor are they well suited for four-corner mapping.

The algorithm treats an image as a collection of interior layers. Informally, the layers are extracted in a manner similar to peeling an onion. A radial path emanates from each boundary point, crossing interior layers until the innermost layer, the skeleton, is reached. Assuming correspondences may be established between the boundary points of the source and target images, the warping problem is reduced to mapping between radial paths in both images. Note that the layers and the radial paths actually comprise a sampling grid.

This algorithm uses a generalization of polar coordinates. The extension lies in that radial paths are not restricted to terminate at a single point. Rather, a fully connected skeleton obtained from a thinning operation may serve as terminators of radial paths directed from the boundary. This permits the processing of arbitrary shapes.

The 1-D resampling operations are introduced in three stages. First, the radial paths in the source image must be resampled so that they all take on the same length. Then these normalized lists, which comprise the columns in our intermediate image, are resampled in the horizontal direction. This serves to put them in direct correspondence to their counterparts in the target image. Finally, each column is resampled to lengths that match those of the radial paths in the target image. In general, these lengths will vary due to asymmetric image boundaries.

The final image is generated by wrapping the resampled radial paths onto the target shape. This procedure is identical to the previous peeling operation except that values are now deposited onto the traversed pixels.

7.6.3. General 2-Pass Algorithm: Wolberg and Boult, 1989

Sampling an arbitrary forward mapping function yields a 2-D *spatial lookup table*. This specifies the output coordinates for all input pixels. A separable technique to implement this utility is of great practical importance. The chief complications arise from the bottleneck and foldover problems described earlier. These difficulties are addressed in [Wolberg 89b].

Wolberg and Boult propose a 2-pass algorithm for realizing arbitrary warps that are specified by spatial lookup tables. It is based on the solution of the three main difficulties of the Catmull-Smith algorithm: bottlenecking, foldovers, and the need for a closed-form inverse. In addition, it addresses some of the errors caused by filtering, especially those caused by insufficient resolution in the sampling of the mapping function. Through careful attention to efficiency and graceful degradation, the method is no more costly than the Catmull-Smith algorithm when bottlenecking and foldovers are not present. However when these problems do surface, they are resolved at a cost proportional to their manifestation. Since the underlying data structures continue to facilitate pipelining, this method offers a promising hardware solution to the implementation of arbitrary spatial mapping functions. The details of this method are given in the next section.

7.7. SEPARABLE IMAGE WARPING

In this section, we describe an algorithm introduced by Wolberg and Boult that addresses the problems that are particular to 2-pass methods [Wolberg 89b]. The result is a separable approach that is general, accurate, and efficient, with graceful degradation for transformations of arbitrary complexity.

The goal of this work is to realize an arbitrary warp with a separable algorithm. The proposed technique is an extension of the Catmull-Smith approach where attention has been directed toward solutions to the bottleneck and foldover problems, as well as to the removal of any need for closed-form inverses. Consequently, the advantages of 1-D resampling are more fully exploited.

Conceptually, the algorithm consists of four stages: intensity resampling, coordinate resampling, distortion measurement, and compositing. Figure 7.30 shows the interaction of these components. Note that bold arrows represent the flow of images through a stage, and thin arrows denote those images that act upon the input. The subscripts x and y are appended to images that have been resampled in the horizontal and vertical directions, respectively.

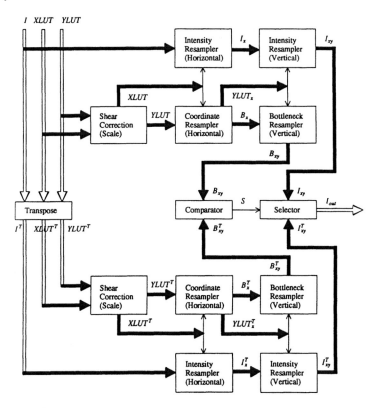

Figure 7.30: Block diagram of the Wolberg-Boult algorithm.

The intensity resampler applies a 2-pass algorithm to the input image. Since the result may suffer bottleneck problems, the identical process is repeated with the transpose of the image. This accounts for the vertical symmetry of Fig. 7.30. Pixels which suffer excessive bottlenecking in the natural processing can be recovered in the transposed processing. In the actual implementation, transposition is realized as a 90° clockwise rotation so as to avoid the need to reorder pixels left to right.

The coordinate resampler computes spatial information necessary for the intensity resampler. It warps the spatial lookup table $Y(u,v)$ so that the second pass of the intensity resampler can access it without the need for an inverse function.

Local measures of shearing, perspective distortion, and bottlenecking are computed to indicate the amount of information lost at each point. This information, together with the transposed and non-transposed results of the intensity resampler, are passed to the compositor. The final output image is generated by the compositor, which samples those pixels from the two resampled images such that information loss is minimized.

7.7.1. Spatial Lookup Tables

Scanline algorithms generally express the coordinate transformation in terms of forward mapping functions X and Y. Sampling X and Y over all input points yields two new real-valued images, $XLUT$ and $YLUT$, specifying the point-to-point mapping from each pixel in the input image onto the output images. $XLUT$ and $YLUT$ are referred to as *spatial lookup tables* since they can be viewed as 2-D tables that express a spatial transformation.

In addition to $XLUT$ and $YLUT$ a mechanism is also provided for the user to specify $ZLUT$, which associates a z-coordinate value with each pixel. This allows warping of planar textures onto non-planar surfaces and is useful in dealing with foldovers. The z-coordinates are assumed to be from a particular point of view that the user determines before supplying $ZLUT$ to the system.

The motivation for introducing spatial lookup tables is generality. The goal is to find a serial warp equivalent to any given parallel warp. Thus, it is impossible to retain the mathematical elegance of closed-form expressions for the mapping functions F, G, and the auxiliary function, H. Therefore, assuming the forward mapping functions, X and Y, have closed-form expressions seems overly restrictive. Instead, the authors assume that the parallel warp is defined by the samples that comprise the spatial lookup tables. This provides a general means of specifying arbitrary mapping functions.

For each pixel (u,v) in input image I, spatial lookup tables $XLUT$, $YLUT$, and $ZLUT$ are indexed at location (u,v) to determine the corresponding (x,y,z) position of the input point after warping. This new position is orthographically projected onto the output image. Therefore, (x,y) is taken to be the position in the output image. (Of course, a perspective projection may be included as part of the warp). The z-coordinate will only be used to resolve foldovers. This straightforward indexing applies only if the dimensions of I, $XLUT$, $YLUT$, and $ZLUT$ are all identical. If this is not the case, then the smaller images are upsampled (magnified) to match the largest dimensions.

7.7.2. Intensity Resampling

The spatial lookup tables determine how much compression and stretching each pixel undergoes. The actual intensity resampling is implemented by using a technique similar to that proposed in [Fant 86]. As described earlier, this method exploits the benefits of operating in scanline order. As a result, it is well-suited for hardware implementation and remains compatible with spatial lookup tables.

The 1-D intensity resampler is applied to the image in two passes, each along orthogonal directions. The first pass resamples horizontal scanlines, warping pixels along a row in the intermediate image. Its purpose is to deposit them into the proper columns for vertical resampling. At that point, the second pass is applied to all columns in the intermediate image, generating the output image.

In Fig. 7.30, input image I is shown warped according to $XLUT$ to generate intermediate image I_x. In order to apply the second pass, $YLUT$ is warped alongside I, yielding $YLUT_x$. This resampled spatial lookup table is applied to I_x in the second pass as a collection of 1-D vertical warps. The result is output image I_{xy}.

The intensity resampling stage must handle multiple output values to be defined in case of foldovers. This is an important implementation detail that has impact on the memory requirements of the algorithm. We defer discussion of this aspect of the intensity resampler until Section 7.7.5, where foldovers are discussed in more detail.

7.7.3. Coordinate Resampling

$YLUT_x$ is computed in the coordinate resampling stage depicted in the second row of the block diagram in Fig. 7.30. The ability to resample $YLUT$ for use in the second pass has important consequences: it circumvents the need for a closed-form inverse of the first pass. As briefly pointed out in [Catmull 80], that inverse provides exactly the same information that was available as the first pass was computed, i.e., the u-coordinate associated with a pixel in the intermediate image. Thus, instead of computing the inverse to index into $YLUT$, we simply warp $YLUT$ into $YLUT_x$ allowing direct access in the second pass.

The coordinate resampler is similar to the intensity resampler. It differs only in the notable absence of antialiasing filtering — the output coordinate values in $YLUT_x$ are computed by *point* sampling $YLUT$. Interpolation is used to compute values when no input data are supplied at the resampling locations. However, unlike the intensity resampler, the coordinate resampler neither weighs the result with its area coverage nor does the resampler average it with the coordinate values of other contributions to that pixel. This serves to secure the accuracy of edge coordinates, even when the edge occupies only a partial output pixel.

7.7.4. Distortions and Errors

In forward mapping, input pixels are taken to be squares that map onto arbitrary quadrilaterals in the output image. Although separable mappings greatly simplify resampling by treating pixels as points along scanlines, the measurement of distortion must necessarily revert to 2-D to consider the deviation of each input pixel as it projects onto the output.

As is standard, we treat the mapping of a square onto a general quadrilateral as a combination of translation, scaling, shearing, rotation, and perspective transformations. Inasmuch as separable kernels exist for realizing translations and scale changes, these transformations do not suffer degradation in scanline algorithms and are not considered further. Shear, perspective and rotations, however, offer significant challenges to the 2-pass approach. In particular, excessive shear and perspective contribute to aliasing problems while rotations account for the bottleneck problem.

We first examine the errors introduced by separable filtering. We then address the three sources of geometric distortion for 2-pass scanline algorithms: shear, perspective, and rotation.

7.7.4.1. Filtering Errors

One of the sources of error for scanline algorithms comes from the use of cascaded orthogonal 1-D filtering. Let us ignore rotation for a moment, and assume we process the image left-to-right and top-to-bottom. Then one can easily show that scanline algorithms will, in the first pass, filter a pixel based only on the horizontal coverage of its top segment. In the second pass, they will filter based only on the vertical coverage of the left-hand segment of the input pixel. As a result, a warped pixel generating a quadrilateral at the output pixel is always approximated by a rectangle (Fig. 7.31). Note this can be either an overestimate or underestimate, and the error depends on the direction of processing. This problem is not unique to our approach. It is shared by all scanline algorithms.

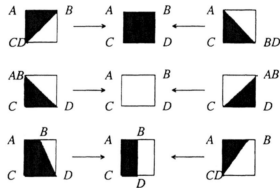

Figure 7.31: Examples of filtering errors.

7.7.4.2. Shear

Figure 7.32 depicts a set of spatial lookup tables that demonstrate horizontal shear. For simplicity, the example includes no scaling or rotation. The figure also shows the result obtained after applying the tables to an image of constant intensity (100). The horizontal shear is apparent in the form of jagged edges between adjacent rows.

Scanline algorithms are particularly sensitive to this form of distortion because proper filtering is applied only *along* scanlines — filtering issues *across* scanlines are not considered. Consequently, horizontal (vertical) shear is a manifestation of aliasing along

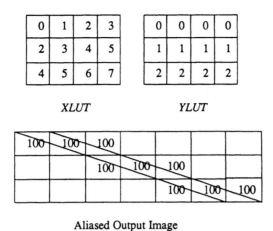

XLUT   YLUT

Aliased Output Image

Figure 7.32: Horizontal shear: Spatial LUTs and output image.

the vertical (horizontal) direction, i.e., between horizontal (vertical) scanlines. The prefiltering stage described below must be introduced to suppress these artifacts *before* the regular 2-pass algorithm is applied.

This problem is a symptom of undersampled spatial lookup tables, and the only proper solution lies in increasing the resolution of the tables by sampling the continuous mapping functions more densely. If the continuous mapping functions are no longer available to us, then new values are computed from the sparse samples by interpolation. In [Wolberg 89], linear interpolation is assumed to be adequate.

We now consider the effect of increasing the spatial resolution of *XLUT* and *YLUT*. The resulting image in Fig. 7.33 is shown to be antialiased, and clearly superior to its counterpart in Fig. 7.32. The values of 37 and 87 reflect the partial coverage of the input slivers at the output. Note that with additional upsampling, these values converge to 25 and 75, respectively. Adjacent rows are now constrained to lie within 1/2 pixel of each other.

The error constraint can be specified by the user and the spatial resolution for the lookup tables can be determined automatically. This offers us a convenient mechanism in which to control error tolerance and address the space/accuracy tradeoff. For the examples herein, both horizontal and vertical shear are restricted to one pixel.

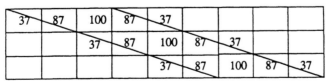

Figure 7.33: Corrected output image.

By now the reader may be wondering if the shear problems might be alleviated, as was suggested in [Catmull 80], by considering a different order of processing. While the problem may be slightly ameliorated by changing processing direction, the fundamental problem lies in undersampling the lookup tables. They are specifying an output configuration (with many long thin slivers) which, because of filtering errors, cannot be accurately realized by separable processing in any order.

7.7.4.3. Perspective

Like shear, perspective distortions may also cause problems by warping a rectangle into a triangular patch which results in significant filtering errors. In fact, if one only considers the warp determined by any three corners of an input pixel, one cannot distinguish shear from perspective projection. The latter requires knowledge of all four corners. The problem generated by perspective warping can also be solved by the same mechanism as for shears: resample the spatial lookup tables to ensure that no long thin slivers are generated. However, unlike shear, perspective also affects the bottleneck problem because, for some orders of processing, the first pass may be contractive while the second pass is expansive. This perspective bottlenecking is handled by the same mechanism as for rotations, as described below.

7.7.4.4. Rotation

In addition to jagginess due to shear and perspective, distortions are also introduced by rotation. Rotational components in the spatial transformation are the *major* source of bottleneck problems. Although all rotation angles contribute to this problem, we consider those beyond 45° to be inadequately resampled by a 2-pass algorithm. This threshold is chosen because 0° and 90° rotations can be performed exactly. If other exact image rotations were available, then the worst case error could be reduced to half the maximum separation of the angles. Local areas whose rotational components exceed 45° are recovered from the transposed results, where they obviously undergo a rotation less than 45°.

7.7.4.5. Distortion Measures

Consider scanning two scanlines jointly, labeling an adjacent pair of pixels in the first row as A, B, and the associated pair in the second row as C and D. Let (x_A, y_A), (x_B, y_B), (x_C, y_C), and (x_D, y_D) be their respective output coordinates as specified by the spatial lookup tables. These points define an output quadrilateral onto which the square input pixel is mapped. From these four points, it is possible to determine the horizontal and vertical scale factors necessary to combat aliasing due to shear and perspective distortions. It is also possible to determine if extensive bottlenecking is present. For convenience, we define

$$\Delta x_{ij} = |x_i - x_j|; \quad \Delta y_{ij} = |y_i - y_j|; \quad s_{ij} = \Delta y_{ij} / \Delta x_{ij}.$$

If AB has not rotated from the horizontal by more than 45°, then its error due to bottlenecking is considered acceptable, and we say that it remains "horizontal." Examples of quadrilaterals that satisfy this case are illustrated in Fig. 7.34. Only the vertical aliasing distortions due to horizontal shearing and/or perspective need to be considered in this case. The vertical scale factor, *vfctr*, for *XLUT* and *YLUT* is given by $vfctr = MAX(\Delta x_{AC}, \Delta x_{BD})$. Briefly, this measures the maximum deviation in the horizontal direction for a unit step in the vertical direction. To ensure an alignment error of at most ε, the image must be rescaled vertically by a factor of $vfctr/\varepsilon$. Note that the maximum *vfctr* computed over the entire image is used to upsample the spatial lookup tables.

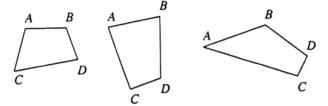

Figure 7.34: Warps where AB remains horizontal.

If AB is rotated by more than 45°, then we say that it has become "vertical" and two possibilities exist: vertical shearing/perspective or rotation. In order to consider vertical shear/perspective, the magnitude of the slope of AC is measured in relation to that of AB. If $s_{AB} \le s_{AC}$, then AC is considered to remain vertical. Examples of this condition are shown in Fig. 7.35. The horizontal scale factor, *hfctr*, for the spatial lookup tables is expressed as $hfctr = MAX(\Delta y_{AB}, \Delta y_{CD})$. Briefly stated, this measures the maximum deviation in the vertical direction for a unit step in the horizontal direction. Again, alignment error can be limited to ε by rescaling the image horizontally by a factor of $hfctr/\varepsilon$.

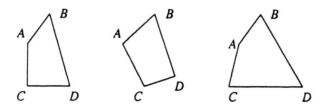

Figure 7.35: AB has rotated while AC remains vertical. Vertical shear.

If, however, angle BAC is also found to be rotated, then the entire quadrilateral $ABCD$ is considered to be bottlenecked because it has rotated and/or undergone a perspective distortion. The presence of the bottleneck problem at this pixel will require contributions to be taken from the transposed result. This case is depicted in Fig. 7.36.

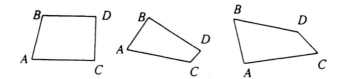

Figure 7.36: Both AB and AC have rotated. Bottleneck problem.

The values for *hfctr* and *vfctr* are computed at each pixel. The maximum values of *hfctr* /ε and *vfctr* /ε are used to scale the spatial lookup tables before they enter the 2-pass resampling stage. In this manner, the output of this stage is guaranteed to be free of aliasing due to undersampled spatial lookup tables.

7.7.4.6. Bottleneck Distortion

The bottleneck problem was described earlier as a many-to-one mapping followed by a one-to-many mapping. The extent to which the bottleneck problem becomes manifest is intimately related to the order in which the orthogonal 1-D transformations are applied. The four possible orders in which a 2-D separable transformation can be implemented are listed in Section 7.4.1.6. Of the four alternatives, we shall only consider variations (1) and (3). Although variations (2) and (4) may have impact on the extent of aliasing in the output image (see Fig. 8 of [Smith 87]), their roles may be obviated by upsampling the spatial lookup tables before they enter the 2-pass resampling stage.

A solution to the bottleneck problem thereby requires us to consider the effects which occur as an image is separably resampled with and without a preliminary image transposition stage. Unlike the Catmull-Smith algorithm that selects only one variation for the entire image, we are operating in a more general domain that may require either of the two variations over arbitrary regions of the image. This leads us to develop a local measure of bottleneck distortion that is used to determine which variation is most suitable at each output pixel. Thus alongside each resampled intensity image, another image of identical dimensions is computed to maintain estimates of the local bottleneck distortion.

A 2-pass method is introduced to compute bottleneck distortion estimates at each point. There are many possible bottleneck metrics that may be considered. The chosen metric must reflect the deviation of the output pixel from the ideal horizontal/vertical orientations that are exactly handled by the separable method. Since the bottleneck problem is largely attributed to rotation (i.e., an affine mapping), only three points are necessary to determine the distortion of each pixel. In particular, we consider points A, B, and C, as shown in the preceding figures. Let θ be the angle between AB and the horizontal axis and let ϕ be the angle between AC and the vertical axis. We wish to minimize $\cos\theta$ and $\cos\phi$ so as to have the transformed input pixels conform to the rectilinear output grid. The function $b = \cos\theta\cos\phi$ is a reasonable measure of accuracy that satisfies this criterion. This is computed over the entire image, generating a bottleneck image B_x. Image B_x reflects the fraction of each pixel in the intermediate image *not* subject to bottleneck distortion in the first pass.

The second pass resamples intermediate image B_x in the same manner as the intensity resampler, thus spreading the distortion estimates to their correct location in the final image. The result is a double-precision bottleneck-distortion image B_{xy}, with values inversely proportional to the bottleneck artifacts. The distortion computation process is repeated for the transpose of the image and spatial lookup tables, generating image B_{xy}^T.

Since the range of values in the bottleneck image are known to lie between 0 and 1, it is possible to quantize the range into N intervals for storage in a lower precision image with $\log_2 N$ bits per pixel. We point out that the measure of area is not exact. It is subject to exactly the same errors as intensity filtering.

7.7.5. Foldover Problem

Up to this point, we have been discussing our warping algorithm as though both passes resulted in only a single value for each point. Unfortunately, this is often not the case — a warped scanline can fold back upon itself.

In [Catmull 80] it was proposed that multiple framebuffers be used to store each level of the fold. While this solution may be viable for low-order warps, as considered in [Catmull 80] and [Smith 87], it may prove to be too costly for arbitrary warps where the number of potential folds may be large. Furthermore, it is often the case that the folded area may represent a small fraction of the output image. Thus, using one frame buffer per fold would be prohibitively expensive, and we seek a solution that degrades more gracefully.

If we are to allow an image to fold upon itself, we must have some means of determining which of the folds are to be displayed. The simplest mechanism, and probably the most useful, is to assume that the user will supply not only *XLUT* and *YLUT*, but also *ZLUT* to specify the output z-coordinates for each input pixel. In the first pass *ZLUT* will be processed in exactly the same way as *YLUT*, so the second pass of the intensity resampler can have access to the z-coordinates.

Given *ZLUT*, we are now faced with the problem of keeping track of the information from the folding. A naive solution might be to use a z-buffer in computing the intermediate and final images. Unfortunately, while z-buffering will work for the output of the second pass, it cannot work for the first pass because some mappings fold upon themselves in the first pass only to have some of the "hidden" part exposed by the second pass of the warp. Thus, we must find an efficient means of incorporating all the data, including the foldovers, in the intermediate image.

7.7.5.1. Representing Foldovers

Our solution is to maintain multiple columns for each column in the intermediate image. The extra columns, or layers, of space are allocated to hold information from foldovers on an as-needed basis. The advantage of this approach is that if a small area of the image undergoes folding, only a small amount of extra information is required. When the warp has folds, the intermediate image has a multi-layered structure, like that in Fig. 7.37.

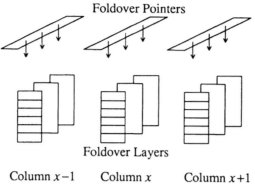

Figure 7.37: Data structure for folded warps.

While this representation is superior to multiple frame buffers, it may still be inefficient unless we allow each layer in the intermediate image to store data from many different folds (assuming that some of them have terminated and new ones were created). Thus, we reuse each foldover layer whenever possible

In addition to the actual data stored in extra layers, we also maintain a number of extra pieces of information (described below), such as various pointers to the layers, and auxiliary information about the last entry in each layer.

7.7.5.2. Tracking Foldovers

It is not sufficient to simply store all the necessary information in some structure for later processing. Given that folds do occur, there is the problem of how to filter the intermediate image. Since filtering requires all the information from one foldover layer to be accessed coherently, it is necessary to track each layer across many rows of the image. For efficiency, we desire to do this tracking by using a purely local match from one row to the next. The real difficulty in the matching is when fold layers are created, terminated, or bifurcated. We note that any "matching" must be a heuristic, since without strong assumptions about the warps, there is *no* procedure to match folds from one row to another. (The approach in [Catmull 80] assumes that the Newton-Raphson algorithm can follow the zeros of the auxiliary function H correctly, which is true only for simple auxiliary functions with limited bifurcations.)

Our heuristic solution to the matching problem uses three types of information: direction of travel when processing the layer (left or right in the row), ordering of folds within a column, and the original u-coordinate associated with each pixel in the intermediate image.

First, we constrain layers to match only those layers where the points are processed in the same order. For instance, matching between two leftward layers is allowed, but matching between leftward and rightward layers is not allowed.

Secondly, we assume the layers within a single column are partially ordered. Within each column, every folded pixel in the current row is assigned a unique number based on the order in which it was added to the foldover lists. The partial order would allow matching pixels 12345 with 1?23??4 (where the symbol ? indicates a match with a null element), but would not allow matching of 12345 with 1?43??2.

Finally, we use the u-coordinate associated with each pixel to define a distance measure between points which satisfies the above constraints. The match is done using a divide-and-conquer technique. Briefly, we first find the best match among all points, i.e., minimum distance. We then subdivide the remaining potential matches to the left and to the right of the best match, thus yielding two smaller subsets on which we reapply the algorithm. For hardware implementation, dynamic programming may be more suitable. This is a common solution for related string matching problems.

Consider a column that previously had foldover layers labeled 123456, with orientation *RLRLRL*, and original u-coordinates of 10,17,25,30,80,95. If two of these layers now disappeared leaving four layers, say *abcd*, with orientation *RLRL* and original u-coordinates of 16,20,78,101, then we would do the matching finding *abcd* matching 1256 respectively.

7.7.5.3. Storing Information from Foldovers

Once the matches are determined, we must rearrange the data so that the intensity resampler can access it in a spatially coherent manner. To facilitate this, each column in the intermediate image has a block of pointers that specify the order of the foldover layers. When the matching algorithm results in a shift in order, a different set of pointers is defined, and the valid range of the previous set is recorded. The advantage of this explicit reordering of pointers is that it allows for efficient access to the folds while processing.

We describe the process from the point of view of a single column in the intermediate image, and note that all columns are processed identically. The first encountered entry for a row goes into the base layer. For each new entry into this column, the fill pointer is advanced (using the block of pointers), and the entry is added at the bottom of the next fold layer. After we compute the "best" match, we move incorrectly stored data, reorder the layers and define a new block of pointers.

Let us continue the example from the end of the last section, where 123456 was matched to 1256. After the matching, we would then move the data, incorrectly stored in columns 3 and 4 into the appropriate location in 5 and 6. Finally, we would reorder the columns and adjust the pointer blocks to reflect the new order 125634. The columns previously labeled 34 would be marked as terminated and would be considered spares to be used in later rows if a new fold layer begins.

7.7.5.4. Intensity Resampling with Foldovers

A final aspect of the foldover problem is how it affects the 2-D intensity resampling process. The discussion above demonstrates that all the intensity values for a given column are collected in such a way that each fold layer is a separate contiguous array of spatially coherent values. Thus, the contribution of each pixel in a fold layer is obtained by standard 1-D filtering of that array.

From the coordinate resampler, we obtain $ZLUT_{xy}$, and thus, merging the foldovers is equivalent to determining which filtered pixels are visible. Given the above information, we implement a simple z-buffer algorithm, which integrates the points in front-to-back order with partial coverage calculations for antialiasing. When the accumulated area coverage exceeds 1, the integration terminates. Note that this z-buffer requires *only* a 1-D accumulator, which can be reused for each column. The result is a single intensity image combining the information from all visible folds.

7.7.6. Compositor

The compositor generates the final output image by selecting the most suitable pixels from I_{xy} and I_{xy}^T as determined by the bottleneck images B_{xy} and B_{xy}^T. A block diagram of the compositor is shown in center row of Fig. 7.30.

Bottleneck images B_{xy} and B_{xy}^T are passed through a comparator to generate bitmap image S. Also known as a *vector mask*, S is initialized according to the following rule.

$$S[x,y] = (B_{xy}[x,y] \leq B_{xy}^T[x,y])$$

Images S, I_{xy}, and I_{xy}^T are sent to the selector where I_{out} is assembled. For each position in I_{out}, the vector mask S is indexed to determine whether the pixel value should be sampled from I_{xy} or I_{xy}^T.

7.7.7. Examples

This section illustrates some examples of the algorithm. Figure 7.38 shows the final result of warping the Checkerboard and Madonna images into 360° circles. This transformation takes each row of the source image and maps it into a radial line. This corresponds directly to a mapping from the Cartesian coordinate system to the polar coordinate system, i.e., $(x, y) \rightarrow (r, \theta)$.

Figure 7.39 illustrates the output of the intensity resampler for the non-transposed and transposed processing. I_{xy} appears in Fig. 7.39a, and I_{xy}^T is shown in Fig. 7.39b. Figure 7.39c shows S, the vector mask image. S selects points from I_{xy} (white) and I_{xy}^T (black) to generate the final output image I_{out}. Gray points in S denote equal bottleneck computations from both sources. Ties are arbitrarily resolved in favor of I_{xy}^T. Finally, in Fig. 7.39d, the two spatial lookup tables $XLUT$ and $YLUT$ that defined the circular warp, are displayed as intensity images, with y increasing top-to-bottom, and x increasing left-to-right. Bright intensity values in the images of $XLUT$ and $YLUT$ denote high coordinate values. Note that if the input were to remain undistorted $XLUT$ and $YLUT$ would be

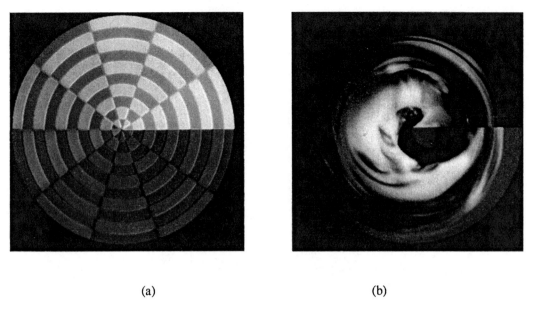

(a)   (b)

Figure 7.38: 360° warps on (a) Checkerboard and (b) Madonna.

ramps. The deviation from the ramp configuration depicts the amount of deformation which the input image undergoes.

Figure 7.40 demonstrates the effect of undersampling the spatial lookup tables. The Checkerboard texture is again warped into a circle. However, *XLUT* and *YLUT* were supplied at lower resolution. The jagginess in the results are now more pronounced.

Figure 7.41a illustrates an example of foldovers. Figure 7.41b shows *XLUT* and *YLUT*. A foldover occurs because *XLUT* is not monotonically increasing from left to right.

In Figs. 7.42a and 7.42b, the foldover regions are shown magnified (with pixel replication) to highlight the results of two different methods of rendering the final image. In Fig. 7.42a, we simply selected the closest pixels. Note that dim pixels appear at the edge of the fold as it crosses the image. This subtlety is more apparent along the fold upon the cheek. The intensity drop is due to the antialiasing filtering that correctly weighted the pixels with their area coverage along the edge. This can be resolved by integrating partially visible pixels in front-to-back order. As soon as the sum of area coverage exceeds unity, no more integration is necessary. The improved result appears in Fig. 7.42b.

Figure 7.43 shows the result of bending horizontal rows. As we scan across the rows in left-to-right order, the row becomes increasingly vertical. This is another example in which the traditional 2-pass method would clearly fail since a wide range of rotation angles are represented. A vortex warp is shown in Fig. 7.44.

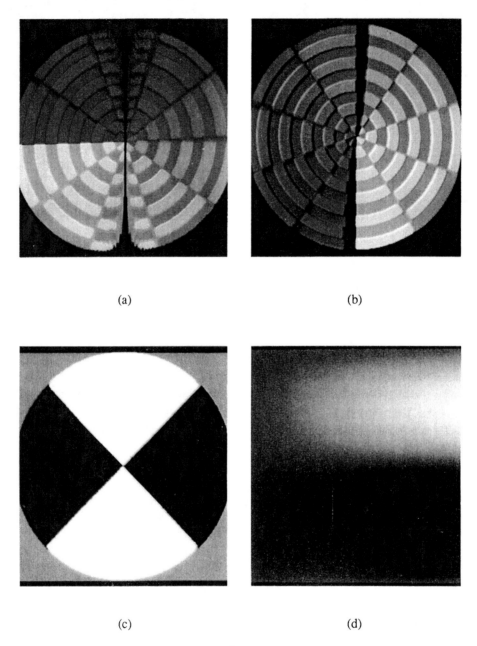

(a)   (b)

(c)   (d)

Figure 7.39: (a) I_{xy}; (b) I_{xy}^{T}; (c) S; (d) *XLUT* and *YLUT*.

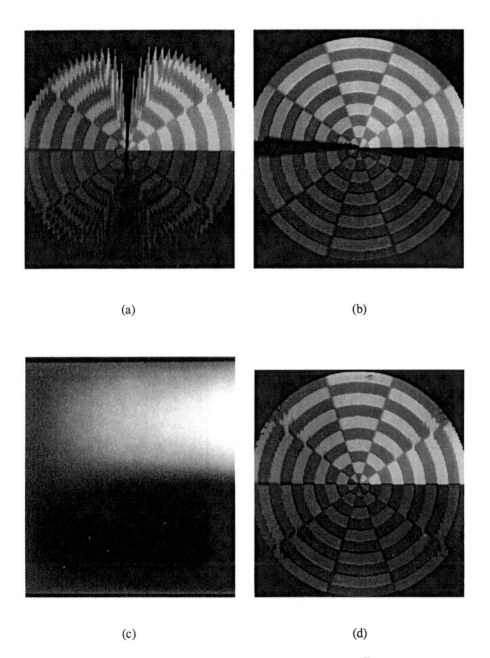

(a)   (b)

(c)   (d)

Figure 7.40: Undersampled spatial lookup tables. (a) I_{xy}; (b) I_{xy}^T; (c) Undersampled LUTs; (d) Output.

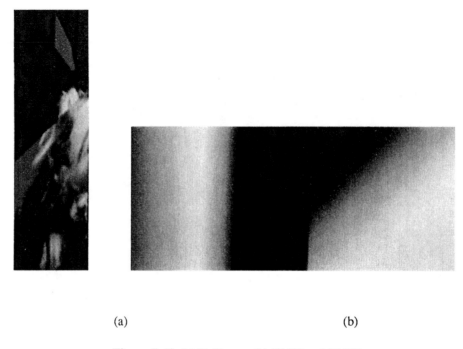

(a)   (b)

Figure 7.41: (a) Foldover; (b) *XLUT* and *YLUT*.

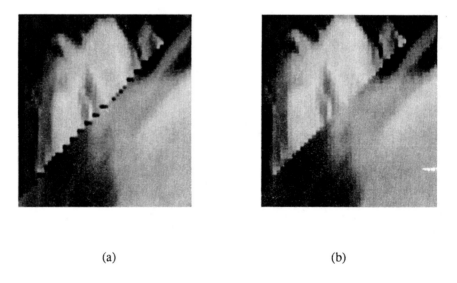

(a)   (b)

Figure 7.42: Magnified foldover. (a) No filtering. (b) Filtered result.

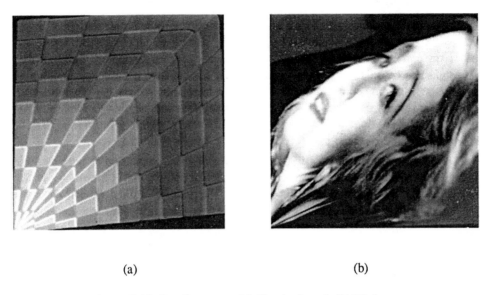

(a)   (b)

Figure 7.43: Bending rows. (a) Checkerboard; (b) Madonna.

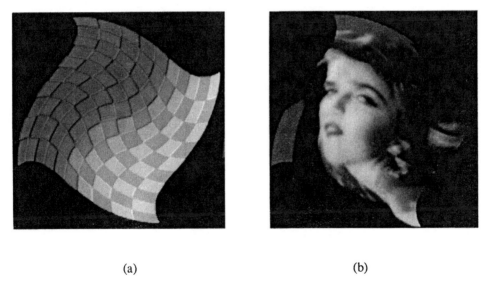

(a)   (b)

Figure 7.44: Vortex warp. (a) Checkerboard; (b) Madonna.

7.8. DISCUSSION

Scanline algorithms all share a common theme: simple interpolation, antialiasing, and data access are made possible when operating along a single dimension. Using a 2-pass transform as an example, the first pass represents a forward mapping. Since the data is assumed to be unidirectional, a single-element accumulator is sufficient for filtering purposes. This is in contrast to a full 2-D accumulator array for standard forward mappings. The second pass is actually a hybrid mapping function, requiring an inverse mapping to allow a new forward mapping to proceed. Namely, auxiliary function H must be solved before G, the second-pass forward mapping, can be evaluated.

A benefit of this approach is that clipping along one dimension is possible. For instance, there is no need to compute H for a particular column that is known in advance to be clipped. This results in some timesavings. The principal difficulty, however, is the bottleneck problem which exists as a form of aliasing. This is avoided in some applications, such as rotation, where it has been shown that no scaling is necessary in any of the 1-D passes. More generally, special attention must be provided to counteract this degradation. This has been demonstrated for the case of arbitrary spatial lookup tables.

8

EPILOGUE

Digital image warping is a subject of widespread interest. It is of practical importance to the remote sensing, medical imaging, computer vision, and computer graphics communities. Typical applications can be grouped into two classes: geometric correction and geometric distortion. Geometric correction refers to distortion compensation of imaging sensors, decalibration, and geometric normalization. This is applied to remote sensing, medical imaging, and computer vision. Geometric distortion refers to texture mapping, a powerful computer graphics tool for realistic image synthesis.

All geometric transformations have three principal components: spatial transformation, image resampling, and antialiasing. They have each received considerable attention. However, due to domain-dependent assumptions and constraints, they have rarely received uniform treatment. For instance, in remote sensing work where there is usually no severe scale change, image reconstruction is more sophisticated than antialiasing. However, in computer graphics where there is often more dramatic image compression, antialiasing plays a more significant role. This has served to obscure the single underlying set of principles that govern all geometric transformations for digital images. The goal of this book has been to survey the numerous contributions to this field, with special emphasis given to the presentation of a single coherent framework.

Various formulations of spatial transformations have been reviewed, including affine and perspective mappings, polynomial transformations, piecewise polynomial transformations, and four-corner mapping. The role of these mapping functions in geometric correction and geometric distortion was discussed. For instance, polynomial transformations were introduced to extend the class of mappings beyond affine transformations. Thus, in addition to performing the common translate, scale, rotate, and shear operations, it is possible to invert pincushion and barrel distortions. For more local control, piecewise polynomial transformations are widespread. It was shown that by establishing several correspondence points, an entire mapping function can be generated through the use of local interpolants. This is actually a surface reconstruction problem. There continues to be a great deal of activity in this area as evidenced by recent papers on multigrid relaxation algorithms to iteratively propagate constraints throughout the surface. Consequently, the tools of this field of mathematics can be applied directly to spatial transformations.

Image resampling has been shown to primarily consist of image reconstruction, an interpolation process. Various interpolation methods have been reviewed, including the (truncated) sinc function, nearest neighbor, linear interpolation, cubic convolution, 2-parameter cubic filters, and cubic splines. By analyzing the responses of their filter kernels in the frequency domain, a comparison of interpolation methods was presented. In particular, the quality of interpolation is assessed by examining the performance of the interpolation kernel in the passbands and stopbands. A review of sampling theory has been included to provide the necessary background for a comprehensive understanding of image resampling and antialiasing.

Antialiasing has recently attracted much attention in the computer graphics community. The earliest antialiasing algorithms were restrictive in terms of the preimage shape and filter kernel that they supported. For example, box filtering over rectangular preimages were common. Later developments obtained major performance gains by retaining these restrictions but permitting the number of computations to be independent of the preimage area. Subsequent improvements offered fewer restrictions at lower cost. In these instances the preimage areas were extended to ellipses and the filter kernels, now stored in lookup tables, were allowed to be arbitrary. The design of efficient filters that operate over an arbitrary input area and accommodate arbitrary filter kernels remains a great challenge.

Development of superior filters used another line of attack: advanced sampling strategies. They include supersampling, adaptive sampling, and stochastic sampling. These techniques draw upon recent results on perception and the human visual system. The suggested sampling patterns that are derived from the blue noise criteria offer promising results. Their critics, however, point to the excessive sampling densities required to reduce noise levels to unobjectionable limits. Determining minimum sampling densities which satisfy some subjective criteria requires additional work.

The final section has discussed various separable algorithms introduced to obtain large performance gains. These algorithms have been shown to apply over a wide range of transformations, including perspective projection of rectangles, bivariate patches, and superquadrics. Hardware products, such as the Ampex ADO and Quantel Mirage, are based on these techniques to produce real-time video effects for the television industry. Recent progress has been made in scanline algorithms that avoid the bottleneck problem, a degradation that is particular to the separable method. These modifications have been demonstrated on the special case of rotation and the arbitrary case of spatial lookup tables.

Despite the relatively short history of geometric transformation techniques for digital images, a great deal of progress has been made. This has been accelerated within the last decade through the proliferation of fast and cost-effective digital hardware. Algorithms which were too costly to consider in the early development of this area, are either commonplace or are receiving increased attention. Future work in the areas of reconstruction and antialiasing will most likely integrate models of the human visual system to achieve higher quality images. This has been demonstrated in a recent study of a family of filters defined by piecewise cubic polynomials, as well as recent work in stochastic sampling. Related problems that deserve attention include new adaptive filtering techniques, irregular sampling algorithms, and reconstruction from irregular samples. In addition, work remains to be done on efficient separable schemes to integrate sophisticated reconstruction and antialiasing filters into a system supporting more general spatial transformations. This is likely to have great impact on the various diverse communities which have contributed to this broad area.

Appendix 1

FAST FOURIER TRANSFORMS

The purpose of this appendix is to provide a detailed review of the Fast Fourier Transform (FFT). Some familiarity with the basic concepts of the Fourier Transform is assumed. The review begins with a definition of the discrete Fourier Transform (DFT) in Section A1.1. Directly evaluating the DFT is demonstrated there to be an $O(N^2)$ process.

The efficient approach for evaluating the DFT is through the use of FFT algorithms. Their existence became generally known in the mid-1960s, stemming from the work of J.W. Cooley and J.W. Tukey. Although they pioneered new FFT algorithms, the original work was actually discovered over 20 years earlier by Danielson and Lanczos. Their formulation, known as the Danielson-Lanczos Lemma, is derived in Section A1.2. Their recursive solution is shown to reduce the computational complexity to $O(N \log_2 N)$.

A modification of that method, the Cooley-Tukey algorithm, is given in Section A1.3. Yet another variation, the Cooley-Sande algorithm, is described in Section A1.4. These last two techniques are also known in the literature as the decimation-in-time and decimation-in-frequency algorithms, respectively. Finally, source code, written in C, is provided to supplement the discussion.

A1.1. DISCRETE FOURIER TRANSFORM

Consider an input function $f(x)$ sampled at N regularly spaced intervals. This yields a list of numbers, f_k, where $0 \le k \le N-1$. For generality, the input samples are taken to be complex numbers, i.e., having real and imaginary components. The DFT of f is defined as

$$F_n = \sum_{k=0}^{N-1} f_k e^{-i 2\pi nk/N} \qquad 0 \le n \le N-1 \qquad (A1.1.1a)$$

$$f_n = \frac{1}{N} \sum_{k=0}^{N-1} F_k e^{i 2\pi nk/N} \qquad 0 \le n \le N-1 \qquad (A1.1.1b)$$

where $i = \sqrt{-1}$. Equations (A1.1.1a) and (A1.1.1b) define the forward and inverse DFTs, respectively. Since both DFTs share the same cost of computation, we shall confine our discussion to the forward DFT and shall refer to it only as the DFT.

The DFT serves to map the N input samples of f into the N frequency terms in F. From Eq. (A1.1.1a), we see that each of the N frequency terms are computed by a linear combination of the N input samples. Therefore, the total computation requires N^2 complex multiplications and $N(N-1)$ complex additions. The straightforward computation of the DFT thereby give rise to an $O(N^2)$ process. This can be seen more readily if we rewrite Eq. (A1.1.1a) as

$$F_n = \sum_{k=0}^{N-1} f_k W^{nk} \qquad 0 \le n \le N-1 \qquad (A1.1.2)$$

where

$$W = e^{-i 2\pi/N} = \cos(-2\pi/N) + i \sin(-2\pi/N) \qquad (A1.1.3)$$

For reasons described later, we assume that

$$N = 2^r$$

where r is a positive integer. That is, N is a power of 2.

Equation (A1.1.2) casts the DFT as a matrix multiplication between the input vector f and the two-dimensional array composed of powers of W. The entries in the 2-D array, indexed by n and k, represent the N equally spaced values along a sinusoid at each of the N frequencies. Since straightforward matrix multiplication is an $O(N^2)$ process, the computational complexity of the DFT is bounded from above by this limit.

In the next section, we show how the DFT may be computed in $O(N \log_2 N)$ operations with the FFT, as originally derived over forty years ago. By properly decomposing Eq. (A1.1.1a), the reduction in proportionality from N^2 to $N \log_2 N$ multiply/add operations represents a significant saving in computation effort, particularly when N is large.

A1.2. DANIELSON-LANCZOS LEMMA

In 1942, Danielson and Lanczos derived a recursive solution for the DFT. They showed that a DFT of length $N^{\dagger}$ can be rewritten as the sum of two DFTs, each of length $N/2$, where N is an integer power of 2. The first DFT makes use of the even-numbered points of the original N; the second uses the odd-numbered points. The following proof is offered.

$$F_n = \sum_{k=0}^{N-1} f_k\, e^{-i 2\pi n k/N} \tag{A1.2.1}$$

$$= \sum_{k=0}^{(N/2)-1} f_{2k}\, e^{-i 2\pi n (2k)/N} + \sum_{k=0}^{(N/2)-1} f_{2k+1}\, e^{-i 2\pi n (2k+1)/N} \tag{A1.2.2}$$

$$= \sum_{k=0}^{(N/2)-1} f_{2k}\, e^{-i 2\pi n k/(N/2)} + W^n \sum_{k=0}^{(N/2)-1} f_{2k+1}\, e^{-i 2\pi n k/(N/2)} \tag{A1.2.3}$$

$$= F_n^e + W^n F_n^o \tag{A1.2.4}$$

Equation (A1.2.1) restates the original definition of the DFT. The summation is expressed in Eq. (A1.2.2) as two smaller summations consisting of the even- and odd-numbered terms, respectively. To properly access the data, the index is changed from k to $2k$ and $2k+1$, and the upper limit becomes $(N/2)-1$. These changes to the indexing variable and its upper limit give rise to Eq. (A1.2.3), where both sums are expressed in a form equivalent to a DFT of length $N/2$. The notation is simplified further in Eq. (A1.2.4). There, F_n^e denotes the nth component of the Fourier Transform of length $N/2$ formed from the even components of the original f, while F_n^o is the corresponding transform derived from the odd components.

The expression given in Eq. (A1.2.4) is the central idea of the Danielson-Lanczos Lemma and the decimation-in-time FFT algorithm described later. It presents a divide-and-conquer solution to the problem. In this manner, solving a problem (F_n) is reduced to solving two smaller subproblems (F_n^e and F_n^o). However, a closer look at the two sums, F_n^e and F_n^o, illustrates a potentially troublesome deviation from the original definition of the DFT: $N/2$ points of f are used to generate N points. (Recall that n in F_n^e and F_n^o is still made to vary from 0 to $N-1$). Since each of the subproblems appears to be no smaller than the original problem, this would thereby seem to be a wasteful approach. Fortunately, there exists symmetries which we exploit to reduce the computational complexity.

The first simplification is found by observing that F_n is periodic in the length of the transform. That is, given a DFT of length N, $F_{n+N} = F_n$. The proof is given below.

---

† This is also known as an N-point DFT.

$$F_{n+N} = \sum_{k=0}^{N-1} f_k e^{-i 2\pi(n+N)k/N} \qquad (A1.2.5)$$

$$= \sum_{k=0}^{N-1} f_k e^{-i 2\pi n k/N} e^{-i 2\pi N k/N}$$

$$= \sum_{k=0}^{N-1} f_k e^{-i 2\pi n k/N}$$

$$= F_n$$

In the second line of Eq. (A1.2.5), the last exponential term drops out because the exponent $-i 2\pi N k/N$ is simply an integer multiple of 2π and $e^{-i 2\pi k} = 1$. Relating this result to Eq. (A1.2.4), we note that F_n^e and F_n^o have period $N/2$. Thus,

$$F_{n+N/2}^e = F_n^e \qquad 0 \le n < N/2 \qquad (A1.2.6)$$

$$F_{n+N/2}^o = F_n^o \qquad 0 \le n < N/2$$

This permits the $N/2$ values of F_n^e and F_n^o to trivially generate the N numbers needed for F_n.

A similar simplification exists for the W^n factor in Eq. (A1.2.4). Since W has period N, the first $N/2$ values can be used to trivially generate the remaining $N/2$ values by the following relation.

$$\cos((2\pi/N)(n+N/2)) = -\cos(2\pi n/N) \qquad 0 \le n < N/2 \qquad (A1.2.7)$$

$$\sin((2\pi/N)(n+N/2)) = -\sin(2\pi n/N) \qquad 0 \le n < N/2$$

Therefore,

$$W^{n+N/2} = -W^n \qquad 0 \le n < N/2 \qquad (A1.2.8)$$

Summarizing the above results, we have

$$F_n = F_n^e + W^n F_n^e \qquad 0 \le n < N/2 \qquad (A1.2.9)$$

$$F_{n+N/2} = F_n^e - W^n F_n^o \qquad 0 \le n < N/2$$

where N is an integer power of 2.

A1.2.1. Butterfly Flow Graph

Equation (A1.2.9) can be represented by the *butterfly* flow graph of Fig. A1.1a, where the minus sign in $\pm W^n$ arises in the computation of $F_{n+N/2}$. The terms along the branches represent multiplicative factors applied to the input nodes. The intersecting node denotes a summation. For convenience, this flow graph is represented by the simplified diagram of Fig. A1.1b. Note that a butterfly performs only *one* complex multiplication ($W^n F_n^o$). This product is used in Eq. (A1.2.9) to yield F_n and $F_{n+N/2}$.

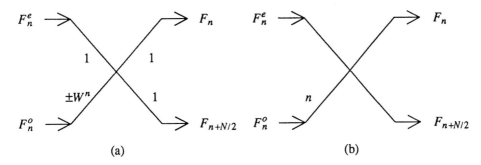

Figure A1.1: (a) Butterfly flow graph; (b) Simplified diagram.

The expansion of a butterfly flow graph in terms of the computed real and imaginary terms is given below. For notational convenience, the real and imaginary components of a complex number are denoted by the subscripts r and i, respectively. We define the following variables.

$$g = F_n^e$$
$$h = F_n^o$$
$$w_r = \cos(-2\pi n/N)$$
$$w_i = \sin(-2\pi n/N)$$

Expanding F_n, we have

$$
\begin{aligned}
F_n &= g + W^n h \qquad\qquad\qquad\qquad\qquad\qquad\text{(A1.2.10)}\\
&= [g_r + ig_i] + [w_r + iw_i][h_r + ih_i]\\
&= [g_r + ig_i] + [w_r h_r - w_i h_i + iw_r h_i + iw_i h_r]\\
&= [g_r + w_r h_r - w_i h_i] + i[g_i + w_r h_i + w_i h_r]
\end{aligned}
$$

The real and imaginary components of $W^n h$ are thus $w_r h_r - w_i h_i$ and $w_r h_i + w_i h_r$, respectively. These terms are isolated in the computation so that they may be subtracted from g_r and g_i to yield $F_{n+N/2}$ without any additional transform evaluations.

A1.2.2. Putting It All Together

The recursive formulation of the Danielson-Lanczos Lemma is demonstrated in the following example. Consider list f of 8 complex numbers labeled f_0 through f_7 in Fig. A1.2. In order to reassign the list entries with the Fourier coefficients F_n, we must evaluate F_n^e and F_n^o. As a result, two new lists are created containing the even and odd components of f. The e and o labels along the branches denote the path of even and odd components, respectively. Applying the same procedure to the newly created lists, successive halving is performed until the lowest level is reached, leaving only one element per list. The result of this recursive subdivision is shown in Fig. A1.2.

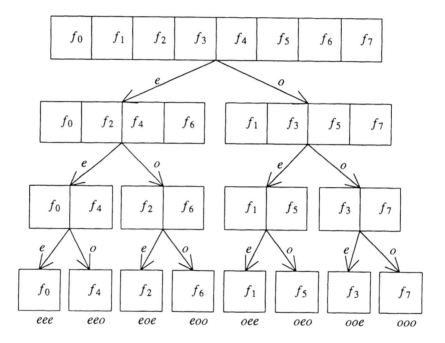

Figure A1.2: Recursive subdivision into even- and odd-indexed lists.

At this point, we may begin working our way back up the tree, building up the coefficients by using the Danielson-Lanczos Lemma given in Eq. (A1.2.9). Figure A1.3 depicts this process by using butterfly flow graphs to specify the necessary complex additions and multiplications. Note that bold lines are used to delimit lists in the figure. Beginning with the 1-element lists, the 1-point DFTs are evaluated first. Since a 1-point DFT is simply an identity operation that copies its one input number into its one output slot, the 1-element lists remain the same.

The 2-point transforms now make use of the 1-point transform results. Next, the 4-point transforms build upon the 2-point results. In this case, N is 4 and the exponent of W is made to vary from 0 to $(N/2)-1$, or 1. In Fig. A1.3, all butterfly flow graphs assume an N of 8 for the W factor. Therefore, the listed numbers are normalized accordingly. For the 4-point transform, the exponents of 0 and 1 (assuming an N of 4) become 0 and 2 to compensate for the implied N value of 8. Finally, the last step is the evaluation of an 8-point transform. In general, we combine adjacent pairs of 1-point transforms to get 2-point transforms, then combine adjacent pairs of pairs to get 4-point transforms, and so on, until the first and second halves of the whole data set are combined into the final transform.

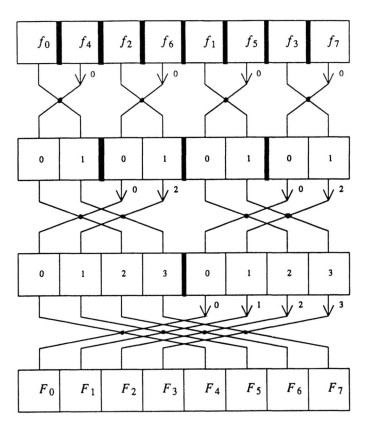

Figure A1.3: Application of the Danielson-Lanczos Lemma.

A1.2.3. Recursive FFT Algorithm

The Danielson-Lanczos Lemma provides an easily programmable method for the DFT computation. It is encapsulated in Eq. (A1.2.9) and presented in the FFT procedure given below.

Procedure FFT(N,f)
1.   **If** N equals 2, **then do**
            Begin
2.   Replace f_0 by $f_0 + f_1$ and f_1 by $f_0 - f_1$.
3.   **Return**
            End
4.   **Else do:**
            Begin
5.   Define g as a list consisting of all points of f which have an even index and h as a list containing the remaining odd points.
6.   Call FFT(N/2, g)
7.   Call FFT(N/2, h)
8.   Replace f_n by $g_n + W^n h_n$ for n=0 to $N-1$.
            End
      End

The above procedure is invoked with two arguments: N and f. N is the number of points being passed in array f. As long as N is greater than 2, f is split into two halves g and h. Array g stores those points of f having an even index, while h stores the odd-indexed points. The Fourier Transforms of these two lists are then computed by invoking the FFT procedure on g and h with length $N/2$. The FFT program will overwrite the contents of the lists with their DFT results. They are then combined in line 8 according to Eq. (A1.2.4).

The successive halving proceeds until N is equal to 2. At that point, as observed in Fig. A1.3, the exponent of W is fixed at 0. Since W^0 is 1, there is no need to perform the multiplication and the results may be determined directly (line 2).

Returning to line 8, the timesavings there arises from using the $N/2$ available elements in g and h to generate the N numbers required. This is a realization of Eq. (A1.2.9), with the real and imaginary terms given in Eq. (A1.2.10). The following segment of C code implements line 8 in the above algorithm. Note that all variables, except $N, N2$, and n, are of type *double*.

```
ang  = 0;                          /* initialize angle */
inc  = -6.2831853 / N;             /* angle increment: 2π/N */
N2   = N / 2;
for(n=0; n<N2; n++) {
        w_r = cos(ang);            /* real part of Wⁿ */
        w_i = sin(ang);            /* imaginary part of Wⁿ */
        ang += inc;                /* next angle in Wⁿ */

        a = w_r*h_r[n] – w_n*h_n[n];   /* real part of Wⁿh (Eq. A1.2.10) */
        f_r[n]    = g_r[n] + a;         /* Danielson-Lanczos Lemma (Eq. A1.2.9) */
        f_r[n+N2] = g_r[n] – a;

        a = w_i*h_r[n] + w_r*h_i[n];   /* imaginary part of Wⁿh (Eq. A1.2.10) */
        f_i[n]    = g_i[n] + a;         /* Danielson-Lanczos Lemma (Eq. A1.2.9) */
        f_i[n+N2] = g_i[n] – a;
}
```

A1.2.4. Cost of Computation

The Danielson-Lanczos Lemma, as given in Eq. (A1.2.9), can be used to calculate the cost of the computation. Let $C(N)$ be the cost for evaluating the transform of N points. Combining the transforms of N points in Eq. (A1.2.9) requires effort proportional to N because of the multiplication of the terms by W^n and the subsequent addition. If c is a constant reflecting the cost of such operations, then we have the following result for $C(N)$.

$$C(N) = 2C\left(\frac{N}{2}\right) + cN \qquad \text{(A1.2.11)}$$

This yields a recurrence relation that is known to result into an $O(N \log N)$ process. Viewed another way, since there are $\log_2 N$ levels to the recursion and cost $O(N)$ at each level, the total cost is $O(N \log_2 N)$.

A1.3. COOLEY-TUKEY ALGORITHM

The Danielson-Lanczos Lemma presented a recursive solution to computing the Fourier Transform. The role of the recursion is to subdivide the original input into smaller lists that are eventually combined according to the lemma. The starting point of the computation thus begins with the adjacent pairing of 1-point DFTs. In the preceding discussion, their order was determined by the recursive subdivision. An alternate method is available to determine their order directly, without the need for the recursive algorithm given above. This result is known as the *Cooley-Tukey*, or *decimation-in-time* algorithm.

To describe the method, we define the following notation. Let F^{ee} be the list of even-indexed terms taken from F^e. Similarly, F^{eo} is the list of odd-indexed terms taken from F^e. In general, the string of symbols in the superscript specifies the path traversed in the tree representing the recursive subdivision of the input data (Fig. A1.2). Note that the height of the tree is $\log_2 N$ and that all leaves denote 1-point DFTs that are actually elements from the input numbers. Thus, for every pattern of e's and o's, numbering $\log_2 N$ in all,

$$F^{eoeeoeo...oee} = f_n \qquad \text{for some } n \qquad (A1.3.1)$$

The problem is now to directly find which value of n corresponds to which pattern of e's and o's in Eq. (A1.3.1). The solution is surprisingly simple: reverse the pattern of e's and o's, then let $e = 0$ and $o = 1$, and the resulting binary string denotes the value of n. This works because the successive subdivisions of the data into even and odd are tests of successive low-order (least significant) bits of n. Examining Fig. A1.2, we observe that traversing successive levels of the tree along the e and o branches corresponds to successively scanning the binary value of index n from the least significant to the most significant bit. The strings appearing under the bottom row designates the traversed path.

The procedure for $N = 8$ is summarized in Table A1.1. There we see the binary indices listed next to the corresponding array elements. The first subdivision of the data into even- and odd-indexed elements amounts to testing the least significant (rightmost) bit. If that bit is 0, an even index is implied; a 1 bit designates an odd index. Subsequent subdivisions apply the same bit tests to successive index bits of higher significance. Observe that in Fig. A1.2, even-indexed lists move down the left branches of the tree. Therefore, the order in which the leaves appear from left to right indicate the sequence of 1s and 0s seen in the index while scanning in *reverse order*, from least to most significant bits.

Original Index	Original Array	Bit-reversed Index	Reordered Array
0 0 0	f_0	0 0 0	f_0
0 0 1	f_1	1 0 0	f_4
0 1 0	f_2	0 1 0	f_2
0 1 1	f_3	1 1 0	f_6
1 0 0	f_4	0 0 1	f_1
1 0 1	f_5	1 0 1	f_5
1 1 0	f_6	0 1 1	f_3
1 1 1	f_7	1 1 1	f_7

Table A1.1: Bit-reversal and array reordering for input into FFT algorithm.

The distinction between the Cooley-Tukey algorithm and the Danielson-Lanczos Lemma is subtle. In the latter, a recursive procedure is introduced in which to compute the DFT. This procedure is responsible for decimating the input signal into a sequence that is then combined, during the traversal back up the tree, to yield the transform output. In the Cooley-Tukey algorithm, though, the recursion is unnecessary since a clever bit-reversal trick is introduced to achieve the same disordered input. Furthermore, directly reordering the input in this way simplifies the bookkeeping necessary in recombining terms. Source code for the Cooley-Tukey FFT algorithm, written in C, is provided in Section A1.5.

A1.3.1. Computational Cost

The computation effort for evaluating the FFT is easily determined from this formulation. First, we observe that there are $\log_2 N$ levels of recursion necessary in computing F_n. At each level, there are $N/2$ butterflies to compute the F_n^e and F_n^o terms (see Fig. A1.3). Since each butterfly requires one complex multiplication and two complex additions, the total number of multiplications and additions is $(N/2)\log_2 N$ and $N \log_2 N$, respectively. This $O(N \log_2 N)$ process represents a considerable savings in computation over the $O(N^2)$ approach of direct evaluation. For example if $N \geq 512$, the number of multiplications is reduced to a fraction of 1 percent of that required by direct evaluation.

A1.4. COOLEY-SANDE ALGORITHM

In the Cooley-Tukey algorithm, the given data sequence is reordered according to a bit-reversal scheme before it is recombined to yield the transform output. The reordering is a consequence of the Danielson-Lanczos Lemma that calls for a recursive subdivision into a sequence of even- and odd-indexed elements.

The *Cooley-Sande* FFT algorithm, also known as the *decimation-in-frequency* algorithm, calls for recursively splitting the given sequence about its midpoint, $N/2$.

$$F_n = \sum_{k=0}^{N-1} f_k e^{-i2\pi nk/N} \tag{A1.4.1}$$

$$= \sum_{k=0}^{(N/2)-1} f_k e^{-i2\pi nk/N} + \sum_{k=N/2}^{N-1} f_k e^{-i2\pi nk/N}$$

$$= \sum_{k=0}^{(N/2)-1} f_k e^{-i2\pi nk/N} + \sum_{k=0}^{(N/2)-1} f_{k+N/2} e^{-i2\pi n(k+N/2)/N}$$

$$= \sum_{k=0}^{(N/2)-1} \left[f_k + f_{k+N/2} e^{-\pi in} \right] e^{-i2\pi nk/N}$$

Noticing that the $e^{-\pi in}$ factor reduces to $+1$ and -1 for even and odd values of n, respectively, we isolate the even and odd terms by changing n to $2n$ and $2n+1$.

$$F_{2n} = \sum_{k=0}^{(N/2)-1} \left[f_k + f_{k+N/2} \right] e^{-i2\pi(2n)k/N} \qquad 0 \le n < N/2 \tag{A1.4.2}$$

$$= \sum_{k=0}^{(N/2)-1} \left[f_k + f_{k+N/2} \right] e^{-i2\pi nk/(N/2)}$$

$$F_{2n+1} = \sum_{k=0}^{(N/2)-1} \left[f_k - f_{k+N/2} \right] e^{-i2\pi(2n+1)k/N} \qquad 0 \le n < N/2 \tag{A1.4.3}$$

$$= \sum_{k=0}^{(N/2)-1} \left[f_k - f_{k+N/2} \right] e^{-i2\pi k/N} e^{-i2\pi nk/(N/2)}$$

Thus, the even- and odd-indexed values of F are given by the DFTs of f_k^e and f_k^o where

$$f_k^e = f_k + f_{k+N/2} \tag{A1.4.4}$$

$$f_k^o = \left[f_k - f_{k+N/2} \right] W^k \tag{A1.4.5}$$

The same procedure can now be applied to f_k^e and f_k^o. This sequence is depicted in Fig. A1.4. The top row represents input list f containing 8 elements. Again, note that lists are delimited by bold lines. Regarding the butterfly notation, the lower left branches denote Eq. (A1.4.4) and the lower right branches denote Eq. (A1.4.5).

Since all the even-indexed values of F need f_k^e, a new list is created for that purpose. This is shown as the left list of the second row. Similarly, the f_k^o list is generated, appearing as the second list on that row. Of course, the list sizes diminish by a factor of two with each level since generating them makes use of f_k and $f_{k+N/2}$ to yield one element in the new list. This process of computing Eqs. (A1.4.4) and (A1.4.5) to generate new lists terminates when $N = 1$, leaving us F, the transform output, in the last row.

In contrast to the decimation-in-time FFT algorithm, in which the input is disordered but the output is ordered, the opposite is true of the decimation-in-frequency FFT algorithm. However, reordering can be easily accomplished by reversing the binary representation of the location index at the end of computation. The advantage of this algorithm is that the values of f are entered in the input array sequentially.

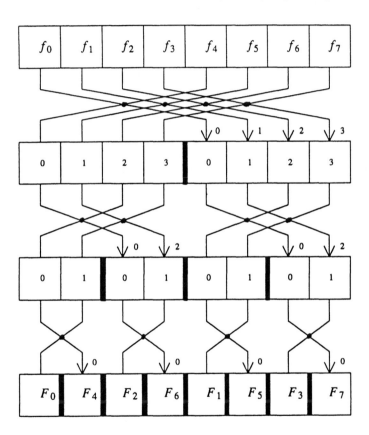

Figure A1.4: Decimation-in-frequency FFT algorithm.

A1.5. SOURCE CODE

This section provides source code for the recursive FFT procedure given in Section A1.2, as well as code for the Cooley-Tukey algorithm described in Section A1.3. The programs are written in C and make use of some library routines described below.

The data is passed to the functions in *quads*. A quad is an image control block, containing information about the image. Such data includes the image dimensions (*height* and *width*), pointers to the uninterleaved image channels (*buf* [0] ... *buf* [15]), and other necessary information. Since the complex numbers have real and imaginary components, they occupy 2 channels in the input and output quads (channels 0 and 1). A brief description of the library routines included in the listing is given below.

1) *cpqd* (q1,q2) simply copies quad q1 into q2.

2) *cpqdinfo* (q1,q2) copies the header information of q1 into q2.

3) *NEWQD* allocates a quad header. The image memory is allocated later when the dimensions are known.

4) *getqd* (h,w,*type*) returns a quad containing sufficient memory for an image with dimensions $h \times w$ and channel datatypes *type*. Note that *FFT_TYPE* is defined as 2 channels of type *float*.

5) *freeqd* (q) frees quad q, leaving it available for any subsequent *getqd* call.

6) *divconst* (q1,*num*,q2) divides the data in q1 by *num* and puts the result in q2. Note that *num* is an array of numbers used to divide the corresponding channels in q1.

7) Finally, PI2 is defined to be 2π, where $\pi = 3.141592653589793$.

A1.5.1. Recursive FFT Algorithm

```
fft1D(q1,dir,q2)        /* Fast Fourier Transform (1D)    */
int dir;                /* dir=0: forward; dir=1: inverse */
qdP q1, q2;             /* q1=input; q2=output            */
{
        int i, N, N2;
        float *r1, *i1, *r2, *i2, *ra, *ia, *rb, *ib;
        double FCTR, fctr, a, b, c, s, num[2];
        qdP qa, qb;

        cpqdinfo(q1, q2);
        N  = q1->width;
        r1 = (float *) q1->buf[0];
        i1 = (float *) q1->buf[1];
        r2 = (float *) q2->buf[0];
        i2 = (float *) q2->buf[1];

        if(N == 2) {                    /* F(0)=f(0)+f(1); F(1)=f(0)-f(1) */
                a = r1[0] + r1[1];      /* a,b needed when r1=r2 */
                b = i1[0] + i1[1];
                r2[1] = r1[0] - r1[1];
                i2[1] = i1[0] - i1[1];
                r2[0] = a;
                i2[0] = b;
        } else {
                N2 = N / 2;
                qa = getqd(1, N2, FFT_TYPE);
                qb = getqd(1, N2, FFT_TYPE);
                ra = (float *) qa->buf[0];    ia = (float *) qa->buf[1];
                rb = (float *) qb->buf[0];    ib = (float *) qb->buf[1];

                /* split list into 2 halves: even and odd */
                for(i=0; i<N2; i++) {
                        ra[i] = *r1++;          ia[i] = *i1++;
                        rb[i] = *r1++;          ib[i] = *i1++;
                }

                /* compute fft on both lists */
                fft1D(qa, dir, qa);
                fft1D(qb, dir, qb);

                /* build up coefficients */
                if(!dir)        /* forward */
                        FCTR = -PI2 / N;
                else    FCTR =  PI2 / N;
                for(fctr=i=0; i<N2; i++,fctr+=FCTR) {
                        c = cos(fctr);
                        s = sin(fctr);
                        a = c*rb[i] - s*ib[i];
```

```
            r2[i]    = ra[i] + a;
            r2[i+N2] = ra[i] - a;

            a = s*rb[i] + c*ib[i];
            i2[i]    = ia[i] + a;
            i2[i+N2] = ia[i] - a;
        }
        freeqd(qa);
        freeqd(qb);
    }
    if(dir) {        /* inverse: divide by log N */
        num[0] = num[1] = 2;
        divconst(q2, num, q2);
    }
}
```

A1.5.2. Cooley-Tukey FFT Algorithm

```
fft1D(q1, dir, q2)          /* Fast Fourier Transform (1D)        */
int dir;                    /* dir=1: forward;  dir= -1: inverse   */
qdP q1, q2;                 /* Uses bit reversal to avoid recursion */
{                           /* and trig recurrence for sin and cos */
        int i, j, logN, N, N1, NN, NN2, itr, offst;
        unsigned int a, b, msb;
        float *r1, *r2, *i1, *i2;
        double wr, wi, wpr, wpi, wtemp, theta, tempr, tempi, num[2];
        qdP qsrc;

        if(q1 == q2) {
                qsrc = NEWQD;
                cpqd(q1, qsrc);
        } else  qsrc = q1;

        cpqdinfo(q1, q2);
        r1 = (float *) qsrc->buf[0];
        i1 = (float *) qsrc->buf[1];
        r2 = (float *) q2->buf[0];
        i2 = (float *) q2->buf[1];

        N  = q1->width;
        N1 = N - 1;
        for(logN=0,i=N/2; i; logN++,i/=2);  /* # of bits sig digits in N */
        msb = LSB << (logN-1);
        for(i=1; i<N1; i++) {            /* swap all nums; ends remain fixed */
                a = i;
                b = 0;
                for(j=0; a && j<logN; j++) {
                        if(a & LSB) b |= (msb>>j);
                        a >>= 1;
                }
                /* swap complex numbers: [i] <--> [b] */
                r2[i] = r1[b];          i2[i] = i1[b];
                r2[b] = r1[i];          i2[b] = i1[i];
        }
        /* copy elements 0 and N1 since they don't swap */
        r2[0]  = r1[0];                 i2[0]  = i1[0];
        r2[N1] = r1[N1];                i2[N1] = i1[N1];

        /* NN denotes the number of points in the transform.
           It grows by a power of 2 with each iteration.
           NN2 denotes NN/2 which is used to trivially generate
           NN points from NN2 complex numbers.

           Computation of the sines and cosines of multiple
           angles is made through recurrence relations.
           wr is the cosine for the real terms; wi is sine for
```

```
        the imaginary terms.
        */
        NN = 1;
        for(itr=0; itr<logN; itr++) {
                NN2 = NN;
                NN  <<= 1;    /* NN *= 2 */

                theta = -PI2 / NN * dir;
                wtemp = sin(.5*theta);
                wpr = -2 * wtemp * wtemp;
                wpi = sin(theta);
                wr  = 1.;
                wi  = 0.;

                for(offst=0; offst<NN2; offst++) {
                        for(i=offst; i<N; i+=NN) {
                                j = i + NN2;
                                tempr = wr*r2[j] - wi*i2[j];
                                tempi = wi*r2[j] + wr*i2[j];
                                r2[j] = r2[i] - tempr;
                                r2[i] = r2[i] + tempr;
                                i2[j] = i2[i] - tempi;
                                i2[i] = i2[i] + tempi;
                        }
                        /* trigonometric recurrence */
                        wr = (wtemp=wr)*wpr - wi*wpi + wr;
                        wi = wi*wpr + wtemp*wpi + wi;
                }
        }
        if(dir == -1) {          /* inverse transform: divide by N */
                num[0] = num[1] = N;
                divconst(q2, num, q2);
        }
        if(qsrc != q1) freeqd(qsrc);
}
```

Appendix 2

INTERPOLATING CUBIC SPLINES

The purpose of this appendix is to review the fundamentals of interpolating cubic splines. We begin by defining a cubic spline in Section A2.1. Since we are dealing with interpolating splines, constraints are imposed to guarantee that the spline actually passes through the given data points. These constraints are described in Section A2.2. They establish a relationship between the known data points and the unknown coefficients used to completely specify the spline. Due to extra degrees of freedom, the coefficients may be solved in terms of the first or second derivatives. Both derivations are given in Section A2.3. Once the coefficients are expressed in terms of either the first or second derivatives, these unknown derivatives must be determined. Their solution, using one of several end conditions, is given in Section A2.4. Finally source code, written in C, is provided in Section A2.5 to implement cubic spline interpolation for uniformly and nonuniformly spaced data points.

A2.1. DEFINITION

A cubic spline $f(x)$ interpolating on the partition $x_0 < x_1 < \cdots < x_{n-1}$ is a function for which $f(x_k) = y_k$. It is a piecewise polynomial function that consists of $n-1$ cubic polynomials f_k defined on the ranges $[x_k, x_{k+1}]$. Furthermore, f_k are joined at x_k $(k=1,...,n-2)$ such that f'_k and f''_k are continuous. An example of a cubic spline passing through n data points is illustrated in Fig. A2.1.

The kth polynomial piece, f_k, is defined over the fixed interval $[x_k, x_{k+1}]$ and has the cubic form

$$f_k(x) = A_3(x - x_k)^3 + A_2(x - x_k)^2 + A_1(x - x_k) + A_0 \qquad (A2.1.1)$$

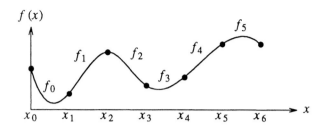

Figure A2.1: Cubic spline.

A2.2. CONSTRAINTS

Given only the data points (x_k, y_k), we must determine the polynomial coefficients, A, for each partition such that the resulting polynomials pass through the data points and are continuous in their first and second derivatives. These conditions require f_k to satisfy the following constraints

$$y_k = f_k(x_k) = A_0 \qquad \text{(A2.2.1)}$$
$$y_{k+1} = f_k(x_{k+1}) = A_3 \Delta x_k^3 + A_2 \Delta x_k^2 + A_1 \Delta x_k + A_0$$

$$y_k' = f_k'(x_k) = A_1 \qquad \text{(A2.2.2)}$$
$$y_{k+1}' = f_k'(x_{k+1}) = 3A_3 \Delta x_k^2 + 2A_2 \Delta x_k + A_1$$

$$y_k'' = f_k''(x_k) = 2A_2 \qquad \text{(A2.2.3)}$$
$$y_{k+1}'' = f_{k+1}''(x_k) = 6A_3 \Delta x_k + 2A_2$$

Note that these conditions apply at the data points (x_k, y_k). If the x_k's are defined on a regular grid, they are equally spaced and $\Delta x_k = x_{k+1} - x_k = 1$. This eliminates all of the Δx_k terms in the above equations. Consequently, Eqs. (A2.2.1) through (A2.2.3) reduce to

$$y_k = A_0 \qquad\qquad (\text{A2.2.4})$$

$$y_{k+1} = A_3 + A_2 + A_1 + A_0$$

$$y'_k = A_1 \qquad\qquad (\text{A2.2.5})$$

$$y'_{k+1} = 3A_3 + 2A_2 + A_1$$

$$y''_k = 2A_2 \qquad\qquad (\text{A2.2.6})$$

$$y''_{k+1} = 6A_3 + 2A_2$$

In the remainder of this appendix, we will refrain from making any simplifying assumptions about the spacing of the data points in order to treat the more general case.

A2.3. SOLVING FOR THE SPLINE COEFFICIENTS

The conditions given above are used to find A_3, A_2, A_1, and A_0 which are needed to define the cubic polynomial piece f_k. Isolating the coefficients, we get

$$A_0 = y_k \qquad\qquad (\text{A2.3.1})$$

$$A_1 = y'_k$$

$$A_2 = \frac{1}{\Delta x_k} \left[3\frac{\Delta y_k}{\Delta x_k} - 2y'_k - y'_{k+1} \right]$$

$$A_3 = \frac{1}{\Delta x_k^2} \left[-2\frac{\Delta y_k}{\Delta x_k} + y'_k + y'_{k+1} \right]$$

In the expressions for A_2 and A_3, $k = 0,...,n-2$ and $\Delta y_k = y_{k+1} - y_k$.

A2.3.1. Derivation of A_2

From (A2.2.1),

$$A_2 = \frac{y_{k+1} - A_3\Delta x_k^3 - y'_k\Delta x_k - y_k}{\Delta x_k^2} \qquad\qquad (\text{A2.3.2a})$$

From (A2.2.2),

$$2A_2 = \frac{y'_{k+1} - 3A_3\Delta x_k^2 - y'_k}{\Delta x_k} \qquad\qquad (\text{A2.3.2b})$$

Finally, A_2 is derived from (A2.3.2a) and (A2.3.2b)

$$\left[3 \times (A\,2.3.2a) \right] - \left[\frac{\Delta x_k}{\Delta x_k} \times (A\,2.3.2b) \right] = A_2$$

A2.3.2. Derivation of A_3

From (A2.2.1),

$$A_3 = \frac{y_{k+1} - A_2 \Delta x_k^2 - y_k' \Delta x_k - y_k}{\Delta x_k^3} \tag{A2.3.2c}$$

From (A2.2.2),

$$3A_3 = \frac{y_{k+1}' - 2A_2 \Delta x_k - y_k'}{\Delta x_k^2} \tag{A2.3.2d}$$

Finally, A_3 is derived from (A2.3.2c) and (A2.3.2d)

$$\left[\frac{\Delta x_k}{\Delta x_k} \times (A\,2.3.2d) \right] - \left[2 \times (A\,2.3.2c) \right] = A_3$$

The equations in (A2.3.1) express the coefficients of f_k in terms of x_k, y_k, x_{k+1}, y_{k+1}, (known) and y_k', y_{k+1}' (unknown). Since the expressions in Eqs. (A2.2.1) through (A2.2.3) present six equations for the four A_i coefficients, the A terms could alternately be expressed in terms of second derivatives, instead of the first derivatives given in Eq. (A2.3.1). This yields

$$A_0 = y_k \tag{A2.3.3}$$

$$A_1 = \frac{\Delta y_k}{\Delta x_k} - \frac{\Delta x_k}{6} \left[y_{k+1}'' + 2y_k'' \right]$$

$$A_2 = \frac{y_k''}{2}$$

$$A_3 = \frac{1}{6\Delta x_k} \left[y_{k+1}'' - y_k'' \right]$$

A2.3.3. Derivation of A_1 and A_3

From (A2.2.1),

$$A_1 = \frac{y_{k+1} - A_3 \Delta x_k^3 - \dfrac{y_k''}{2}\Delta x_k^2 - y_k}{\Delta x_k} \tag{A2.3.4a}$$

From (A2.2.3),

$$A_3 = \frac{y_{k+1}'' - y_k''}{6\Delta x_k} = \frac{\Delta y_k''}{6\Delta x_k} \tag{A2.3.4b}$$

Plugging Eq. (A2.3.4b) into (A2.3.4a),

$$A_1 = \frac{\Delta y_k}{\Delta x_k} - \frac{\Delta x_k}{6} \left[y_{k+1}'' - y_k'' \right] - \frac{y_k''}{2}\Delta x_k = \frac{\Delta y_k}{\Delta x_k} - \frac{\Delta x_k}{6} \left[y_{k+1}'' + 2y_k'' \right] \tag{A2.3.4c}$$

A2.4. EVALUATING THE UNKNOWN DERIVATIVES

Having expressed the cubic polynomial coefficients in terms of data points and derivatives, the unknown derivatives still remain to be determined. They are typically not given explicitly. Instead, we may evaluate them from the given constraints. Although the spline coefficients require *either* the first derivatives or the second derivatives, we shall derive both forms for the sake of completeness.

A2.4.1. First Derivatives

We begin by deriving the expressions for the first derivatives using Eqs. (A2.2.1) through (A2.2.3). Recall that the A coefficients expressed in terms of y' made use of Eqs. (A2.2.1) and (A2.2.2). We therefore use the remaining constraint, given in Eq. (A2.2.3), to express the desired relation. Constraint Eq. (A2.2.3) defines the second derivative of f_k at the endpoints of its interval. By establishing that $f''_{k-1}(x_k) = f''_k(x_k)$, we enforce the continuity of the second derivative across the intervals and give rise to a relation for the first derivatives.

$$6A_3^{k-1} \Delta x_{k-1} + 2A_2^{k-1} = 2A_2^k \qquad (A2.4.1)$$

Note that the superscripts refer to the interval of the coefficient. Plugging Eq. (A2.3.1) into Eq. (A2.4.1) yields

$$\frac{1}{\Delta x_{k-1}}\left[-12\frac{\Delta y_{k-1}}{\Delta x_{k-1}} + 6y'_{k-1} + 6y'_k\right] + \frac{1}{\Delta x_{k-1}}\left[6\frac{\Delta y_{k-1}}{\Delta x_{k-1}} - 4y'_{k-1} - 2y'_k\right] =$$

$$\frac{1}{\Delta x_k}\left[6\frac{\Delta y_k}{\Delta x_k} - 4y'_k - 2y'_{k+1}\right]$$

$$\frac{1}{\Delta x_{k-1}}\left[-6\frac{\Delta y_{k-1}}{\Delta x_{k-1}} + 2y'_{k-1} + 4y'_k\right] = \frac{1}{\Delta x_k}\left[6\frac{\Delta y_k}{\Delta x_k} - 4y'_k - 2y'_{k+1}\right]$$

After collecting the y' terms on one side, we have Eq. (A2.4.2):

$$y'_{k-1}\left[\frac{1}{\Delta x_{k-1}}\right] + y'_k\left[2\left[\frac{1}{\Delta x_{k-1}} + \frac{1}{\Delta x_k}\right]\right] + y'_{k+1}\left[\frac{1}{\Delta x_k}\right] = 3\left[\frac{\Delta y_{k-1}}{\Delta x_{k-1}^2} + \frac{\Delta y_k}{\Delta x_k^2}\right]$$

Equation (A2.4.2) yields a matrix of $n-2$ equations in n unknowns. We can reduce the need for division operations by multiplying both sides by $\Delta x_{k-1}\Delta x_k$. This gives us the following system of equations, with $1 \le k \le n-2$. For notational convenience, we let $h_k = \Delta x_k$ and $r_k = \Delta y_k / \Delta x_k$.

$$\begin{bmatrix} h_1 & 2(h_0+h_1) & h_0 & & & \\ & h_2 & 2(h_1+h_2) & h_1 & & \\ & & \cdot & \cdot & & \\ & & & \cdot & \cdot & \\ & & & h_{n-2} & 2(h_{n-3}+h_{n-2}) & h_{n-3} \end{bmatrix} \begin{bmatrix} y'_0 \\ y'_1 \\ y'_2 \\ \cdot \\ \cdot \\ y'_{n-2} \\ y'_{n-1} \end{bmatrix} = \begin{bmatrix} 3(r_0h_1 + r_1h_0) \\ 3(r_1h_2 + r_2h_1) \\ \cdot \\ \cdot \\ 3(r_{n-3}h_{n-2} - r_{n-2}h_{n-3}) \end{bmatrix}$$

When the two *end* tangent vectors y_0' and y_{n-1}' are specified, then the system of equations becomes determinable. One of several boundary conditions described later may be selected to yield the remaining two equations in the matrix.

A2.4.2. Second Derivatives

An alternate, but equivalent, course of action is to determine the spline coefficients by solving for the unknown second derivatives. This procedure is virtually identical to the approach given above. Note that while there is no particular benefit in using second derivatives rather than first derivatives, it is presented here for generality.

As before, we note that the A coefficients expressed in terms of y'' made use of Eqs. (A2.2.1) and (A2.2.3). We therefore use the remaining constraint, given in Eq. (A2.2.2), to express the desired relation. Constraint Eq. (A2.2.2) defines the first derivative of f_k at the endpoints of its interval. By establishing that $f'_{k-1}(x_k) = f'_k(x_k)$ we enforce the continuity of the first derivative across the intervals and give rise to a relation for the second derivatives.

$$3A_3^{k-1} \Delta x_{k-1}^2 + 2A_2^{k-1} \Delta x_{k-1} + A_1^{k-1} = A_1^k \qquad (A2.4.3)$$

Again, the superscripts refer to the interval of the coefficient. Plugging Eq. (A2.3.3) into Eq. (A2.4.3) yields

$$\frac{\Delta x_{k-1}}{2} \left[y_k'' - y_{k-1}'' \right] + y_{k-1}'' \Delta x_{k-1} + \left[\frac{\Delta y_{k-1}}{\Delta x_{k-1}} - \frac{\Delta x_{k-1}}{6} \left(y_k'' + 2y_{k-1}'' \right) \right] = $$
$$\left[\frac{\Delta y_k}{\Delta x_k} - \frac{\Delta x_k}{6} \left(y_{k+1}'' + 2y_k'' \right) \right]$$

After collecting the y'' terms on one side, we have

$$y_{k-1}'' \left[\Delta x_{k-1} \right] + y_k'' \left[2\Delta x_{k-1} + \Delta x_k \right] + y_{k+1}'' \left[\Delta x_k \right] = 6 \left(\frac{\Delta y_k}{\Delta x_k} - \frac{\Delta y_{k-1}}{\Delta x_{k-1}} \right) \quad (A2.4.4)$$

Equation (A2.4.4) yields the following matrix of $n-2$ equations in n unknowns. Again, for notational convenience we let $h_k = \Delta x_k$ and $r_k = \Delta y_k / \Delta x_k$.

$$\begin{bmatrix} h_0 & 2(h_0+h_1) & h_1 & & & \\ & h_1 & 2(h_1+h_2) & h_2 & & \\ & & \cdot & \cdot & & \\ & & & \cdot & \cdot & \\ & & & h_{n-3} & 2(h_{n-3}+h_{n-2}) & h_{n-2} \end{bmatrix} \begin{bmatrix} y_0'' \\ y_1'' \\ y_2'' \\ \cdot \\ \cdot \\ y_{n-2}'' \\ y_{n-1}'' \end{bmatrix} = \begin{bmatrix} 6(r_1 - r_0) \\ 6(r_2 - r_1) \\ \cdot \\ \cdot \\ 6(r_{n-2} - r_{n-3}) \end{bmatrix}$$

The system of equations becomes determinable once the boundary conditions are specified.

A2.4.3. Boundary Conditions: Free-end, Cyclic, and Not-A-Knot

A trivial choice for the boundary condition is achieved by setting $y_0'' = y_{n-1}'' = 0$. This is known as the *free-end* condition that results in *natural spline interpolation*. Since $y_0'' = 0$, we know from Eq. (A2.2.6) that $A_2 = 0$. As a result, we derive the following expression from Eq. (A2.3.1).

$$y_0' + \frac{y_1}{2} = \frac{3\Delta y_0}{2\Delta x_0} \tag{A2.4.5}$$

Similarly, since $y_{n-1}'' = 0$, $6A_3 + 2A_2 = 0$, and we derive the following expression from Eq. (A2.3.1).

$$2y_{n-2}' + 4y_{n-1}' = 6\frac{\Delta y_{n-2}}{\Delta x_{n-2}} \tag{A2.4.6}$$

Another condition is called the *cyclic* condition, where the derivatives at the end-points of the span are set equal to each other.

$$y_0' = y_{n-1}' \tag{A2.4.7}$$

$$y_0'' = y_n''$$

The boundary condition that we shall consider is the *not-a-knot* condition. This requires y''' to be continuous across x_1 and x_{n-2}. In effect, this extrapolates the curve from the adjacent interior segments [de Boor 78]. As a result, we get

$$A_3^0 = A_3^1 \tag{A2.4.8}$$

$$\frac{1}{\Delta x_0^2}\left[-2\frac{\Delta y_0}{\Delta x_0} + y_0' + y_1' \right] = \frac{1}{\Delta x_1^2}\left[-2\frac{\Delta y_1}{\Delta x_1} + y_1' + y_2' \right]$$

Replacing y_2' with an expression in terms of y_0' and y_1' allows us to remain consistent with the structure of a tridiagonal matrix already derived earlier. From Eq. (A2.4.2), we isolate y_2' and get

$$y_2' = 3\Delta x_1\left[\frac{\Delta y_0}{\Delta x_0^2} + \frac{\Delta y_1}{\Delta x_1^2} \right] - y_0'\frac{\Delta x_1}{\Delta x_0} - 2y_1'\left[\frac{\Delta x_1 + \Delta x_0}{\Delta x_0} \right] \tag{A2.4.9}$$

Substituting this expression into Eq. (A2.4.8) yields

$$y_0'\Delta x_1\left[\Delta x_0 + \Delta x_1 \right] + y_1'\left[\Delta x_0 + \Delta x_1 \right]^2 = \frac{\Delta y_0}{\Delta x_0}\left[3\Delta x_0\Delta x_1 + 2\Delta x_1^2 \right] + \frac{\Delta y_1}{\Delta x_1}\left[\Delta x_0^2 \right]$$

Similarly, the last row is derived to be

$$y_{n-2}'\left[\Delta x_{n-3} + \Delta x_{n-2} \right]^2 + y_{n-1}'\Delta x_{n-3}\left[\Delta x_{n-3} + \Delta x_{n-2} \right] =$$

$$\frac{\Delta y_{n-3}}{\Delta x_{n-3}}\left[\Delta x_{n-2}^2 \right] + \frac{\Delta y_{n-2}}{\Delta x_{n-2}}\left[3\Delta x_{n-3}\Delta x_{n-2} + 2\Delta x_{n-3}^2 \right]$$

The version of this boundary condition expressed in terms of second derivatives is left to the reader as an exercise.

Thus far we have placed no restrictions on the spacing between the data points. Many simplifications are possible if we assume that the points are equispaced, i.e., $\Delta x_k = 1$. This is certainly the case for image reconstruction, where cubic splines can be used to compute image values between regularly spaced samples. The not-a-knot boundary condition used in conjunction with the system of equations given in Eq. (A2.4.2) is shown below. To solve for the polynomial coefficients, the column vector containing the first derivatives must be solved and then substituted into Eq. (A2.3.1).

$$
\begin{bmatrix}
2 & 4 & & & & & \\
1 & 4 & 1 & & & & \\
& 1 & 4 & 1 & & & \\
& & & \cdot & & & \\
& & & & \cdot & & \\
& & & & 1 & 4 & 1 \\
& & & & & 4 & 2
\end{bmatrix}
\begin{bmatrix}
y_0' \\
y_1' \\
y_2' \\
\cdot \\
\cdot \\
y_{n-2}' \\
y_{n-1}'
\end{bmatrix}
=
\begin{bmatrix}
-5y_0 + 4y_1 + y_2 \\
3(y_2 - y_0) \\
3(y_3 - y_1) \\
\cdot \\
\cdot \\
3(y_{n-1} - y_{n-3}) \\
-y_{n-3} - 4y_{n-2} + 5y_{n-1}
\end{bmatrix}
\qquad \text{(A2.4.10)}
$$

A2.5. SOURCE CODE

Below we include two C programs for interpolating cubic splines. The first program, called *ispline*, assumes that the supplied data points are equispaced. The second program, *ispline_gen*, addresses the more general case of irregularly spaced data.

A2.5.1. Ispline

The function *ispline* takes $Y1$, a list of $len1$ numbers in double-precision, and passes an interpolating cubic spline through that data. The spline is then resampled at $len2$ equal intervals and stored in list $Y2$. It begins by computing the unknown first derivatives at each interval endpoint. It invokes the function *getYD*, which returns the first derivatives in the list YD. Along the way, function *tridiag* is called to solve the tridiagonal system of equations shown in Eq. (A2.4.10). Since each derivative is coupled only to its adjacent neighbors on both sides, the equations can be solved in linear time, i.e., $O(n)$. Once YD is initialized, it is used together with $Y1$ to compute the spline coefficients. In the interest of speed, the cubic polynomials are evaluated by using Horner's rule for factoring polynomials. This requires three additions and three multiplications per evaluated point.

```
/***********************************************************************
        Interpolating cubic spline function for equispaced points
        Input: Y1 is a list of equispaced data points with len1 entries
        Output: Y2 <- cubic spline sampled at len2 equispaced points
 ***********************************************************************/
ispline(Y1,len1,Y2,len2)
double *Y1, *Y2;
int len1, len2;
{
        int i, ip, oip;
        double *YD, A0, A1, A2, A3, x, p, fctr;

        /* compute 1st derivatives at each point -> YD */
        YD = (double *) calloc(len1, sizeof(double));
        getYD(Y1,YD,len1);

        /*
         * p is real-valued position into spline
         * ip is interval's left endpoint (integer)
         * oip is left endpoint of last visited interval
         */
        oip = -1;               /* force coefficient initialization */
        fctr = (double) (len1-1) / (len2-1);
        for(i=p=ip=0; i < len2; i++) {
                /* check if in new interval */
                if(ip != oip) {
                        /* update interval */
                        oip = ip;

                        /* compute spline coefficients */
                        A0 = Y1[ip];
                        A1 = YD[ip];
                        A2 = 3.0*(Y1[ip+1]-Y1[ip]) - 2.0*YD[ip] - YD[ip+1];
                        A3 = -2.0*(Y1[ip+1]-Y1[ip]) + YD[ip] + YD[ip+1];
                }
                /* use Horner's rule to calculate cubic polynomial */
                x = p - ip;
                Y2[i] = ((A3*x + A2)*x + A2)*x + A0;

                /* increment pointer */
                ip = (p += fctr);
        }
        cfree((char *) YD);
}
```

A2.5.2. Ispline_gen

The function *ispline_gen* takes the data points in $(X1, Y1)$, two lists of *len* 1 numbers, and passes an interpolating cubic spline through that data. The spline is then resampled at *len* 2 positions and stored in $Y2$. The resampling locations are given by $X2$. The function assumes that $X2$ is monotonically increasing and lies withing the range of numbers in $X1$.

As before, we begin by computing the unknown first derivatives at each interval endpoint. The function *getYD_gen* is then invoked to return the first derivatives in the list *YD*. Along the way, function *tridiag_gen* is called to solve the tridiagonal system of equations given in Eq. (A2.4.2). Once *YD* is initialized, it is used together with $Y1$ to compute the spline coefficients. Note that in this general case, additional consideration must now be given to determine the polynomial interval in which the resampling point lies.

```
/******************************************************************
        Interpolating cubic spline function for irregularly-spaced points
        Input:  Y1 is a list of irregular data points (len1 entries)
                Their x-coordinates are specified in X1
        Output: Y2 <- cubic spline sampled according to X2 (len2 entries)
                Assume that X1,X2 entries are monotonically increasing
 ******************************************************************/
ispline_gen(X1,Y1,len1,X2,Y2,len2)
double *X1, *Y1, *X2, *Y2;
int len1, len2;
{
        int i, j;
        double *YD, A0, A2, A2, A3, x, dx, dy, p1, p2, p3;

        /* compute 1st derivatives at each point -> YD */
        YD = (double *) calloc(len1, sizeof(double));
        getYD_gen(X1,Y1,YD,len1);

        /* error checking */
        if(X2[0] < X1[0] || X2[len2-1] > X1[len1-1]) {
                fprintf(stderr,"ispline_gen: Out of range0);
                exit();
        }

        /*
         * p1 is left endpoint of interval
         * p2 is resampling position
         * p3 is right endpoint of interval
         * j  is input index for current interval
         */
        p3 = X2[0] - 1;         /* force coefficient initialization */
        for(i=j=0; i < len2; i++) {
                /* check if in new interval */
                p2 = X2[i];
```

```
if(p2 > p3) {
        /* find the interval which contains p2 */
        for(; j<len1 && p2>X1[j]; j++);
        if(p2 < X1[j]) j--;
        p1 = X1[j];          /* update left  endpoint */
        p3 = X1[j+1];        /* update right endpoint */

        /* compute spline coefficients */
        dx = 1.0 / (X1[j+1] - X1[j]);
        dy = (Y1[j+1] - Y1[j]) * dx;
        A0 = Y1[j];
        A2 = YD[j];
        A2 = dx * (3.0*dy - 2.0*YD[j] - YD[j+1]);
        A3 = dx*dx * (-2.0*dy + YD[j] + YD[j+1]);
    }
    /* use Horner's rule to calculate cubic polynomial */
    x = p2 - p1;
    Y2[i] = ((A3*x + A2)*x + A1)*x + A0;
    }
    cfree((char *) YD);
}
```

```
/*********************************************************************
        YD <- Computed 1st derivative of data in X,Y (len entries)
        The not-a-knot boundary condition is used
*********************************************************************/
getYD_gen(X,Y,YD,len)
double *X, *Y, *YD;
int len;
{
        int i;
        double h0, h1, r0, r1, *A, *B, *C;

        /* allocate memory for tridiagonal bands A,B,C */
        A = (double *) calloc(len, sizeof(double));
        B = (double *) calloc(len, sizeof(double));
        C = (double *) calloc(len, sizeof(double));

        /* init first row data */
        h0 = X[1] - X[0];           h1 = X[2] - X[1];
        r0 = (Y[1] - Y[0]) / h0;    r1 = (Y[2] - Y[1]) / h1;
        B[0] = h1 * (h0+h1);
        C[0] = (h0+h1) * (h0+h1);
        YD[0] = r0*(3*h0*h1 + 2*h1*h1) + r1*h0*h0;

        /* init tridiagonal bands A, B, C, and column vector YD */
        /* YD will later be used to return the derivatives */
        for(i = 1; i < len-1; i++) {
                h0 = X[i] - X[i-1];          h1 = X[i+1] - X[i];
                r0 = (Y[i] - Y[i-1]) / h0;   r1 = (Y[i+1] - Y[i]) / h1;
                A[i] = h1;
                B[i] = 2 * (h0+h1);
                C[i] = h0;
                YD[i] = 3 * (r0*h1 + r1*h0);
        }

        /* last row */
        A[i] = (h0+h1) * (h0+h1);
        B[i] = h0 * (h0+h1);
        YD[i] = r0*h1*h1 + r1*(3*h0*h1 + 2*h0*h0);

        /* solve for the tridiagonal matrix: YD=YD*inv(tridiag matrix) */
        tridiag_gen(A,B,C,YD,len);

        cfree((char *) A);
        cfree((char *) B);
        cfree((char *) C);
}

/*********************************************************************
        Gauss Elimination with backsubstitution for general
        tridiagonal matrix with bands A,B,C and column vector D.
```

```
·····························································································/
tridiag_gen(A,B,C,D,len)
double *A, *B, *C, *D;
int len;
{
        int i;
        double b, *F;

        F = (double *) calloc(len, sizeof(double));

        /* Gauss elimination; forward substitution */
        b = B[0];
        D[0] = D[0] / b;
        for(i = 1; i < len; i++) {
                F[i] = C[i-1] / b;
                b = B[i] - A[i]*F[i];
                if(b == 0) {
                        fprintf(stderr,"getYD_gen: divide-by-zero0);
                        exit();
                }
                D[i] = (D[i] - D[i-1]*A[i]) / b;
        }

        /* backsubstitution */
        for(i = len-2; i >= 0; i--) D[i] -= (D[i+1] * F[i+1]);

        cfree((char *) F);
}
```

Appendix 3

FORWARD DIFFERENCE METHOD

The method of forward differences is used to simplify the computation of polynomials. It basically extends the incremental evaluation, as used in scanline algorithms, to higher-order functions, e.g., quadratic and cubic polynomials. We find use for this in Chapter 7 where, for example, perspective mappings are approximated without costly division operations. In this appendix, we derive the forward differences for quadratic and cubic polynomials. The method is then generalized to polynomials of arbitrary degree.

The (first) forward difference of a function $f(x)$ is defined as

$$\Delta f(x) = f(x+\delta) - f(x), \qquad \delta > 0 \qquad (A3.1)$$

It is used together with $f(x)$, the value at the current position, to determine $f(x+\delta)$, the value at the next position. In our application, $\delta = 1$, denoting a single pixel step size. We shall assume this value for δ in the discussion that follows.

For simplicity, we begin by considering the forward difference method for linear interpolation. In this case, the first forward difference is simply the slope of the line passing through two supplied function values. That is, $\Delta f(x) = a_1$ for the function $f(x) = a_1 x + a_0$. We have already seen it used in Section 7.2 for Gouraud shading, whereby the intensity value at position $x+1$ along a scanline is computed by adding $\Delta f(x)$ to $f(x)$. Surprisingly, this approach readily lends itself to higher-order interpolants. The only difference, however, is that $\Delta f(x)$ is itelf subject to update. That update is driven by a second increment, known as the second forward difference. The extent to which these increments are updated is based on the degree of the polynomial being evaluated. In general, a polynomial of degree N requires N forward differences.

We now describe forward differencing for evaluating quadratic polynomials of the form

$$f(x) = a_2 x^2 + a_1 x + a_0 \qquad (A3.2)$$

The first forward difference for $f(x)$ is expressed as

$$\Delta f(x) = f(x+1) - f(x) \tag{A3.3}$$

$$= \left[a_2(x+1)^2 + a_1(x+1) + a_0 \right] - \left[a_2 x^2 + a_1 x + a_0 \right]$$

$$= a_2(2x+1) + a_1$$

Thus, $\Delta f(x)$ is a linear expression. If we apply forward differences to $\Delta f(x)$, we get

$$\Delta^2 f(x) = \Delta(\Delta f(x)) \tag{A3.4}$$

$$= \Delta f(x+1) - \Delta f(x)$$

$$= \left[a_2(2[x+1]+1) + a_1 \right] - \left[a_2(2x+1) + a_1 \right]$$

$$= 2a_2$$

Since $\Delta^2 f(x)$ is a constant, there is no need for further terms. The second forward difference is used at each iteration to update the first forward difference which, in turn, is added to the latest result to compute the new value. Each loop in the iteration can be rewritten as

$$f(x+1) = f(x) + \Delta f(x) \tag{A3.5}$$

$$\Delta f(x+1) = \Delta f(x) + \Delta^2 f(x)$$

If computation begins at $x = 0$, then the basis for the iteration is given by f, Δf, and $\Delta^2 f$ evaluated at $x = 0$. Given these three values, the second-degree polynomial can be evaluated from 0 to *lastx* using the following C code.

```
for(x = 0; x < lastx; x++) {
    f[x+1] = f[x] + Δf;        /* compute next point */
    Δf += Δ²f;                 /* update 1st forward difference */
}
```

Notice that Δf is subject to update by $\Delta^2 f$, but the latter term remains constant throughout the iteration.

A similar derivation is given for cubic polynomials. However, an additional forward difference constant must be incorporated due to the additional polynomial degree. For a third-degree polynomial of the form

$$f(x) = a_3 x^3 + a_2 x^2 + a_1 x + a_0 \tag{A3.6}$$

the first forward difference is

$$\Delta f(x) = f(x+1) - f(x) \tag{A3.7}$$

$$= \left[a_3(t+1)^3 + a_2(t+1)^2 + a_1(t+1) + a_0 \right] - \left[a_3 x^3 + a_2 x^2 + a_1 x + a_0 \right]$$

$$= 3a_3 x^2 + (3a_3 + 2a_2)x + a_3 + a_2 + a_1$$

Since $\Delta f(x)$ is a second-degree polynomial, two more forward difference terms are derived. They are

$$\Delta^2 f(x) = \Delta(\Delta f(x)) = 6a_3 x + 6a_3 + 2a_2 \tag{A3.8}$$

$$\Delta^3 f(x) = \Delta(\Delta^2 f(x)) = 6a_3$$

The use of forward differences to evaluate a cubic polynomial between 0 and *lastx* is demonstrated in the following C code.

```
for(x = 0; x < lastx; x++) {
    f[x+1] = f[x] + Δf;        /* compute next point */
    Δf   += Δ²f;                /* update 1st forward difference */
    Δ²f += Δ³f;                 /* update 2nd forward difference */

}
```

In contrast to the earlier example, this case has an additional forward difference term that must be updated, i.e., $\Delta^2 f$. Even so, this method offers the benefit of computing a third-degree polynomial with only three additions per value. An alternate approach, using Horner's rule for factoring polynomials, requires three additions *and* three multiplications. This makes forward differences the method of choice for evaluating polynomials.

The forward difference approach for cubic polynomials is depicted in Fig. A3.1. The basis of the entire iteration is shown in the top row. For consistency with our discussion of this method in Chapter 7, texture coordinates are used as the function values. Thus, we begin with u_0, Δu_0, and $\Delta^2 u_0$ defined for position x_0, where the subscripts refer to the position along the scanline.

In order to compute our next texture coordinate at $x = 1$, we add Δu_0 to u_0. This is denoted by the arrows that are in contact with u_0. Note that diagonal arrows denote addition, while vertical arrows refer to the computed sum. Therefore, u_1 is the result of adding Δu_0 to u_0. The following coordinate, u_2, is the sum of u_1 and Δu_1. The latter term is derived from Δu_0 and $\Delta^2 u_0$. This regular structure collapses nicely into a compact iteration, as demonstrated by the short programs given earlier.

Higher-order polynomials are handled by adding more forward difference terms. This corresponds to augmenting Fig. A3.1 with additional columns to the right. The order of computation is from the left to right. That is, the summations corresponding to the diagonal arrows are executed beginning from the left column. This gives rise to the adjacent elements directly below. Those elements are then combined in similar fashion. This cycle continues until the last diagonal is reached, denoting that the entire span of points has been evaluated.

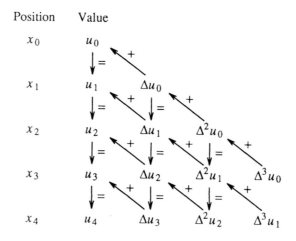

Figure A3.1: Forward difference method.

REFERENCES

[Abdou 82]   Abdou, Ikram E. and Kwan Y. Wong, "Analysis of Linear Interpolation Schemes for Bi-Level Image Applications," *IBM J. Res. Develop.*, vol. 26, no. 6, pp. 667-680, November 1982.

[Abram 85]   Abram, Greg, Lee Westover, and Turner Whitted, "Efficient Alias-free Rendering Using Bit-Masks and Look-Up Tables," *Computer Graphics*, (SIGGRAPH '85 Proceedings), vol. 19, no. 3, pp. 53-59, July 1985.

[Akima 78]   Akima, H., "A Method of Bivariate Interpolation and Smooth Surface Fitting for Irregularly Distributed Data Points," *ACM Trans. Math. Software*, vol. 4, pp. 148-159, 1978.

[Akima 84]   Akima, H., "On Estimating Partial Derivatives for Bivariate Interpolation of Scattered Data," *Rocky Mountain J. Math.*, vol. 14, pp. 41-52, 1984.

[Anderson 90] Anderson, Scott E. and Mark A.Z. Dippe, "A Hybrid Approach to Facial Animation," ILM Technical Memo #1026, Computer Graphics Department, Lucasfilm Ltd., 1990.

[Andrews 76] Andrews, Harry C. and Claude L. Patterson III, "Digital Interpolation of Discrete Images," *IEEE Trans. Computers*, vol. C-25, pp. 196-202, 1977.

[Antoniou 79] Antoniou, Andreas, *Digital Filters: Analysis and Design*, McGraw-Hill, New York, 1979.

[Atteia 66]   Atteia, M., "Existence et determination des fonctions splines a plusieurs variables," *C. R. Acad. Sci. Paris*, vol. 262, pp. 575-578, 1966.

[Ballard 82] Ballard, Dana and Christopher Brown, *Computer Vision*, Prentice-Hall, Englewood Cliffs, NJ, 1982.

[Barnhill 77] Barnhill, Robert E., "Representation and Approximation of Surfaces," *Mathematical Software III*, Ed. by J.R. Rice, Academic Press, London, pp. 69-120, 1977.

[Bennett 84a]   Bennett, Phillip P. and Steven A. Gabriel, "Spatial Transformation System Including Key Signal Generator," U.S. Patent 4,463,372, Ampex Corp., July 31, 1984.

[Bennett 84b]   Bennett, Phillip P. and Steven A. Gabriel, "System for Spatially Transforming Images," U.S. Patent 4,472,732, Ampex Corp., Sep. 18, 1984.

[Bergland 69]   Bergland, Glenn D., "A Guided Tour of the Fast Fourier Transform," *IEEE Spectrum*, vol. 6, pp. 41-52, July 1969.

[Bernstein 71]   Bernstein, Ralph and Harry Silverman, "Digital Techniques for Earth Resource Image Data Processing," *Proc. Amer. Inst. Aeronautics and Astronautics 8th Annu. Meeting*, vol. C21, AIAA paper no. 71-978, October 1971.

[Bernstein 76]   Bernstein, Ralph, "Digital Image Processing of Earth Observation Sensor Data," *IBM J. Res. Develop.*, vol. 20, pp. 40-57, January 1976.

[Bier 86]   Bier, Eric A. and Ken R. Sloan, Jr., "Two-Part Texture Mappings," *IEEE Computer Graphics and Applications*, vol. 6, no. 9, pp. 40-53, September 1986.

[Bizais 83]   Bizais, Y., I.G. Zubal, R.W. Rowe, G.W. Bennett, and A.B. Brill, "2-D Fitting and Interpolation Applied to Image Distortion Analysis," *Pictorial Data Analysis*, Ed. by R.M. Haralick, NATO ASI Series, vol. F4, 1983, pp. 321-333.

[Blake 87]   Blake, Andrew and Andrew Zisserman, *Visual Reconstruction*, MIT Press, Cambridge, MA, 1987.

[Blinn 76]   Blinn, James F. and Martin E. Newell, "Texture and Reflection in Computer Generated Images," *Comm. ACM*, vol. 19, no. 10, pp. 542-547, October 1976.

[Boult 86a]   Boult, Terrance E., "Visual Surface Reconstruction Using Sparse Depth Data," *Proc. IEEE Conference on Computer Vision and Pattern Recognition*, pp. 68-76, June 1986.

[Boult 86b]   Boult, Terrance E., *Information Based Complexity: Applications in Nonlinear Equations and Computer Vision*, Ph.D. Thesis, Dept. of Computer Science, Columbia University, NY, 1986.

[Braccini 80]   Braccini, Carlo and Giuseppe Marino, "Fast Geometrical Manipulations of Digital Images," *Computer Graphics and Image Processing*, vol. 13, pp. 127-141, 1980.

[Bracewell 86]   Bracewell, Ron, *The Fourier Transform and Its Applications*, McGraw-Hill, NY, 1986.

[Briggs 74]   Briggs, I.C., "Machine Contouring Using Minimum Curvature," *Geophysics*, vol. 39, pp. 39-48, 1974.

[Brigham 88]   Brigham, E. Oran, *The Fast Fourier Transform and Its Applications*, Prentice-Hall, Englewood Cliffs, NJ, 1988.

[Burt 88]   Burt, Peter J., "Moment Images, Polynomial Fit Filters, and the Problem of Surface Interpolation," *Proc. IEEE Conference on Computer Vision and Pattern Recognition*, pp. 144-152, June 1988.

[Butler 87]   Butler, David A. and Patricia K. Pierson, "Correcting Distortion in Digital Images," *Proc. Vision '87*, pp. 6-49-6-69, June 1987.

[Caelli 81]   Caelli, Terry, *Visual Perception: Theory and Practice*, Pergamon Press, Oxford, 1981.

[Casey 71]   Casey, R.G. and M.A. Wesley, "Parallel Linear Transformations on Two-Dimensional Binary Images," *IBM Technical Disclosure Bulletin*, vol. 13, no. 11, pp. 3267-3268, April 1971.

[Catmull 74]   Catmull, Edwin, *A Subdivision Algorithm for Computer Display of Curved Surfaces*, Ph.D. Thesis, Dept. of Computer Science, University of Utah, Tech. Report UTEC-CSc-74-133, December 1974.

[Catmull 80]   Catmull, Edwin and Alvy Ray Smith, "3-D Transformations of Images in Scanline Order," *Computer Graphics*, (SIGGRAPH '80 Proceedings), vol. 14, no. 3, pp. 279-285, July 1980.

[Chen 88]   Chen, Yu-Tse, P. David Fisher, and Michael D. Olinger, "The Application of Area Antialiasing on Raster Image Displays," *Graphics Interface '88*, pp. 211-216, 1988.

[Clough 65]   Clough, R.W. and J.L. Tocher, "Finite Element Stiffness Matrices for Analysis of Plates in Bending," *Proc. Conf. on Matrix Methods in Structural Mechanics*, pp. 515-545, 1965.

[Cochran 67]   Cochran, W.T., Cooley, J.W., *et al.*, "What is the Fast Fourier Transform?," *IEEE Trans. Audio and Electroacoustics*, vol. AU-15, no. 2, pp. 45-55, 1967.

[Cook 84]   Cook, Robert L., Tom Porter, and Loren Carpenter, "Distributed Ray Tracing," *Computer Graphics*, (SIGGRAPH '84 Proceedings), vol. 18, no. 3, pp. 137-145, July 1984.

[Cook 86]   Cook, Robert L., "Stochastic Sampling in Computer Graphics," *ACM Trans. on Graphics*, vol. 5, no. 1, pp. 51-72, January 1986.

[Cooley 65]   Cooley, J.W. and Tukey, J.W., "An Algorithm for the Machine Calculation of Complex Fourier Series," *Math. Comp.*, vol. 19, pp. 297-301, April 1965.

[Cooley 67a]   Cooley, J.W., Lewis, P.A.W., and Welch P.D., "Historical Notes on the Fast Fourier Transform," *IEEE Trans. Audio and Electroacoustics*, vol. AU-15, no. 2, pp. 76-79, 1967.

[Cooley 67b]   Cooley, J.W., Lewis, P.A.W., and Welch P.D., "Application of the Fast Fourier Transform to Computation of Fourier Integrals," *IEEE Trans.*

Audio and *Electroacoustics*, vol. AU-15, no. 2, pp. 79-84, 1967.

[Cooley 69]   Cooley, J.W., Lewis, P.A.W., and Welch P.D., "The Fast Fourier Transform and Its Applications," *IEEE Trans. Educ.*, vol. E-12, no. 1, pp. 27-34, 1969.

[Crow 77]   Crow, Frank C., "The Aliasing Problem in Computer-Generated Shaded Images," *Comm. ACM*, vol. 20, no. 11, pp. 799-805, November 1977.

[Crow 81]   Crow, Frank C., "A Comparison of Antialiasing Techniques," *IEEE Computer Graphics and Applications*, vol. 1, no. 1, pp. 40-48, January 1981.

[Crow 84]   Crow, Frank C., "Summed-Area Tables for Texture Mapping," *Computer Graphics*, (SIGGRAPH '84 Proceedings), vol. 18, no. 3, pp. 207-212, July 1984.

[Cutrona 60]   Cutrona, L.J., E.N. Leith, C.J. Palermo, and L.J. Porcello, "Optical Data Processing and Filtering Systems," *IRE Trans. Inf. Theory*, vol. IT-6, pp. 386-400, 1960.

[Danielson 42]   Danielson, G.C. and Lanczos, C., "Some Improvements In Practical Fourier Analysis and Their Application to X-Ray Scattering from Liquids," *J. Franklin Institute*, vol. 233, pp. 365-380 and 435-452, 1942.

[de Boor 78]   de Boor, Carl, *A Practical Guide to Splines*, Springer-Verlag, NY, 1978.

[De Floriani 87]De Floriani, Leila, "Surface Representations Based on Triangular Grids," *Visual Computer*, vol. 4, no. 3, pp. 27-50, 1987.

[Dippe 85a]   Dippe, Mark A.Z., *Antialiasing in Computer Graphics*, Ph.D. Thesis, Dept. of CS, U. of California at Berkeley, 1985.

[Dippe 85b]   Dippe, Mark A.Z. and Erling H. Wold, "Antialiasing Through Stochastic Sampling," *Computer Graphics*, (SIGGRAPH '85 Proceedings), vol. 19, no. 3, pp. 69-78, July 1985.

[Dongarra 79]   Dongarra, J.J., et al., *LINPACK User's Guide*, Society for Industrial and Applied Mathematics (SIAM), Philadelphia, PA, 1979.

[Duchon 76]   Duchon, J., "Interpolation des Fonctions de Deux Variables Suivant le Principe de la Flexion des Plaques Minces," *R.A.I.R.O Analyse Numerique*, vol. 10, pp. 5-12, 1976.

[Duchon 77]   Duchon, J., "Splines Minimizing Rotation-Invariant Semi-Norms in Sobolev Spaces," *Constructive Theory of Functions of Several Variables*, A. Dodd and B. Eckmann (ed.), Springer-Verlag, Berlin, pp. 85-100, 1977.

[Dungan 78]   Dungan, W. Jr., A. Stenger, and G. Sutty, "Texture Tile Considerations for Raster Graphics," *Computer Graphics*, (SIGGRAPH '78 Proceedings), vol. 12, no. 3, pp. 130-134, August 1978.

[Edelsbrunner 87]

Edelsbrunner, Herbert, *Algorithms in Combinatorial Geometry*,

Springer-Verlag, New York, 1987.

[Fant 86] Fant, Karl M., "A Nonaliasing, Real-Time Spatial Transform Technique," *IEEE Computer Graphics and Applications*, vol. 6, no. 1, pp. 71-80, January 1986. See also "Letters to the Editor" in vol. 6, no. 3, pp. 66-67, March 1986 and vol. 6, no. 7, pp. 3,8, July 1986.

[Fant 89] Fant, Karl M., "Nonaliasing Real-Time Spatial Transform Image Processing System," U.S. Patent 4,835,532, Honeywell Inc., May 30, 1989.

[Faux 79] Faux, Ivor D. and Michael J. Pratt, *Computational Geometry for Design and Manufacture*, Ellis Horwood Ltd., Chichester, England, 1979.

[Feibush 80] Feibush, Elliot A., Marc Levoy, and Robert L. Cook, "Synthetic Texturing Using Digital Filters," *Computer Graphics*, (SIGGRAPH '80 Proceedings), vol. 14, no. 3, pp. 294-301, July 1980.

[Fisher 88] Fisher, Timothy E. and Richard D. Juday, "A Programmable Video Image Remapper," *Proc. SPIE Digital and Optical Shape Representation and Pattern Recognition*, vol. 938, pp. 122-128, 1988.

[Fiume 83] Fiume, Eugene and Alain Fournier, "A Parallel Scan Conversion Algorithm with Anti-Aliasing for a General-Purpose Ultracomputer," *Computer Graphics*, (SIGGRAPH '83 Proceedings), vol. 17, no. 3, pp. 141-150, July 1983.

[Fiume 87] Fiume, Eugene, Alain Fournier, and V. Canale, "Conformal Texture Mapping," *Eurographics '87*, pp. 53-64, September 1987.

[Floyd 75] Floyd, R.W. and L. Steinberg, "Adaptive Algorithm for Spatial Grey Scale," *SID Intl. Sym. Dig. Tech. Papers*, pp. 36-37, 1975.

[Foley 90] Foley, James D., Andries Van Dam, Steven K. Feiner, and John F. Hughes, *Computer Graphics: Principles and Practice*, 2nd Ed., Addison-Wesley, Reading, MA, 1990.

[Fournier 88] Fournier, Alain and Eugene Fiume, "Constant-Time Filtering with Space-Variant Kernels," (SIGGRAPH '88 Proceedings), vol. 22, no. 4, pp. 229-238, August 1988.

[Franke 79] Franke, R., "A Critical Comparison of Some Methods for Interpolation of Scattered Data," *Naval Postgraduate School Technical Report*, NPS-53-79-003, 1979.

[Fraser 85] Fraser, Donald, Robert A. Schowengerdt, and Ian Briggs, "Rectification of Multichannel Images in Mass Storage Using Image Transposition," *Computer Vision, Graphics, and Image Processing*, vol. 29, no. 1, pp. 23-36, January 1985.

[Frederick 90] Frederick, C. and E.L. Schwartz, "Conformal Image Warping," *IEEE Computer Graphics and Applications*, vol. 10, no. 2, pp. 54-61, March 1990.

[Gabriel 84]   Gabriel, Steven A. and Michael A. Ogrinc, "Controller for System for Spatially Transforming Images, U.S. Patent 4,468,688, Ampex Corp., Aug. 28, 1984.

[Gangnet 82]   Gangnet, M., D. Perny, P. Coueignoux, "Perspective Mapping of Planar Textures," *Eurographics '82*, pp. 57-71, September 1982.

[Glassner 86]   Glassner, Andrew, "Adaptive Precision in Texture Mapping," (SIG-GRAPH '86 Proceedings), vol. 20, no. 4, pp. 297-306, July 1986.

[Gonzalez 87]   Gonzalez, Rafael C. and Paul Wintz, *Digital Image Processing*, Addison-Wesley, Reading, MA, 1987.

[Goshtasby 86]   Goshtasby, Ardeshir, "Piecewise Linear Mapping Functions for Image Registration," *Pattern Recognition*, vol. 19, no. 6, pp. 459-466, 1986.

[Goshtasby 87]   Goshtasby, Ardeshir, "Piecewise Cubic Mapping Functions for Image Registration," *Pattern Recognition*, vol. 20, no. 5, pp. 525-533, 1987.

[Goshtasby 88]   Goshstasby, Ardeshir, "Image Registration by Local Approximation Methods," *Image and Vision Computing*, vol. 6, no. 4, pp. 255-261, Nov. 1988.

[Gouraud 71]   Gouraud, Henri, "Continuous Shading of Curved Surfaces," *IEEE Trans. Computers*, vol. 20, no. 6, pp., 623-628, 1971.

[Graf 87]   Graf, Carl P., Kim M. Fairchild, Karl M. Fant, George W. Rusler, and Michael O. Schroeder, "Computer Generated Synthesized Imagery," U.S. Patent 4,645,459, Honeywell Inc., February 24, 1987.

[Green 78]   Green, P.J. and R. Sibson, "Computing Dirichlet Tessellations in the Plane," *Computer Journal*, vol. 21, pp. 168-173, 1978.

[Green 89]   Green, William B., *Digital Image Processing: A Systems Approach*, Van Nostrand Reinhold Co., NY, 1989.

[Greene 86]   Greene, Ned, and Paul Heckbert, "Creating Raster Omnimax Images from Multiple Perspective Views Using the Elliptical Weighted Average Filter," *IEEE Computer Graphics and Applications*, vol. 6, no. 6, pp. 21-27, June 1986.

[Grimson 81]   Grimson, W.E.L., *From Images to Surfaces: A Computational Study of the Human Early Visual System*, MIT Press, Cambridge, MA, 1981.

[Grimson 83]   Grimson, W.E.L., "An Implementation of a Computational Theory of Visual Surface Interpolation," *Computer Vision, Graphics, and Image Processing*, vol. 22, pp. 39-69, 1983.

[Gupta 81]   Gupta, Satish. and Robert F. Sproull, "Filtering Edges for Gray-Scale Displays," *Computer Graphics*, (SIGGRAPH '81 Proceedings), vol. 15, no. 3, pp. 1-5, August 1981.

[Haralick 76]   Haralick, Robert M., "Automatic Remote Sensor Image Processing," *Topics in Applied Physics, Vol. 11: Digital Picture Analysis*, Ed. by A. Rosenfeld, Springer-Verlag, 1976, pp. 5-63.

[Harder 72]   Harder, R.L. and R.N. Desmarais, "Interpolation Using Surface Splines," *J. Aircraft*, vol. 9, pp. 189-191, 1972.

[Hardy 71]   Hardy, R.L., "Multiquadratic Equations of Topography and Other Irregular Surfaces," *J. Geophysical Research*, vol. 76, pp. 1905-1915, 1971.

[Hardy 75]   Hardy, R.L., "Research Results in the Application of Multiquadratic Equations to Surveying and Mapping Problems," *Surveying and Mapping*, vol. 35, pp. 321-332, 1975.

[Heckbert 83]   Heckbert, Paul, "Texture Mapping Polygons in Perspective," Tech. Memo No. 13, NYIT Computer Graphics Lab, April 1983.

[Heckbert 86a]   Heckbert, Paul, "Filtering by Repeated Filtering," *Computer Graphics*, (SIGGRAPH '86 Proceedings), vol. 20, no. 4, pp. 315-321, July 1986.

[Heckbert 86b]   Heckbert, Paul, "Survey of Texture Mapping," *IEEE Computer Graphics and Applications*, vol. 6, no. 11, pp. 56-67, November 1986.

[Heckbert 89]   Heckbert, Paul, *Fundamentals of Texture Mapping and Image Warping*, Masters Thesis, Dept. of EECS, U. of California at Berkeley, Technical Report No. UCB/CSD 89/516, June 1989.

[Holzmann 88]   Holzmann, Gerard J., *Beyond Photography — The Digital Darkroom*, Prentice-Hall, Englewood Cliffs, NJ, 1988.

[Horner 87]   Horner, James L., *Optical Signal Processing*, Academic Press, NY, 1987.

[Hou 78]   Hou, Hsieh S. and Harry C. Andrews, "Cubic Splines for Image Interpolation and Digital Filtering," *IEEE Trans. Acoust., Speech, Signal Process.*, vol. ASSP-26, pp. 508-517, 1987.

[Hu 90]   Hu, Lincoln, *Personal Communication*, 1990.

[IMSL 80]   *IMSL Library Reference Manual*, ed. 8 (IMSL Inc., 7500 Bellaire Boulevard, Houston TX 77036), 1980.

[Jain 89]   Jain, Anil K., *Fundamentals of Digital Image Processing*, Prentice-Hall, Englewood Cliffs, NJ, 1989.

[Janesick 87]   Janesick, J.R., T. Elliott, S. Collins, M.M. Blouke, and J. Freeman, "Scientific Charge-Coupled Devices," *Optical Engineering*, vol. 26, no. 8, pp. 692-714, August 1987.

[Jarvis 76]   Jarvis, J.F., C.N. Judice, and W.H. Ninke, "A Survey of Techniques for the Display of Continuous-Tone Pictures on Bilevel Displays," *Computer Graphics and Image Processing*, vol. 5, pp. 13-40, 1976.

[Jensen 86]   Jensen, J.R., *Introductory Image Processing*, Prentice-Hall, Englewood Cliffs, NJ, 1986.

[Jou 89]   Jou, J.-Y. and Alan C. Bovik, "Improved Initial Approximation and Intensity-Guided Discontinuity Detection in Visible-Surface Reconstruction," *Computer Vision, Graphics, and Image Processing*, vol. 47, pp. 292-326, 1989.

[Joy 88]   Joy, Kenneth I., Charles W. Grant, Nelson L. Max, and Lansing Hatfield, *Computer Graphics: Image Synthesis*, IEEE Computer Society Press, Los Alamitos, CA, 1988.

[Juday 89]   Juday, Richard D. and David S. Loshin, "Quasi-Conformal Remapping for Compensation of Human Visual Field Defects: Advances in Image Remapping for Human Field Defects," *Proc. SPIE Optical Pattern Recognition*, vol. 1053, pp. 124-130, 1989.

[Kajiya 86]   Kajiya, James T., "The Rendering Equation," *Computer Graphics*, (SIGGRAPH '86 Proceedings), vol. 20, no. 4, pp. 143-150, July 1986.

[Keys 81]   Keys, Robert G., "Cubic Convolution Interpolation for Digital Image Processing," *IEEE Trans. Acoust., Speech, Signal Process.*, vol. ASSP-29, pp. 1153-1160, 1981.

[Klucewicz 78]   Klucewicz, I.M., "A Piecewise C^1 Interpolant to Arbitrarily Spaced Data," *Computer Graphics and Image Processing*, vol. 8, pp. 92-112, 1978.

[Knowlton 72]   Knowlton, K. and L. Harmon, "Computer-Produced Grey Scales," *Computer Graphics and Image Processing*, vol. 1, pp. 1-20, 1972.

[Lawson 77]   Lawson, C.L., "Software for C^1 Surface Interpolation," *Mathematical Software III*, Ed. by J.R. Rice, Academic Press, London, pp. 161-194, 1977.

[Leckie 80]   Leckie, D.G., "Use of Polynomial Transformations for Registration of Airborne Digital Line Scan Images," *14th Intl. Sym. Remote Sensing of Environment*, pp. 635-641, 1980.

[Lee 80]   Lee, D.T. and B.J. Schachter, "Two Algorithms for Constructing a Delaunay Triangulation," *Intl. J. Computer Info. Sci.*, vol. 9, pp. 219-242, 1980.

[Lee 83]   Lee, Chin-Hwa, "Restoring Spline Interpolation of CT Images," *IEEE Trans. Medical Imaging*, vol. MI-2, no. 3, pp. 142-149, September 1983.

[Lee 85]   Lee, Mark, Richard A. Redner, and Samuel P. Uselton, "Statistically Optimized Sampling for Distributed Ray Tracing," *Computer Graphics*, (SIGGRAPH '85 Proceedings), vol. 19, no. 3, pp. 61-67, July 1985.

[Lee 87]   Lee, David, Theo Pavlidis, and Greg W. Wasilkowski, "A Note on the Trade-off Between Sampling and Quantization in Signal Processing," *J. Complexity*, vol. 3, no. 4, pp. 359-371, December 1987.

[Lien 87]   Lien, S.-L., M. Shantz, and V. Pratt, "Adaptive Forward Differencing for Rendering Curves and Surfaces," *Computer Graphics*, (SIGGRAPH '87 Proceedings), vol. 21, no. 4, pp. 111-118, July 1987.

[Lillestrand 72] Lillestrand, R.L., "Techniques for Changes in Urban Development from Aerial Photography," *IEEE Trans. Computers*, vol. C-21, pp. 546-549, 1972.

[Limb 69]   Limb, J.O., "Design of Dither Waveforms for Quantized Visual Signals," *Bell System Tech. J.*, vol. 48, pp. 2555-2582, 1969.

[Limb 77]   Limb, J.O., "Digital Coding of Color Video Signals — A Review," *IEEE Trans. Comm.*, vol. COMM-25, no. 11, pp. 1349-1382, November 1977.

[Ma 88]   Ma, Song De and Hong Lin, "Optimal Texture Mapping," *Eurographics '88*, pp. 421-428, September 1988.

[Maeland 88]   Maeland, Einar, "On the Comparison of Interpolation Methods," *IEEE Trans. Medical Imaging*, vol. MI-7, no. 3, pp. 213-217, September 1988.

[Markarian 71] Markarian, H., R. Bernstein, D.G. Ferneyhough, L.E. Gregg, and F.S. Sharp, "Implementation of Digital Techniques for Correcting High Resolution Images," *Proc. Amer. Inst. Aeronautics and Astronautics 8th Annu. Meeting*, vol. C21, AIAA paper no. 71-326, pp. 285-304, October 1971.

[Marvasti 87] Marvasti, Farokh A., *A Unified Approach to Zero-Crossings and Nonuniform Sampling*, Nonuniform Publ. Co., Oak Park, IL, 1987.

[Massalin 90] Massalin, Henry, *Personal Communication*, 1990.

[Meinguet 79a] Meinguet, Jean, "An Intrinsic Approach to Multivariate Spline Interpolation at Arbitrary Points," *Polynomial and Spline Approximation: Theory and Applications*, B.N. Sahney (ed.), Reidel Dordrecht, Holland, pp. 163-190, 1979.

[Meingeut 79b] Meinguet, Jean, "Multivariate Interpolation at Arbitrary Points Made Simple," *J. Applied Math. and Physics (ZAMP)*, vol. 30, pp. 292-304, 1979.

[Mertz 34]   Mertz, Pierre and Frank Gray, "A Theory of Scanning and its Relation to the Characteristics of the Transmitted Signal in Telephotography and Television," *Bell System Tech. J.*, vol. 13, pp. 464-515, July 1934.

[Mitchell 87] Mitchell, Don P., "Generating Antialiased Images at Low Sampling Densities," *Computer Graphics*, (SIGGRAPH '87 Proceedings), vol. 21, no. 4, pp. 65-72, July 1987.

[Mitchell 88] Mitchell, Don P. and Arun N. Netravali, "Reconstruction Filters in Computer Graphics," *Computer Graphics*, (SIGGRAPH '88 Proceedings), vol. 22, no. 4, pp. 221-228, August 1988.

[Nack 77]   Nack, M.L., "Rectification and Registration of Digital Images and the Effect of Cloud Detection," *Proc. Machine Processing of Remotely Sensed Data*, pp. 12-23, 1977.

[NAG 80]   *NAG Fortran Library Manual Mark* 8, (NAG Central Office, 7 Banbury Road, Oxford OX26NN, U.K.), 1980.

[Nagy 83]   Nagy, George, "Optical Scanning Digitizers," *IEEE Computer*, vol. 16, no. 5, pp. 13-24, May 1983.

[Naiman 87]   Naiman, A. and A. Fournier, "Rectangular Convolution for Fast Filtering of Characters," *Computer Graphics*, (SIGGRAPH '87 Proceedings), vol. 21, no. 4, pp. 233-242, July 1987.

[Netravali 80]   Netravali, A.N. and J.O. Limb, "Picture Coding: A Review," *Proc. IEEE* vol. 68, pp. 366-406, 1980.

[Netravali 88]   Netravali, A.N. and B.G. Haskell, *Digital Pictures : Representation and Compression*, Plenum Press, New York, 1988.

[Nielson 83]   Nielson, G.M. and R. Franke, "Surface Construction Based Upon Triangulations," *Surfaces in Computer Aided Geometric Design*, Ed. by R.E. Barnhill and W. Boehm, North-Holland Publishing Co., Amsterdam, pp. 163-177, 1983.

[Norton 82]   Norton, A., A.P. Rockwood, and P.T. Skolmoski, "Clamping: A Method of Antialiasing Textured Surfaces by Bandwidth Limiting in Object Space," *Computer Graphics*, (SIGGRAPH '82 Proceedings), vol. 16, no. 3, pp. 1-8, July 1982.

[Oakley 90]   Oakley, J.P. and M.J. Cunningham, "A Function Space Model for Digital Image Sampling and Its Application in Image Reconstruction," *Computer Vision, Graphics, and Image Processing*, vol. 49, pp. 171-197, 1990.

[Oka 87]   Oka, M., K. Tsutsui, A. Ohba, Y. Kurauchi, and T. Tagao, "Real-Time Manipulation of Texture-Mapped Surfaces," *Computer Graphics*, (SIGGRAPH '87 Proceedings), vol. 21, no. 4, pp. 181-188, July 1987.

[Paeth 86]   Paeth, Alan W., "A Fast Algorithm for General Raster Rotation," *Graphics Interface '86*, pp. 77-81, May 1986.

[Park 82]   Park, Stephen K. and Robert A. Schowengerdt, "Image Sampling, Reconstruction, and the Effect of Sample-Scene Phasing," *Applied Optics*, vol. 21, no. 17, pp. 3142-3151, September 1982.

[Park 83]   Park, Stephen K. and Robert A. Schowengerdt, "Image Reconstruction by Parametric Cubic Convolution," *Computer Vision, Graphics, and Image Processing*, vol. 23, pp. 258-272, 1983.

[Parker 83]   Parker, J. Anthony, Robert V. Kenyon, and Donald E. Troxel, "Comparison of Interpolating Methods for Image Resampling," *IEEE Trans. Medical Imaging*, vol. MI-2, no. 1, pp. 31-39, March 1983.

[Pavlidis 82]   Pavlidis, Theo, *Algorithms for Graphics and Image Procesing*, Computer Science Press, Rockville, MD, 1982.

[Penna 86]   Penna, M.A. and R.R. Patterson, *Projective Geometry and Its Applications to Computer Graphics*, Prentice-Hall, Englewood Cliffs, NJ, 1986.

[Percell 76]   Percell, P., "On Cubic and Quartic Clough-Tocher Finite Elements," *SIAM J. Numerical Analysis*, vol. 13, pp. 100-103, 1976.

[Perlin 85]   Perlin, K., "Course Notes," *SIGGRAPH '85 State of the Art in Image Synthesis Seminar Notes*, July 1985.

[Pitteway 80]   Pitteway, M.L.V and D.J. Watkinson, "Bresenham's Algorithm with Grey Scale," *Comm. ACM*, vol. 23, no. 11, pp. 625-626, November 1980.

[Pohlmann 89]   Pohlmann, Ken C., *Principles of Digital Audio*, Howard W. Sams & Company, Indianapolis, IN, 1989.

[Porter 84]   Porter, T. and T. Duff, "Compositing Digital Images," *Computer Graphics*, (SIGGRAPH '84 Proceedings), vol. 18, no. 3, July 1984, pp. 253-259.

[Powell 77]   Powell, M.J.D. and M.A Sabin, "Piecewise Quadratic Approximations on Triangles," *ACM Trans. Mathematical Software*, vol. 3, pp. 316-325, 1977.

[Press 88]   Press, W.H., Flannery, B.P., Teukolsky, S.A., and Vetterling, W.T., *Numerical Recipes in C*, Cambridge University Press, Cambridge, 1988.

[Ramirez 85]   Ramirez, R.W., *The FFT: Fundamentals and Concepts*, Prentice-Hall, Englewood Cliffs, NJ, 1985.

[Ratzel 80]   Ratzel, J.N., "The Discrete Representation of Spatially Continuous Images," Ph.D. Thesis, Dept. of EECS, MIT, 1980.

[Reichenbach 89]
                 Reichenbach, Stephen E., and Stephen K. Park, "Two-Parameter Cubic Convolution for Image Reconstruction," *Proc. SPIE Visual Communications and Image Processing*, vol. 1199, pp. 833-840, 1989.

[Rifman 74]   Rifman, S.S. and D.M. McKinnon, "Evaluation of Digital Correction Techiques for ERTS Images — Final Report, Report 20634-6003-TU-00, TRW Systems, Redondo Beach, Calif., July 1974.

[Robertson 87]   Robertson, Philip K., "Fast Perspective Views of Images Using One-Dimensional Operations," *IEEE Computer Graphics and Applications*, vol. 7, no. 2, pp. 47-56, February 1987.

[Robertson 89]   Robertson, Philip K., "Spatial Transformations for Rapid Scan-Line Surface Shadowing," *IEEE Computer Graphics and Applications*, vol. 9, no. 2, pp. 30-38, March 1989.

[Rogers 76]   Rogers, D.F. and J.A. Adams, *Mathematical Elements for Computer Graphics*, McGraw-Hill, NY, 1976.

[Rosenfeld 82]   Rosenfeld, A. and A.C. Kak, *Digital Picture Processing, Vol. 2*, Academic Press, Orlando, FL, 1982.

[Sakrison 77]   Sakrison, David J., "On the Role of the Observer and a Distortion Measure in Image Transmission," *IEEE Trans. Comm.*, vol. COMM-25, no. 11, pp. 1251-1267, November 1977.

[Samek 86]   Samek, Marcel, Cheryl Slean, and Hank Weghorst, "Texture Mapping and Distortion in Digital Graphics," *Visual Computer*, vol. 3, pp. 313-320, 1986.

[Schafer 73]   Schafer, Ronald W. and Lawrence R. Rabiner, "A Digital Signal Processing Approach to Interpolation," *Proc. IEEE*, vol. 61, pp. 692-702, June 1973.

[Schowengerdt 83]
                Schowengerdt, Robert A., *Techniques for Image Processing and Classification in Remote Sensing*, Academic Press, Orlando, FL, 1983.

[Schreiber 85] Schreiber, William F., and Donald E. Troxel, "Transformation Between Continuous and Discrete Representations of Images: A Perceptual Approach," *IEEE Trans. Pattern Analysis and Machine Intelligence*, vol. PAMI-7, no. 2, pp. 178-186, March 1985.

[Schreiber 86] Schreiber, William F., *Fundamentals of Electronic Imaging Systems*, Springer-Verlag, Berlin, 1986.

[Shannon 48]   Shannon, Claude E., "A Mathematical Theory of Communication," *Bell System Tech. J.*, vol. 27, pp. 379-423, July 1948, and vol. 27, pp. 623-656, October 1948.

[Shannon 49]   Shannon, Claude E., "Communication in the Presence of Noise," *Proc. Inst. Radio Eng.*, vol. 37, no. 1, pp. 10-21, January 1949.

[Simon 75]   Simon, K.W., "Digital Image Reconstruction and Resampling for Geometric Manipulation," *Proc. IEEE Symp. on Machine Processing of Remotely Sensed Data*, pp. 3A-1-3A-11, 1975.

[Singh 79]   Singh, M., W. Frei, T. Shibita, G.H. Huth, and N.E. Telfer, "A Digital Technique for Accurate Change Detection in Nuclear Medicine Images — with Application to Myocardial Perfussion Studies Using Thallium 201," *IEEE Trans. Nuclear Science*, vol. NS-26, pp. 565-575, February 1979.

[Smith 83]   Smith, Alvy Ray, "Digital Filtering Tutorial for Computer Graphics," parts 1 and 2, *SIGGRAPH '83 Introduction to Computer Animation seminar notes*, pp. 244-272, July 1983.

[Smith 87]   Smith, Alvy Ray, "Planar 2-Pass Texture Mapping and Warping," *Computer Graphics*, (SIGGRAPH '87 Proceedings), vol. 21, no. 4, pp. 263-272, July 1987.

[Smythe 90]   Smythe, Douglas B., "A Two-Pass Mesh Warping Algorithm for Object Transformation and Image Interpolation," ILM Technical Memo #1030, Computer Graphics Department, Lucasfilm Ltd., 1990.

[Stead 84]   Stead, S.E., "Estimation of Gradients from Scattered Data," *Rocky Mountain J. Math.*, vol. 14, pp. 265-279, 1984.

[Steiner 77]   Steiner, D. and M.E. Kirby, "Geometrical Referencing of Landsat Images by Affine Transformation and Overlaying of Map Data,"

Photogrametria, vol. 33, pp. 41-75, 1977.

[Stoffel 81]   Stoffel, J.C. and J.F. Moreland, "A Survey of Electronic Techniques for Pictorial Image Reproduction," *IEEE Trans. Comm.*, vol. COMM-29, no. 12, pp. 1898-1925, December 1981.

[Strang 80]   Strang, Gilbert, *Linear Algebra and Its Applications*, 2nd ed., Academic Press, NY, 1980.

[Tabata 86]   Tabata, Kuniaki and Haruo Takeda, "Processing Method for the Rotation of an Image," U.S. Patent 4,618,991, Hitachi Ltd., October 21, 1986.

[Tanaka 86]   Tanaka, A., M. Kameyama, S. Kazama, and O. Watanabe, "A Rotation Method for Raster Image Using Skew Transformation," *Proc. IEEE Conference on Computer Vision and Pattern Recognition*, pp. 272-277, June 1986.

[Tanaka 88]   Tanaka, Atsushi and Masatoshi Kameyama, "Image Rotating System By an Arbitrary Angle," U.S. Patent 4,759,076, Mitsubishi Denki Kabushiki Kaisha, July 19, 1988.

[Terzopoulos 83]
Terzopoulos, Demetri, "Multilevel Computational Processes for Visual Surface Reconstruction," *Computer Vision, Graphics, and Image Processing*, vol. 24, pp. 52-96, 1983.

[Terzopoulos 84]
Terzopoulos, Demetri, *Multiresolution Computation of Visible-Surface Representations*, Ph.D. Thesis, Dept. of EECS, MIT, 1984.

[Terzopoulos 85]
Terzopoulos, Demetri, "Computing Visible Surface Representations," AI Lab, Cambridge, MA, AI Memo 800, 1985.

[Terzopoulos 86]
Terzopoulos, Demetri, "Regularization of Inverse Visual Problems Involving Discontinuities," *IEEE Trans. Pattern Analysis and Machine Intelligence*, vol. PAMI-8, no. 4, pp. 413-424, 1986.

[Tikhonov 77]   Tikhonov, A.N. and V.A. Arsenin, *Solutions of Ill−Posed Problems*, Winston and Sons, Washington, D.C., 1977.

[Turkowski 82]   Turkowski, Ken, "Anti-Aliasing Through the Use of Coordinate Transformations," *ACM Trans. on Graphics*, vol. 1, no. 3, pp. 215-234, July 1982.

[Turkowski 88a] Turkowski, Ken, "Several Filters for Sample Rate Conversion," Technical Report No. 9, Apple Computer, Cupertino, CA, May 1988.

[Turkowski 88b] Turkowski, Ken, "The Differential Geometry of Texture Mapping," Technical Report No. 10, Apple Computer, Cupertino, CA, May 1988.

[Ulichney 87]   Ulichney, Robert, *Digital Halftoning*, MIT Press, Cambridge, MA, 1987.

[Van Wie 77]   Van Wie, Peter and Maurice Stein, "A Landsat Digital Image Rectification System," *IEEE Trans. Geoscience Electronics*, vol. GE-15, pp. 130-17, 1977.

[Volder 59]   Volder, Jack E., "The CORDIC Trigonometric Computing Technique," *IRE Trans. Electron. Comput.*, vol. EC-8, no. 3, pp. 330-334, September 1959.

[Ward 89]   Ward, Joseph and David R. Cok, "Resampling Algorithms for Image Resizing and Rotation," *Proc. SPIE Digital Image Processing Applications*, vol. 1075, pp. 260-269, 1989.

[Weiman 79]   Weiman, Carl F.R. and George M. Chaikin, "Logarithmic Spiral Grids for Image Processing and Display," *Computer Graphics and Image Processing*, vol. 11, pp. 197-226, 1979.

[Weiman 80]   Weiman, Carl F.R., "Continuous Anti-Aliased Rotation and Zoom of Raster Images," *Computer Graphics*, (SIGGRAPH '80 Proceedings), vol. 14, no. 3, pp. 286-293, July 1980.

[Whitted 80]   Whitted, Turner, "An Improved Illumination Model for Shaded Display," *Comm. ACM*, vol. 23, no. 6, pp. 343-349, June 1980.

[Williams 83]   Williams, Lance, "Pyramidal Parametrics," *Computer Graphics*, (SIGGRAPH '83 Proceedings), vol. 17, no. 3, pp. 1-11, July 1983.

[Wolberg 88]   Wolberg, George, "Image Warping Among Arbitrary Planar Shapes," *New Trends in Computer Graphics* (Proc. Computer Graphics Intl. '88), Ed. by N. Magnenat-Thalmann and D. Thalmann, Springer-Verlag, pp. 209-218, 1988.

[Wolberg 89a]   Wolberg, George, "Skeleton-Based Image Warping," *Visual Computer*, vol. 5, pp. 95-108, 1989.

[Wolberg 89b]   Wolberg, George and Terrance E. Boult, "Image Warping with Spatial Lookup Tables," *Computer Graphics*, (SIGGRAPH '89 Proceedings), vol. 23, no. 3, pp. 369-378, July 1989.

[Wolberg 90]   Wolberg, George, *Separable Image Warping: Implications and Techniques*, Ph.D. Thesis, Dept. of Computer Science, Columbia University, NY, 1990.

[Wong 77]   Wong, Robert Y., "Sensor Transformation," *IEEE Trans. Syst. Man Cybern.*, vol. SMC-7, pp. 836-840, Dec. 1977.

[Yellott 83]   Yellott, John I. Jr., "Spectral Consequences of Photoreceptor Sampling in the Rhesus Retina," *Science*, vol. 221, pp. 382-385, 1983.

INDEX

GEORGE WOLBERG

BIOGRAPHICAL SKETCH

George Wolberg was born on February 25, 1964, in Buenos Aires, Argentina. He received the B.S. and M.S. degrees in electrical engineering from Cooper Union, New York, NY, in 1985, and the Ph.D. degree in computer science from Columbia University, New York, NY, in 1990.

He is currently an Assistant Professor in the Computer Science department at the City College of New York / CUNY, and an Adjunct Assistant Professor at Columbia University. He has worked at AT&T Bell Laboratories, Murray Hill, NJ, and at IBM T.J. Watson Research Center, Yorktown Heights, NY, during the summers of 1983/4 and 1985/9, respectively. His research at these labs centered on image restoration, image segmentation, graphics algorithms, and texture mapping. From 1985 to 1988, he served as an image processing consultant to Fantastic Animation Machine, New York, NY, and between 1986 and 1989, he had been an Instructor of Computer Science at Columbia University. He spent the summer of 1990 at the Electrotechnical Laboratory in Tsukuba, Ibaraki, Japan, as a selected participant in the Summer Institute in Japan, a research program sponsored by the U.S. National Science Foundation and by the Science and Technology Agency of Japan.

Dr. Wolberg was the recipient of a National Science Foundation Graduate Fellowship. His research interests include image processing, computer graphics, and computer vision. He is a member of Tau Beta Pi, Eta Kappa Nu, and the IEEE Computer Society.

IEEE Computer Society

Other IEEE Computer Society Press Texts

Monographs

Analyzing Computer Architecture
Written by J.C. Huck and M.J. Flynn
(ISBN 0-8186-8857-2); 206 pages

Desktop Publishing for the Writer: Designing, Writing, Developing
Written by Richard Ziegfeld and John Tarp
(ISBN 0-8186-8840-8); 380 pages

Integrating Design and Test: Using CAE Tools for ATE Programming
Written by K.P. Parker
(ISBN 0-8186-8788-6 (case)); 160 pages

JSP and JSD: The Jackson Approach to Software Development (Second Edition)
Written by J.R. Cameron
(ISBN 0-8186-8858-0); 560 pages

National Computer Policies
Written by Ben G. Matley and Thomas A. McDannold
(ISBN 0-8186-8784-3); 192 pages

Physical Level Interfaces and Protocols
Written by Uyless Black
(ISBN 0-8186-8824-6); approximately 272 pages

Protecting Your Proprietary Rights in the Computer and High Technology Industries
Written by Tobey B. Marzouk, Esq.
(ISBN 0-8186-8754-1); 224 pages

Tutorials

Advanced Computer Architecture
Edited by D.P. Agrawal
(ISBN 0-8186-0667-3); 400 pages

Advanced Microprocessors and High-Level Language Computer Architectures
Edited by V. Milutinovic
(ISBN 0-8186-0623-1); 608 pages

Advances in Distributed System Reliability
Edited by Suresh Rai and Dharma P. Agrawal
(ISBN 0-8186-8907-2); 352 pages

Computer Architecture
Edited by D.D. Gajski, V.M. Milutinovic, H. Siegel, and B.P. Furht
(ISBN 0-8186-0704-1); 602 pages

Computer Communications: Architectures, Protocols, and Standards (Second Edition)
Edited by William Stallings
(ISBN 0-8186-0790-4); 448 pages

Computer Graphics (2nd Edition)
Edited by J.C. Beatty and K.S. Booth
(ISBN 0-8186-0425-5); 576 pages

Computer Graphics Hardware: Image Generation and Display
Edited by H.K. Reghbati and A.Y.C. Lee
(ISBN 0-8186-0753-X); 384 pages

Computer Graphics: Image Synthesis
Edited by Kenneth Joy, Max Nelson, Charles Grant, and Lansing Hatfield
(ISBN 0-8186-8854-8); 384

Computer and Network Security
Edited by M.D. Abrams and H.J. Podell
(ISBN 0-8186-0756-4); 448 pages

Computer Networks (4th Edition)
Edited by M.D. Abrams and I.W. Cotton
(ISBN 0-8186-0568-5); 512 pages

Computer Text Recognition and Error Correction
Edited by S.N. Srihari
(ISBN 0-8186-0579-0); 364 pages

Computers for Artificial Intelligence Applications
Edited by B. Wah and G.-J. Li
(ISBN 0-8186-0706-8); 656 pages

Database Management
Edited by J.A. Larson
(ISBN 0-8186-0714-9); 448 pages

Digital Image Processing and Analysis: Volume 1: Digital Image Processing
Edited by R. Chellappa and A.A. Sawchuk
(ISBN 0-8186-0665-7); 736 pages

Digital Image Processing and Analysis: Volume 2: Digital Image Analysis
Edited by R. Chellappa and A.A. Sawchuk
(ISBN 0-8186-0666-5); 670 pages

Digital Private Branch Exchanges (PBXs)
Edited by E.R. Coover
(ISBN 0-8186-0829-3); 400 pages

Distributed Computing Network Reliability
Edited by Suresh Rai and Dharma P. Agrawal
(ISBN 0-8186-8908-0); 368 pages

Distributed Control (2nd Edition)
Edited by R.E. Larson, P.L. McEntire, and J.G. O'Reilly
(ISBN 0-8186-0451-4); 382 pages

Distributed Database Management
Edited by J.A. Larson and S. Rahimi
(ISBN 0-8186-0575-8); 580 pages

Distributed-Software Engineering
Edited by S.M. Shatz and J.-P. Wang
(ISBN 0-8186-8856-4); 304 pages

DSP-Based Testing of Analog and Mixed-Signal Circuits
Edited by M. Mahoney
(ISBN 0-8186-0785-8); 272 pages

Fault-Tolerant Computing
Edited by V.P. nelson and B.D. Carroll
(ISBN 0-8186-0677-0 (paper) 0-8186-8667-4 (case)); 432 pages

Gallium Arsenide Computer Design
Edited by V.M. Milutinovic and D.A. Fura
(ISBN 0-8186-0795-5); 368 pages

Human Factors in Software Development (2nd Edition)
Edited by B. Curtis
(ISBN 0-8186-0577-4); 736 pages

Integrated Services Digital Networks (ISDN) (Second Edition)
Edited by W. Stallings
(ISBN 0-8186-0823-4); 404 pages

For Further Information:

IEEE Computer Society, 10662 Los Vaqueros Circle, P.O. Box 3014,
Los Alamitos, CA 90720-1264

IEEE Computer Society, 13, Avenue de l'Aquilon, 2,
B-1200 Brussels, BELGIUM

IEEE Computer Society,
Ooshima Building, 2-19-1 Minami-Aoyama,
Minato-ku, Tokyo 107, JAPAN